MANAGEMENT FOR QUALITY IN HIGH-TECHNOLOGY ENTERPRISES

MANAGEMENT FOR QUALITY IN HIGH-TECHNOLOGY ENTERPRISES

Yefim Fasser
Donald Brettner

WILEY-INTERSCIENCE
A JOHN WILEY & SONS, INC., PUBLICATION

This book is printed on acid-free paper. ∞

Copyright © 2002 by John Wiley & Sons, Inc. All rights reserved.

Published simultaneously in Canada.

No part of this publication may be reproduced, stored in a retrieval system or transmitted in any form or by any means, electronic, mechanical, photocopying, recording, scanning or otherwise, except as permitted under Sections 107 or 108 of the 1976 United States Copyright Act, without either the prior written permission of the Publisher, or authorization through payment of the appropriate per-copy fee to the Copyright Clearance Center, 222 Rosewood Drive, Danvers, MA 01923, (978) 750-8400, fax (978) 750-4744. Requests to the Publisher for permission should be addressed to the Permissions Department, John Wiley & Sons, Inc., 605 Third Avenue, New York, NY 10158-0012, (212) 850-6011, fax (212) 850-6008, E-Mail: PERMREQ @ WILEY.COM.

For ordering and customer service, call 1-800-CALL-WILEY.

Library of Congress Cataloging-in-Publication Data is available.

ISBN: 0-471-20958-9

Printed in the United States of America.

10 9 8 7 6 5 4 3 2 1

To my wife, Ida; my daughter, Elina; and my granddaughter, Natasha. For their understanding and encouragement.
Y.F.

To my wife Amy, for her constant understanding and support.
D.B.

Acknowledgment

We gratefully acknowledge the help and support of the people who in one way or another helped us to work on this book. Our deep appreciation to:

Chuck Anderson
Don Bottarini
Tina Chow
Roy Devaraj
Yuthana Hemungkorn
Susan Ho
Jim Hooper
E.K. Khoo
Melissa Lee
P.C. Loh
Adam Mah
Ajay Marathe

Heros Mardirossian
Raj Master
M.S. Maung
Yusuke Mizukami
Penny Ong
John Owen
Debi Roybal
Chuck Stiteler
Fred Stillger
T.S. Tan
C.H. Teoh

Contents

- **Introduction**   xvii
- **Part I: A Systemic Approach to Organizational Transformation**   1

Chapter 1   **A Systems View of Organization**   3
- 1.1 What is an Organization?   4
- 1.2 Seeing an Organization from Different Perspectives   5
  - 1.2.1 Thinking of an Organization as a Machine   6
  - 1.2.2 Thinking of an Organization as a Living Entity   7
  - 1.2.3 Comparing the Mechanistic and Living Entity Metaphors   8
  - 1.2.4 Thinking of an Organization as a Symphony Orchestra   11
  - 1.2.5 Thinking of an Organization as a Community   12
  - 1.2.6 The Web as a Metaphor of Organizational Architecture   13
- 1.3 The Lifetime of an Organizational System   14
- 1.4 Growth and Development   15
- 1.5 Shaping the Organization's Architecture   17
  - 1.5.1 Thinking Differently about Organizational Architecture   17
  - 1.5.2 Three Major Principles that Underpin the Organization's Architecture   18

Chapter 2	**Systems: A General Concept**	**21**
2.1	The Meaning of System	22
2.2	The Control System	23
2.3	The Planning and Control System	26
2.4	Analytical and Systems Thinking	27
	2.4.1 Analytical Thinking	28
	2.4.2 Systems Thinking	28
	2.4.3 Summary	29
2.5	Deming's View of a System	30
Chapter 3	**The Total Continuous Process of Improvement and Innovation (TCPI^2) Macro System**	**33**
3.1	The Evolution of the Macro System	33
3.2	A Description of the Macro System	35
3.3	The System of Profound Knowledge in Action	38
	3.3.1 What is Profound Knowledge?	39
	3.3.2 Deming's Philosophy on Profound Knowledge	41
3.4	The Description of the System of Profound Knowledge in Action	44
	3.4.1 The Theory and Application of Systems	46
	3.4.2 The Theory of Knowledge	47
	3.4.3 The Theory and Practice of Globalization	48
	3.4.4 The Theory and Knowledge of Variation	49
	3.4.5 Organizational Psychology	56

- **Part II: Managing a Knowledge-Based Organization** — **63**

Chapter 4	**Organizational Learning**	**65**
4.1	What is Learning and When Does it Occur?	65
4.2	Learning How to Learn	67
	4.2.1 How We Usually Work and Learn	68
	4.2.2 How We React to Problems	68
	4.2.3 Two Scenarios of Learning	69
4.3	Single-Loop and Double-Loop Learning	71
4.4	Other Classifications of Learning	74
	4.4.1 Maintenance Learning	74
	4.4.2 Shock Learning	75
	4.4.3 Anticipatory Learning	75

	4.5	Emotional Learning	76
	4.6	Learning from Partners	77
	4.7	What Form of Learning is Preferable?	78
Chapter 5		**Systemic Problem Solving (SPS) as an Effective Way of Learning**	**81**
	5.1	The Problem of Problems	81
	5.2	Rapid Change Requires Fast Learning	82
	5.3	A Holistic View of Problem Solving	83
	5.4	The Anatomy of a Problem	84
	5.5	Why Problems Repeat Themselves	86
	5.6	Three Forms of Problems	88
		5.6.1 An Example to Demonstrate the Three Types of Problem Resolution	89
	5.7	Systemic Problem Solving (SPS)	90
	5.8	The Current Reality Tree	93
		5.8.1 A Glimpse of the Theory	93
		5.8.2 Current Reality Tree	95
		5.8.3 A Manufacturing Example	96
Chapter 6		**Knowledge-Based Innovation**	**101**
	6.1	Creating an Innovative Environment	101
	6.2	Characteristics of Knowledge-Based Innovation	105
		6.2.1 The Lead Time of Knowledge-Based Innovation	106
		6.2.2 The Convergences of Knowledge-Based Innovation	107
	6.3	Requirements for Creating Knowledge-Based Innovation	108
		6.3.1 Conduct Careful Analysis	109
		6.3.2 Develop a Clear Focus on the Strategic Position	110
		6.3.3 Practice Entrepreneurial Management	112
	6.4	The Life Span of an Innovation	112
	6.5	Dr. Deming on Innovation	114
	6.6	Incrementalism and Innovation: Can They Coexist?	115
		6.6.1 Motorola's "Obsession" with Quality	116
		6.6.2 Tom Peters' View of Innovation and Incrementalism	117
		6.6.3 Friend or Enemy?	118
	6.7	Creative Thinking: A Way Toward Innovation	120
	6.8	Knowledge and Mental Models	122

Chapter 7 Knowledge Managers and Knowledge Workers — 126

- 7.1 Time and the Meaning of Management — 127
- 7.2 The Knowledge Manager — 127
- 7.3 The Knowledge Worker — 128
- 7.4 The T Shape of Knowledge: A Requisite for the Knowledge Worker — 129
- 7.5 Recruiting Knowledge Workers — 132
- 7.6 Managing Knowledge Workers — 133
- 7.7 The Relationship Between the Organization and the Knowledge Worker — 134

Chapter 8 Knowledge Transfer and Knowledge Management — 137

- 8.1 Managing Knowledge Transfer — 137
  - 8.1.1 Informal Knowledge Transfer — 138
  - 8.1.2 Structured (Formal) Knowledge Transfer — 141
  - 8.1.3 International Conferences and Symposia: A Form of Sharing — 142
  - 8.1.4 The Culture of Knowledge Transfer — 142
- 8.2 The System of Knowledge Management — 144

• Part III: Managing in a Global Environment — 147

Chapter 9 On the Road to Globalization — 149

- 9.1 The Trend Toward Globalization — 149
- 9.2 The Driving Forces of Globalization — 154
- 9.3 Technical Growth — 155
- 9.4 The Evolution of Globalization — 157
  - 9.4.1 Going Beyond Classic Globalization — 158
  - 9.4.2 The Levels of International Development — 164
- 9.5 The Meaning of Globalization in the Twenty-First Century — 165

Chapter 10 Managing Mergers, Acquisititions, and Other Strategic Alliances — 171

- 10.1 Mergers and Acquisitions — 172
  - 10.1.1 The Acquisition-Integration Process — 173
  - 10.1.2 Merger and Acquisition Due Diligence — 175
  - 10.1.3 The Cultural Aspect of Due Diligence — 176
  - 10.1.4 The Integration Manager — 178
- 10.2 Other Strategic Alliances — 180
  - 10.2.1 Joint Ventures — 180

	10.2.2	Supplier Agreements	182
	10.2.3	Licensing Technology	182
10.3	Some Forms of Global Competitiveness		184
	10.3.1	High Quality: The Entry Ticket into the Global Market	185
	10.3.2	A Large Variety of Products and Services	185
	10.3.3	Customer Convenience in Dealing with the Global Supplier	185
	10.3.4	On-Time Innovative Products	186
	10.3.5	Competitive Cost	187
	10.3.6	Global Mindset	188

Chapter 11   Globalization and Culture — 190

11.1	Cultural Conditioning		190
11.2	Cultural Strategies		192
	11.2.1	Ignoring Cultural Differences	193
	11.2.2	Minimizing Cultural Differences	194
	11.2.3	Making Cultural Differences a Core Competence	196
11.3	The Global Manager		198
11.4	Shaping the Culture in a Global Environment		200
11.5	Cultural Influences on Mergers and Acquisitions		201
	11.5.1	Cultural Pluralism	202
	11.5.2	Cultural Blending	203
	11.5.3	Cultural Takeover	203
	11.5.4	Cultural Resistance	203
11.6	Changing the Culture During the Merger		203
	11.6.1	Behavioral Change	205
	11.6.2	Justifying Behavioral Change	206
	11.6.3	Communicating the Cultural Change	206
	11.6.4	Hiring and Socialization	207
	11.6.5	Removal of Deviants	207

- **Part IV: Some Aspects of Managing Quality** — 210

Chapter 12   Some Fundamental Concepts of Managing Quality — 211

12.1	Quality and Customer Satisfaction	212
12.2	A Dualistic Look at Quality and Cost	214
12.3	Managing the Intangible Part of the Product	215

xii CONTENTS

12.4	Listening to the Customer's Voice	216
12.5	Creating a Quality Culture	219
12.6	Continuous Improvement and Innovation as Core Components of a Quality Culture	223
12.7	Quality Culture Means Focus on Quality	225
12.8	Measuring Quality in a Contemporary Organization	226
	12.8.1 Applying the Malcolm Baldrige Criteria for Measuring Quality	227
	12.8.2 The Baldrige Award Criteria	228
	12.8.2.1 Criteria Purpose	228
	12.8.2.2 Core Values and Concepts	228
	12.8.2.3 The Baldrige Criteria for Performance Excellence Framework: A Systems Perspective	233
	12.8.2.4 What Baldrige Criteria Score is Required to Become a World-Class Organization?	233
	12.8.3 Global Criteria for Measuring Quality	237
	12.8.4 Concluding Thoughts on the Baldrige Criteria	237

Chapter 13 Managing Variation: A Requisite for Quality **240**

13.1	Managing Process Variation	240
	13.1.1 The Peculiarities of Managing Variation in a High-Technology Enterprise	242
	13.1.2 Creating a Sense of Urgency	243
13.2	Taguchi's Quality Philosophy	245
	13.2.1 Beyond Conforming to Specifications	245
	13.2.2 What a Manager Should Know about Taguchi's Loss Function Concept	246
	13.2.3 Concluding Thoughts on Taguchi's Quality Philosophy	247
13.3	Creating a Low ppm Environment	248
13.4	Challenging and Changing Our Assumptions about Quality	251
13.5	Defining the Boundaries Between Low ppm and Perfectionism	252
13.6	The Importance of Double-Loop Learning in Quality Improvement	254
13.7	Do We Need Statistical Process Control for Automated Equipment?	256
13.8	Motorola's Six Sigma Methodology	257

Chapter 14	Some Major Quality Initiatives		262
14.1	Two Types of Improvement		263
14.2	The Kaizen Approach to Quality Improvement		264
14.3	Total Quality Control (TQC)		265
14.4	Understanding the Concept of TQM		268
14.5	Comparing TQC and TQM		271
14.6	The Six Sigma Philosophy		272
14.7	Reengineering and Beyond		275
	14.7.1	The Concept of Reengineering	275
	14.7.2	Comparing Reengineering with Other Quality Programs	276
	14.7.3	Beyond Reengineering	279
	14.7.4	Creating Holonic Business Systems	281
		14.7.4.1 Holon and Holonic Networks	281
		14.7.4.2 The Portuguese Man-Of-War as a Metaphor of a Holonic Network	283
14.8	Concluding Thoughts on Quality Initiatives		284

Chapter 15	Achieving High Quality Through Transformational Changes		288
15.1	Understanding the Concept of Change		289
15.2	Do We Love or Hate Change?		291
15.3	Recognizing the Cycles of Change		293
	15.3.1	Janssen's Four-Room Apartment	294
	15.3.2	Three Phases of Change Management	298
	15.3.3	Second-Order Change	306
15.4	Force Field Analysis		309
15.5	Some Strategies for Cultural Change		312
	15.5.1	The Meaning of Cultural Change	312
	15.5.2	The Substitutability of Strategy and Culture	313
	15.5.3	Methods and Forms for Cultural Change	314

- **Part V: Reshaping the Organizational Culture** — 321

Chapter 16	The System of the Organizational Culture	323
16.1	Organizational Culture: What is It?	323
16.2	The Structure of the Organizational Culture System	324

Chapter 17		**Managing the Core of the Organizational System**	**328**
	17.1	The Power of Vision	328
	17.2	Vision and Transformational Leadership	330
	17.3	Purpose—The Guiding Star of an Organization	331
		17.3.1 The Purpose of an Organization	331
		17.3.2 Purpose and Vision	333
		17.3.3 Purpose and Core Values	334
		17.3.4 Purpose from a System Perspective	335
		17.3.5 Creating Partnership Through the Exchange of Purpose	336
		17.3.6 The Personal Purpose	337
		17.3.7 Being True to Our Purpose	337
	17.4	The Mission of an Organization	338
		17.4.1 The Mission Statement of a Contemporary Organization	338
		17.4.2 Mission and Purpose	340
		17.4.3 Mission and Vision	341
		17.4.4 Testing Your Mission Statement	341
		17.4.5 The Process of Forming a Mission Statement	343
	17.5	Setting Motivational Goals	345
		17.5.1 Goals and Tension	345
		17.5.2 A Goal from a Systems Perspective	346
		17.5.3 Hard and Soft Goals	347
		17.5.4 How to Make the Goal Motivational	348
		17.5.5 Goals in a Multicultural Organization	349
		17.5.6 Creating Learning Goals	351
	17.6	Creating a Great Strategy	352
	17.7	The Moon Vision: A Case Study	354
		17.7.1 A Vision Statement that Became a Model	354
		17.7.2 How Realistic Should a Vision Be?	355
		17.7.3 Looking for Answers by Analyzing a Higher System	355
		17.7.4 Building an Environment for Risk Taking	357
		17.7.5 Pulling the Resources Together	357
		17.7.6 Strategy and Culture	358
		17.7.7 How Do We Create a "Moon" Vision?	359
		17.7.8 What Can We "Take Home?"	359

Chapter 18	Values, Behavioral Standards, and Business Ethics	364
18.1	A General Concept of Values	364
	18.1.1 Values and Beliefs	364
	18.1.2 Recognizing Different Values	365
18.2	Building Shared Values in an Organization	366
18.3	Core Values	367
	18.3.1 Where Do Core Values Come From?	367
18.4	Operational Values	369
18.5	Espoused Values	370
18.6	Behavioral Standards	371
	18.6.1 What Comes First: Values or Behavior?	371
	18.6.2 Developing Core Behavioral Standards	372
18.7	Business Ethics	376
	18.7.1 Understanding the Concept of Ethics	376
	18.7.2 The "Testron" Story	377
	18.7.3 The Meaning of Morality	378
	18.7.4 Where Do Moral Standards Come From?	379
	18.7.5 The Meaning of Ethics	379
	18.7.6 The Meaning of Business Ethics	380
	18.7.7 The Distinction Between Values and Ethics	381
	18.7.8 Codes of Ethics	381

Chapter 19	Symbols, Symbolic Actions, and Metaphors	383
19.1	The Founder and Leaders as Symbols of the Organization	383
19.2	Stories and Myths	384
19.3	The Organization's History	384
19.4	The Organization's Rituals in Working Life	385
	19.4.1 Meeting with Leaders	386
	19.4.2 National Rituals	386
	19.4.3 Groundbreaking Ceremonies	387
	19.4.4 An Organization's Anniversary	387
	19.4.5 Humor as a Form of Symbolic Action	387
19.5	Interpretation of Symbols	388
19.6	A Note of Caution	389
19.7	A Metaphor: The Tree of Culture	389

Chapter 20	Understanding an Organization's Behavior	393
20.1	Should Organizational Values Change?	395
20.2	Should We Prioritize Values?	395
20.3	Culture and Change	396
20.4	Learning from IBM's Experience	397
20.4.1	The Timeline of IBM's Leaders	397
20.4.2	A Little Bit of History	399
20.4.3	The Formation of IBM's Culture	399
20.4.4	Thomas Watson, Sr., a Great Entrepreneur	401
20.4.5	The Second CEO: Thomas Watson, Jr.	405
20.4.6	The Beginning of Gerstner's Era: The Renaissance of IBM	411
20.4.7	IBM: Moving Toward a Strong Strategic Vision	418
20.4.8	How to Measure the Success of Cultural Transformation	419
20.4.9	What Type of Leaders Have Led IBM?	420
20.4.10	Studying IBM's Pattern of Cultural Change	423
20.4.11	Ten Major Lessons We Can Learn from IBM's Story	427

Introduction

This book, *Management for Quality In High-Technology Enterprises*, is not about sampling plans, performance to specifications, or other specific issues related to quality products and services. The contents of this book are based on our observations of what a contemporary manager needs to successfully manage for quality. Management for quality has a broad definition in this book. It is all the activities that managers and leaders are involved in to fulfill all stakeholder needs, with a strong focus on customers.

Here the term "focus on customers" also has a broad definition. Today the customers want not only a quality product, but they also want it for a reasonable price, and they want it now. To accomplish this, managers need to know how to create the most effective systems and processes and how to motivate people to delight the customers and foresee their future desires. Customer focus also means continuously putting new products and services that the customers could not have dreamed of before on the market. And this needs to be done before your competitor does it. Because of this, managers of high-technology enterprises need to know how to think systemically and how to create an environment for creativity and innovation. Customer focus also means satisfying the global customers. Because of this, we need managers with a global mind-set who know how to work in a global environment, how to manage the multinational workforce, and how to communicate with people from all over the world.

In the past, a manager could survive with having only hard skills related to the process of making products. Today as never before, managers are also involved in forming mergers and acquisitions, joint ventures, technology transfers, licensing and other alliances. In addition, managers also need to be knowledgeable in a number of soft skills. They are responsible for the creation and development of the organization's purpose, vision, and mission and are involved in shaping the organization's culture. They are obligated to be a model in living the corporate values and helping people follow their example. Also, they should be process oriented and understand how to manage variation. They need to be familiar with all the major quality initiatives. In these turbulent times, organizations are going through a continuous process of change. Managing change is another skill for a contemporary manager.

We think that all this relates to the subject of management for quality because if any of these elements are not considered, the customer will feel it. These are not easy tasks for management, especially in high-technology enterprises. In these organizations the processes and products are becoming more and more sophisticated, the global competition is continuously growing, the lifetime of products is getting shorter and shorter, and communications are reaching all over the world. Because of this, continuously learning and creating knowledge are probably the most important responsibilities of management.

It was difficult to put all these components of management in one book. However, as the contents of the book suggest, we have tried to include the main topics that we believe a manager should know to be able to cope with the continuously growing requirements of high-technology enterprises.

The book is structured in five parts, which contain 20 chapters. Part I, A Systemic Approach to Organizational Transformation, has three chapters. Chapter 1 gives a systems view of an organization. A metaphorical approach is then used to help the reader see an organization from different perspectives. Chapter 2 is dedicated to the concept of systems and systems thinking, which includes Dr. Deming's view of a system. Chapter 3 describes Advanced Micro Devices' (AMD) experiences in developing the Total Continuous Process Improvement and Innovation (TCPI^2) macro system. Dr. Deming's system of profound knowledge is at the core of the TCPI^2 macro system.

Part II, Managing a Knowledge-Based Organization, has five chapters. Chapter 4 brings the reader's attention to the importance of learning and creating knowledge. Here the reader will become more familiar with single-loop and double-loop learning and other interesting concepts related to the subject of learning. Chapter 5 brings up issues related to systemic problem solving. Here we provide a holistic view to problem solving and continue with descriptions of different techniques for systemic problem solving. Chapter 6 is dedicated to the subject of knowledge-based innovation. It starts with a description of how to create an innovative environment and continues with different sections that allow readers to see the process of innovation from different angles. Chapter 7 shows the relationship between knowledge and innovation. Here you will become familiar with the concept of knowledge workers and with some peculiarities of leading this category of workers. Chapter 8 covers the subject of managing knowledge transfer. Two types of transfer are described: informal and structured (formal). Our experiences conducting international conferences and symposia are described as important forms of knowledge sharing and transfer.

Part III, Managing In a Global Environment, has three chapters. Chapter 9 gives a general description of globalization, describes its evolution, and shows the importance of globalization in the twenty-first century. Chapter 10 describes the strategy of globalization and will familiarize you with such concepts as mergers and acquisitions, joint ventures, supplier agreements, and others that will allow you to form a global mind-set. Global competitiveness is a central topic of this chapter. In Chapter 11 you will become familiarized with the relationship between globalization and organizational culture. Different cultural strategies that are in sync with the requirements of globalization are demonstrated. The role of a global manager is also described in some detail.

Part IV, Some Aspects of Managing Quality, consists of four chapters. Chapter 12 contains some fundamental concepts of managing quality. Customer focus, the relationship between quality and cost, and continuous improvement and innovation are just some of the topics. This chapter ends with a description of Malcolm Baldrige criteria as a complex measure of quality that covers almost all aspects of the organization. Chapter 13 relates to the topic of variation. The reader will have the opportunity to become familiar with the peculiarities of managing variation in high-technology enterprises. Motorola's experience in reducing variation is shown as an example of benchmarking. A way of creating a low ppm (parts per million) environment is also demonstrated. Chapter 14 contains a description of a number of quality initiatives, which includes TQC, TQM, the Six Sigma philosophy, and Reengineering. We make a comparison between these initiatives and explain their relationship to the high technology enterprises. Chapter 15 is related to change and change management. Starting with the general concept of change, a number of different techniques are described that can help managers manage changes.

Part V, Reshaping the Organizational Culture, consists of five chapters. Chapter 16 gives a systemic view of organizational culture, starting with a description of what culture is all about and moving to the architecture of a cultural system. Chapter 17 is dedicated to a description of vision, purpose, mission, and goals. Here we show the relationship between these components, which together create the core of the whole organizational system. Chapter 18 concerns the concepts of values, behavioral standards, and business ethics and the relationship between these concepts. Chapter 19 demonstrates that symbols, symbolic actions, and metaphors are important parts of the organizational culture. Stories, myths, rituals, and humor are also described as components of the cultural system. Chapter 20 shows the relationship between culture and change. We demonstrate the origin of the power of culture and show the importance of adapting and shaping the culture over time by using a real example from IBM's history.

This book was based mainly on experiences gained in one of AMD's largest groups—the Manufacturing Services Group (MSG). In September 2001, to achieve a greater focus in technological development, AMD was restructured to give the Memory Group and Computation Products Group direct responsibility for the front-end and back-end manufacturing and technology. This realignment required that MSG be reorganized into three separate divisions. However, the accumulated experience, knowledge, and culture of MSG will continue to develop because all the people and their executive management are the same.

In addition to using MSG's experiences, the authors used materials that came from organizations such as IBM, Motorola, Intel, General Electric, Hewlett-Packard, and others. This allowed us to better reflect the peculiarities of the high-technology industry.

The 20 chapters of this book touch on topics that are different and yet strongly interconnected. We hope that this book will demonstrate a way to achieve organizational excellence, expand your knowledge in managing for quality, and develop your interest in reading more about these subjects. The references at the end of each chapter may help you find the right books for further reading. The main intent of writing this book was to put together a body of knowledge that would be interesting for leaders and managers at all levels of the organization. We think that this book might also be of interest to students as reference material.

PART I

A Systemic Approach to Organizational Transformation

As organizations grow more complex, the need for skilled leadership becomes greater. One of the important skills managers will need to develop more than ever before is the skill of applying systems. Seeing organizations as complex systems, learning how to think systemically, will help managers comprehend the complexity of organizations and make their management process more effective. This part of the book is dedicated to the theory and practice of systems and systems thinking.

PART TWO

A Systemic Approach to
Operational Persistence

Chapter 1

A Systems View of Organization

As an introduction to this chapter, we would like to use the old Indian story about the blind men and the elephant. The first blind man touched the elephant's side and said, "I think it is a wall." The second blind man touched the elephant's long trunk and said, "Oh, it is just like a snake!" The third blind man touched the elephant's smooth ivory tusk and said, "It's as sharp as a spear!" All three blind men perceived the elephant differently; depending on what part of the elephant they touched. We can say the same thing about an organization, especially when it grows large. Management and employees sometimes have difficulty comprehending the whole organization, and they mainly see it through the prism of their workplace. As an organization grows and becomes more global, more specialization occurs and it becomes more difficult to embrace and see the "whole elephant."

Management needs to make sure that every employee, regardless of what part of the organization he or she works in, can see the whole organization and feel that he or she is an important part of a larger whole. This will allow the employees to recognize that the purpose of their work is in context with and a part of the purpose of the whole organization, that their vision and goals are derived from and form a part of the organization's vision and goals. Obviously, this will positively impact on the organization's performance and make its employees more participative.

Management must learn how to see the whole picture of the organization, which is continuously changing. This will have an impact on the leadership style and the way the organization operates. In this chapter, we want to introduce some metaphors that can be used as prisms through which to see an organization. This may help you to better understand the meaning of an organization and to develop new forms of management.

1.1 WHAT IS AN ORGANIZATION?

You probably are familiar with the frog experiment frequently described by psychologists. When a frog was dropped in a bowl of boiling water, the frog immediately jumped out. But when the frog was placed in a bowl of room-temperature water that was gradually heated up to boiling, the frog was not aware of the dangerous situation. This experiment can be used as an analogy for the process of organizational changes that occur over a number of years. Because the process is gradual, we usually don't react to these changes quickly enough. The organization today is not what it was yesterday, and tomorrow's organization may be absolutely different. Knowledge managers need to periodically review their way of thinking about an organization and have a deep understanding of it to build an extraordinary organization.

It may sound naive to ask a person who has spent ten or twenty years in an organization, "What precisely is an organization? How does it work?" However, ask these questions and you will see that the answers will be different and contradictory; they will show that the majority of people think too simply about the concept of an organization. Because of this, there are a lot of chronic problems in organizations that are not getting solved, even though we work on them very hard. "Organization" has become an everyday term, but we don't always recognize that its meaning is continuously changing. Some authors have even suggested that because of continuous increases in business complexity, organizations will not work in the future. In the introduction to the book *Organization of the Future*, Peter Drucker wrote, "A good many writers, seeing all these changes and all this turmoil, are writing of 'the end of organizations.' That, however, is the one thing we can predict with certainty will not happen...."[1] Certainly there will be a strong need for organizations in the future, but the writing about "the end of organizations" should be perceived by management as a strong signal that the meaning and purpose of "organization" will be going through a lot of changes.

To better understand how to shape today's organizations and ensure that they meet our future needs, it is not as important to seek the best definition as it is to find a way to understand organizations from different perspectives. As a result, this will allow us to become aware of new insights and to create a wide and varied range of organizational improvement.

In his book *Images of Organization*,[2] Gareth Morgan offers an interesting approach to the study of organizations by using metaphors. We at AMD adapted this approach as a way to analyze and improve our understanding of the concept of an organization. The study and application of Morgan's approach have allowed us to become more open-minded in the process of organizational transformation. In this chapter, we describe our view of the metaphorical concept of learning about "organization" and share some of our practices in this direction.

For most people a metaphor is regarded as a tool that helps us enhance the way we speak, but its importance is much greater than this. Metaphors greatly influence the way we think, the way we see things, and the way we act. In their book *Metaphors We Live By*, George Lakoff and Mark Johnson wrote, "We have found... that metaphor is pervasive in everyday life, not just in language but in thought and action. Our ordinary conceptual system, in terms of which we both think and act, is fundamentally metaphorical in nature."[3] Because of this, it is very important to understand how metaphor works and where its power comes from.

We usually use metaphor whenever we attempt to understand one element of experience in terms of another. For example, when we say, "This manager is a tiger," we place emphasis on the tigerlike aspects of the manager. The metaphor makes us see the manager in a distinctive, yet partial, way. In other words, by highlighting certain qualities of the manager, we tend to force other qualities into the background. As in our example, by drawing attention to the tigerlike strength or bravery, the metaphor makes us neglect the fact that the same manager may well also be a sloppy pig, a stubborn mule, etc. Our ability to comprehend a broader picture of the manager depends on our skill to see how these different qualities of the same manager can coexist and complement or contradict each other.

This way of thinking can be applied not only to individuals but also to an organization. For example, for many years we used to think about an organization as a machine. This metaphorical concept focuses our attention on an organization that is designed to achieve predetermined objectives and to run smoothly and efficiently like an oiled and well-tuned machine. Because of this kind of mind-set, we have developed a whole set of actions and behaviors that are influenced by the mechanistic way of thinking, moving the human part of an organization into a secondary role. The application of different metaphors to an organization will allow us to enhance the way we think and to broaden our understanding of the complexity and paradoxical character of an organization. Below we will see how different metaphors generate different ways of thinking, which in turn allows us to improve our way of designing and managing contemporary organizations.

1.2 SEEING AN ORGANIZATION FROM DIFFERENT PERSPECTIVES

In 1999 we held a series of seminars on leadership for AMD employees, with a total attendance of more than 500 people from different countries. One of the workshops conducted at these seminars was designed to demonstrate the influence of metaphors on the way we think. To do this, we divided the participants into small groups, supplied them with a list of metaphors, and asked them to think of an organization by using a metaphor selected from the list or one of their own. It was interesting to see how the selected metaphor rapidly formed the group's way of thinking.

Each group started to think about an organization with categories that came from the metaphor they selected. For example, if they selected the metaphor "organization as a machine," their way of thinking about the organization was absolutely different than those groups who selected the metaphor "organization as an organism," etc. The effect of the metaphor on the way people perceive the concept of an organization was demonstrated with the same strength in all five countries where we conducted the seminars.

Without pretending to make any fundamental conclusions from the data we collected, we believe that the results of this mini-experiment suggest that if an organization accepts a particular metaphor, whatever it should be, the metaphor will influence the way people in the organization think and act. For example, if an organization sees itself as a community, it will live and act as a community, using all elements inherent in a community. Because of this, it is very important to make sure that organizations use metaphors that conform to the internal and external environments of an enterprise and match the needs of the people who work for them.

A large variety of metaphors can be applied to different aspects of organization. However, here we will offer the reader a selected number of metaphors that we feel deserve more attention (see Fig. 1-1). Two metaphors—"organization as a machine" and "organization as a living entity"—will be described in more detail and compared with each other. We will only give a brief description of some other metaphors such as "organization as a symphony orchestra" and "organization as a community."

1.2.1 Thinking of an Organization as a Machine

When we design, buy, or use a machine, we always know what we want it to do for us. We have a purpose, and we want the machine to serve our purpose. We tune it up, lubricate it properly, plug it in, and let it run. If something goes wrong, we fix it. If a part wears out, we replace it. If the machine starts giving us frequent

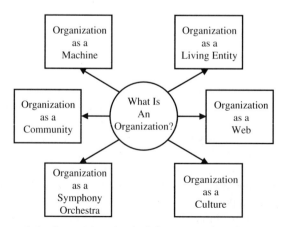

Figure 1-1 Some Metaphorical Concepts of an Organization

problems, we may throw it out and buy a new one. The same thing applies when we think about an organization as a machine. We design the product, establish the process, form the structure, hire and train people, and require them to perform according to policies and procedures. We know the input and expected output. We try to control the overall business processes, and if something goes wrong, we investigate and correct the problem. If a person is not performing properly, we try to "fix the problem," and if this does not work, we replace the person. The purpose of the organization is to make money.

This is a simplified description of the mechanistic way of thinking that has driven many organizations since the Industrial Revolution. Russell L. Ackoff called this period "the machine age." In his book *Creating the Corporate Future*, he wrote, "I believe we are leaving an age that can be called the *Machine Age*. In the Machine Age the universe was believed to be *a machine that was created by God to do His work*. Man, as part of that machine, was expected to serve God's purposes, to do His will...."[4] This metaphorical concept—the universe as a machine—strongly influenced our way of thinking. We started to mold our world in accordance with mechanistic thinking. We started to see organizations as machines and people who work there as its parts. Gradually, our work became dehumanized. This created many problems that organizations have faced and are trying to resolve without recognizing that the source of most of these problems came from the mechanistic thinking.

This is not to say that the metaphorical concept "organization as a machine" should not be used. This is one of the oldest concepts, which has brought enormous benefits, increasing organizational capabilities a thousand fold. The problem is that by becoming mechanistically oriented, we act like machines and lose our focus on the human aspects of the organization.

To make a point here, we intentionally "purified" the image of the metaphor "organization as a machine." In real life, organizations use a combination of different metaphors that make the whole organization less mechanistic. McDonald's is a typical example of a mechanistic organization. All the facilities look the same. The product made in these facilities is also similar. Every franchise has the same set of policies and procedures, and they perform like a great machine. However, this does not mean that McDonald's is not innovative and does not take care of its people. McDonald's is also a great example that shows that it is possible in one organization to apply the mechanistic model in some of its functions and at the same time apply different metaphorical models in its other functions. The art of contemporary leadership is to tap the strength of mechanistic thinking and, at the same time, recognize its weaknesses and seek new ways to organize enterprises.

1.2.2 Thinking of an Organization as a Living Entity

Accepting the metaphorical concept "an organization as a living entity" introduces an absolutely new way of thinking that is in contrast with the mechanistic view. It is normal to think that someone owns a machine, but these days it is difficult to imagine that a living organization has an owner. Living beings communicate with and learn from each other. They have their own purpose, which is linked with the organization's purpose, and they have their own vision, which is aligned with the organization's vision. The employees in a living entity grow together with the organization because they *are* the organization. Employees in a living organization are empowered to do their job. They cannot be controlled like machines. They also participate in the decision-making process and have their own identity.

The metaphorical concept of a living organization will lead you to think about it as a company that has members who subscribe to a set of common values and who believe that the goal of the company supports them and gives them conditions to achieve their own individual goals. Both the employees and the organization have the same driving motive—they want to survive. To survive, they need to grow and expand their potential. We could say that in a living organization there is an unwritten contract between the individual and the company that they will help each other to grow, reach their potential, and expand their working life span.

Using the metaphorical concept "an organization is a living entity" totally changes the viewpoint of the organization's purpose. As Arie De Geus wrote in his book *The Living Company*, "It probably doesn't matter very much whether a company is actually *alive* in a strict biological sense, or whether the 'living company' is simply a useful metaphor.... *Like all organisms, the living company exists primarily for its own survival and improvement: to fulfill its potential and to become as great as it can be*. It does not exist solely to provide customers with goods, or to return investment to shareholders...."[5] As you can see, Arie de Geus is emphasizing *survival* as the major purpose of any living system, including an organization. He says, "After all, you too are a living entity. You exist to survive and thrive; working at your job is a means to that end."[6] Seeing an organization as a living being is obviously a more progressive metaphor—a metaphor of tomorrow's organization that begins today.

1.2.3 Comparing the Mechanistic and Living Entity Metaphors

Now that you have a general understanding of the two major metaphorical concepts-mechanistic and living entity—we can describe and make a comparison of their strengths and weaknesses.

Let's start with the metaphor "organization as a machine." This metaphor can be applied when the following conditions exist:

1. When we need to perform a straightforward task,
2. When we have a relatively stable external environment,
3. When the product has a long life span,
4. When precision is a very important component of the process,
5. When people feel relatively satisfied with performing simple tasks,

6. When the work does not require continuous innovation and a lot of creativity, and
7. When the job does not require fast and continuous learning.

In this kind of environment, the mechanistic approach works just great, assuming that the organization's mind-set is not based only on mechanistic thinking.

However, the environment for most contemporary organizations does not fit the seven points described above. Because of this, the metaphor "organization as a machine" is losing its popularity, and this is why organizations have started to look for other metaphorical concepts.

There are still other reasons why the machine metaphor is losing power. Organizations that have been influenced by this metaphor have difficulties being flexible enough and instituting timely changes under the influence of the external environment. Mechanistic organizations are more bureaucratic and have difficulties introducing innovation and creativity. They usually also have difficulties introducing such democratic concepts as empowerment, teamwork, and other forms of employee involvement. This holds true, especially when the percentage of knowledge workers in the organization is rapidly growing. Knowledge workers do not fit in well with the mechanistic approach to management.

Mechanistically structured organizations can have a dehumanizing effect on the employees, especially for the frontline employees who usually do the work. Such an organization fosters mechanistic thinking and creates an atmosphere of ineffective communication because of the multilayer organizational structure and the standardized procedures that cannot embrace all necessary rules and behavior (in particular circumstances). So what is the alternative?

Now, let's move to the other metaphor: "organization as a living entity." One of the major strengths of the living entity metaphor is that it brings management's attention to the fact that an organization is interconnected with its external environment—that it is an open system. In contrast, the mechanistic metaphor makes us think about an organization as a closed system where the connection with the external environment is ignored. From an open-system point of view, the organization performs as an ongoing process and its purpose is survival. From a closed-system point of view, the organization's purpose is focused on specific operational goals, for example, making money. It is important to emphasize the difference here. Survival is a process, whereas goals are end points to be achieved. Reorienting the management's attention from the end point goals to the survival process, and focusing their attention on the external environment (customers, competitors, suppliers, subcontractors, and other institutions), makes them more flexible and gives them a much broader view of an organization. Changing the management's mind-set from a mechanistic to a living entity view of an organization will allow them to see culture, strategy, structure, technology, and the human part of the organization as the interacting subsystems in a larger system—the whole organization.

In a mechanistic environment, management's attention is more focused on the internal environment, on tight control, on policies and procedures, and on standards and specifications. The human part of the organization is not strongly developed. The living entity approach also allows management to see a larger range of options when it comes to business improvement. This comes from the understanding of the existence of different "species" within the organization, which require different approaches to their well-being and development. Comparing the mechanistic and living entity metaphors from a humanistic, innovation, and creativity point of view, the internal environment of a living entity organization is more team- and project oriented, less hierarchical, and more humanistic than the bureaucratic environment in a mechanistic organization.

While describing the positive part of the living entity metaphor, we must, however, also pay attention to its limitations and weaknesses to make sure that there is no misinterpretation or overuse of this metaphor. One of the main limitations of the living entity metaphor is that it influences us to think about organizations and their environments in an excessively concrete way. Organisms live in a natural environment, and that environment determines the life of these organisms. The same cannot be said about an organization and its environment. People create organizations. They are products of people's vision, mission, values, and beliefs.

This suggests that the shape and structure of the organization is not constant and is more fragile than the material structure of an organism. The environment of an organization can also be seen as being a product of the employees' innovation and creativity because it is made through the actions of those who work in the organization. Because of this, it is incorrect to suggest that the organization alone needs to adapt to its environment. The view, derived from the living entity metaphor, that organizations depend on forces operating in the external environment undermines the power of organizations and their employees to create their own future.

A second limitation of the living entity metaphor is related to the assumption of "functional unity." If we observe organisms, we can see their functional interdependence where every element of the whole system works for all the other elements. For example, in a human body, all its parts—the heart, the lungs, the brain, etc.—normally work together to preserve the homeostatic functioning of the whole. In other words, the human system is unified and shares a common life. However, if we look at an organization, we can observe times when the different parts are not always as functionally unified as those of organisms. The living entity metaphor leads us to think and believe that unity and harmony, which can be seen in the world of organisms, can also be achieved in organizations.

One more limitation related to an organism's growth was noted by Russell L. Acroff in his book *Creating the Corporate Future*, where he wrote, "Growth usually occurs in organisms without choice. Nevertheless, purposeful systems [organizations] can deter or accelerate their growth by the choices they make...."[7]

A third limitation must be considered when using the metaphorical concept of a living entity for organizational design. Living entities and organizations are systems that usually have purposes of their own. However, the parts of a living entity do not. For example, the parts of a human system (i.e., the heart, lungs, brain, hands, etc.) do not have a purpose of their own. The parts of an organization do have their own purpose, for example every division, department, and person in an organization has its or his own purpose. When we focus on organizations, we are concerned with three levels of purpose: the purpose of the system, of its parts, and the system of which it is a part, the suprasystem (see Section 17.2).

We can also find other limitations in the living entity metaphor, as we would continue to find when analyzing any metaphor. However, as we see from the comparison of mechanistic and living entity metaphors, the living entity metaphor suits the contemporary organization more precisely. Although there is still a place for the mechanistic metaphor, which can be used if the environment allows, the living entity image of an organization is becoming more and more popular and is establishing its powerful credentials as a form of organization.

Once again, it is important to emphasize the point that an organization should not base its philosophy solely on one "best" metaphorical concept. To better understand the internal and external environments of a particular organization and find the right ways to improve the organization, we need to use a variety of metaphorical concepts that will complement each other.

Below is a brief description of several other metaphorical concepts of organization.

1.2.4 Thinking of an Organization as a Symphony Orchestra

How would an organization look if it were designed based on a symphony orchestra model? In his book *The New Economics*, W. Edwards Deming described the meaning of a system by using a symphony orchestra as a metaphor. He says, "An example of a system, well optimized, is a good orchestra. The players are not there to play solos as prima donnas, each one trying to catch the ear of the listener. They are there to support each other. Individually, they need not be the best players in the country."[8] This comparison can also be applied to an organization, simply because an organization is a system.

A symphony orchestra has many properties that are similar to those of a modern organization. Even while the players compete for a name and a position inside the orchestra, there are strong elements of cooperation and collaboration. Each player has his or her own personal mastery, but the symphony can be performed only by working together as a whole organization. All the players complement each other and have the same interest—to perform the best they can to serve the listener's (customer's) needs. All of them play different instruments, like working on a different process, but the result is one piece of music at any given time. They collaborate and support each other.

The conductor's role is not so much to achieve control as to facilitate the players to help them attain the greatest cooperation and collaboration toward a specific interpretation of the music. The players (workers) have joy in the work they do. A lot of organizations are introducing elements of a symphonic orchestra into manufacturing life. The leader's activity becomes more like that of a conductor, more like the role of a mentor or facilitator. At the same time as growing high-level specialists, organizations introduce a team-based atmosphere in which the results come from all players working together.

Many scientists and practitioners have used a symphony orchestra as a model of an ideal organization. Peter Drucker, who has a significant influence on the way people think, also expressed his interest in a symphony orchestra as a metaphor of an organization. He wrote, "The prototype of the modern organization is the symphony orchestra.... The orchestra performs only because all two hundred fifty musicians have the same score...."[9] For an organization, "having the same score" can be related to having the same shared vision, purpose, values, etc. Of all the characteristics of a symphony orchestra, the most impressive in our opinion is the high level of coordinated effort. The interaction and the immediate feedback between the players are so complicated that they can be easily compared with those of a large corporation. However, the way a symphony orchestra achieves coordination may not be exactly the way a modern organization should act. Symphony orchestras do not access the full range of human potential. During the performance, the musicians are expected to perform only as directed by the conductor. Here it is appropriate to bring in another item related to metaphors—the jazz group. The leader acts as a member of the group, and leadership is often rotated among group members. This provides all the members with the experience of learning to be a leader and a follower—of listening and being listened to. Another peculiarity of a jazz group is that the leader faces the audience (the customers). He and his team constantly receive immediate feedback on how they are performing, so there is a continuous learning process going on. Being constantly in tune with the customer's voice is an important component of a contemporary organization. If, as in a symphony, the attention of the players were mainly directed to the score and the conductor, for an organization this would mean that work would be conducted only according to the policies and procedures and that workers would have to make sure they do what the boss says. There are other peculiarities that differentiate a jazz group metaphor from a that of symphonic orchestra. Both metaphors can be used to form a model of a contemporary organization.

1.2.5 Thinking of an Organization as a Community

What images would be evoked when using the conceptual metaphor of an organization as a community? At the seminar mentioned above, people had almost the same opinion about the mechanistic view of an organization. In contrast, when a dialogue on the metaphor of an organization as a community came up, their reaction was different. Some people talked about the contribution of the organization to the outside world. Others perceived an organization as a community where the element of democracy is largely involved. Still others talked about team spirit and referred to highly initiative opportunities for learning, a higher quality of life, more cooperation than competition, etc. In general, the dialogue generated the greatest interest compared with those for other metaphorical concepts. Why?

In the last decade, organizations have displayed a great interest in such concepts as knowledge creation, self-management, cooperation and collaboration, entrepreneurship and intrapreneurship, and cross-functional teamwork. These and many other concepts related to involvement of people and their concerns about individual growth belong to the characteristics of an organization as a community. This is probably why people are so interested in this timely concept.

Introducing the metaphorical concept of a community certainly does not mean that organizations should ever replace existing "real" communities of interest. An organization will still remain an enterprise with its own purpose. However, this metaphorical concept can help to reshape the organization's structure and make people feel as though they are working in a community.

As Daniel Kemmis nicely describes in his book *Community and the Politics of Place*,[10] local communities were the places where people learned, through personal discipline and practice, the skills of local participation and the meaning of the common good. Organizations can best serve their own interest by becoming "practice fields" for the skills that will lead to more democratic and collaborative behavior than that which has characterized corporate life in the past. Linking community practice with the organization's experience is a way of linking the individual to a larger space of interest. This would help to develop both the business and the society by establishing a larger purpose and shared meaning. An organization as a community is a metaphor that reflects not only the needs of an organization but also the needs of our society.

1.2.6 The Web as a Metaphor of Organizational Architecture

In her book *The Web of Inclusion*, Sally Helgesen used the web as a metaphor to express the architecture of a contemporary organization that fits today's requirements. She wrote, "In architectural terms, the most obvious characteristics of the web are that it builds from the center out, and that this building is a never-ending process. The architect of the web works as the spider does, by ceaselessly spinning new tendrils of connection, while also continually strengthening those that already exist. The architect's tools are not force, not the ability to issue commands, but rather providing access and engaging in constant dialogue."[11]

This metaphorical concept clearly describes an architectural design suitable for a high-performance organization. It also describes the role of the leader as an architect and the tools the leader should use to successfully maintain and continuously develop the organizational "web." In describing the leader's role in architectural terms, Helgesen notes, "... Such an architect recognizes that the periphery and the center are interdependent, parts of a fabric, no seam of which can be rent without tearing the whole. Balance and harmony are essential if the periphery is to hold; if only the center is strong, the edges will quickly fray."[12]

This thought is especially meaningful when we consider the growing tendency of globalization. For example, AMD as a global organization is continuously developing its periphery by building subsidiaries in other countries (see Part III, Managing in a Global Environment). At the same time, AMD enhances its corporate capabilities and maintains the necessary balance and harmony between both parts.

These days, we can clearly observe a gradual tendency toward replacing the old organizational architecture, which was based on the hierarchical concept, with a new architecture that reflects the metaphor of a web. Seeing the organization as architecture is another way of seeing the pattern of the whole. This way of seeing organizations complements the concept of systems thinking and seeing the organization as a complex system, which will be described in Chapter 2.

Conclusion

The metaphors we have just described allow you to see organizations in a different light. You probably have your own metaphors that you use to form your own view of the organization. Which is the best one? There is no best one. Each of them has its good and bad points. An organization is too complex to be related to only one metaphor. As a leader, you can select one major metaphor that most fits your mental model and introduce it to your people to help them understand your way of thinking. This main metaphor can be selected as a basis to create or redesign your organizational model. But because every metaphor has its limitations, and some "bad" metaphors have good elements that you might like to introduce, we would recommend that you use all kinds of metaphors as tools to share your thoughts. Making metaphors a way of expressing thoughts has been proven to be very effective.

1.3 THE LIFETIME OF AN ORGANIZATIONAL SYSTEM

Using a living system as a metaphor of an organization, we should say that its major purpose is survival. To survive it needs to grow, and to grow it needs to adjust itself to the external environment. The form of adjustment also depends on the age of the organization. This is why we can see different types or "species" of organizations that can survive in different environments. Just as we can find a camel in the desert, or a polar bear in arctic regions, we can find different species of organizations that survive and grow in different environments. However, there is at least one major peculiarity that differentiates an organizational system from a living system. A living system independent of how well it adapts itself to the environment has a definite lifetime. Sooner or later it must die. An organization, if well managed, can live forever.

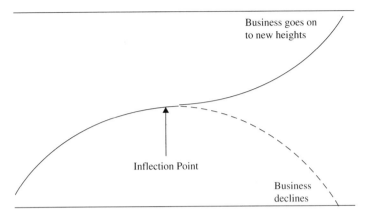

Figure 1-2 The Concept of the Inflection Point
Source: From *Only the Paranoid Survive* by Andrew S. Grove, copyright © 1996 by Andrew S. Grove. Used by permission of Doubleday, a division of Random House, Inc.[13]

This is the right time to introduce the inflection curve concept as described in Andrew S. Grove's book *Only the Paranoid Sur*vive. An inflection point occurs where existing strategy and structure of the organization dissolves. Here an organization has two options. One is to inject new life with the introduction of new products, technologies, structure, and strategy; the other is to let the organization's business gradually decline and finally disappear as an enterprise. In real life, management does not see the inflection point as we see it in Figure 1-2. The art of management is to foresee the time when the peak will occur and act accordingly on time. There are symptoms and signals from the external and internal environments that management can pick up. Management needs to learn how to listen to them, know how to decode them, and be ready to act and take off for new heights. For example, one of AMD's inflection points was reached in 1999. At that time, the corporation was ready to make a significant change in technology, and launched a new microprocessor, AMD Athlon™ microprocessor, which made AMD a sustainable and significant player in the microprocessor field (AMD Athlon™ is a trademark of Advanced Micro Devices, Inc.). It is quite possible that an organization could miss the shared time period when there is a chance to take off for new heights. An organization must always be ready for an "organ transplant" in the existing system. Separate organs (products) may decay and die, but if there is a culture that allows the right replacement of the old products, the organization will continue to live and grow forever.

1.4 GROWTH AND DEVELOPMENT

What is the meaning of development in relation to an organizational system or an individual? Is there any difference between development and growth? What is more important, development or growth? Are they dependent on each other? The aim of this section is to elaborate on these issues.

When we talk about "development" and "growth" we usually use these two terms interchangeably, as though they have the same meaning. Actually these two processes are different and are not dependent on each other. For example, a musician or an engineer can develop without growing.

The importance of understanding the difference between these two terms is more than just a matter of semantics, because taking these two processes as the same thing can negatively influence the way management thinks and acts.

Sometimes organizations take actions toward their development, but in actuality these efforts are directed at organizational growth. For an organization as a system to grow, it means to increase in size or the number of employees. For example, when one organization acquires another organization, it may grow in the number of employees and the size of its revenue. From a career perspective, for an individual to grow usually means the growth of his position in the organizational structure.

In contrast, development is an increase in capability and competence. Development of individuals and organizations is a matter of acquiring new knowledge, and it is more a matter of learning than earning, even though learning and earning are usually interrelated. Becoming a knowledge worker is an example of development. Being given the right to work in a team that is usually involved in serious projects is another example of development.

When AMD came up with its faster microprocessor, AMD Athlon™, that was an example of development. In addition to technological or scientific development, which increases the organization's competencies, development is also reflected in the quality of working life it provides for its employees. This is why, when an organization is included in the list of best companies to work for, it is evidence of organizational development. A well-developed organization can do more with its resources than one that is less developed. There is no shortage of examples where organizations with fewer resources invested in R&D came up with greater results than competitors who spent more on R&D. This is not to say that the amount of resources available is irrelevant. Having the opportunity to invest more money can certainly accelerate the development of an organization.

The point we want to make is that an organization that is better developed can generate greater results from the same amount of money as that spent by a company that is less developed. The same point also applies to individuals. A well-developed individual knows how to spend money more wisely and more effectively.

Growth and development do not conflict with each other. They can reinforce each other. The best evidence that the two processes—growth and development—reinforce each other is when we observe a simultaneous increase in the standards of work and quality of life for the organization and the individuals who work there.

To summarize, we would like to emphasize again that growth and development are two independent phenomena, but at the same time, they can reinforce each other. A lack of resources can limit growth but not development. However, the availability of resources can accelerate development.

1.5 SHAPING THE ORGANIZATION'S ARCHITECTURE

"We shape our buildings: Thereafter they shape us."

Winston Churchill[14]

When we think about continuous improvement, we usually think about changing the manufacturing process, buying new equipment, reorganizing the organization's structure, etc., all of which will improve product quality, reduce cost, and delight the customer. However, another no less important area of improvement is organizational architecture, which influences the relationships of all components in the organization. This section provides another view of the organization and the relationship of its parts.

1.5.1 Thinking Differently about Organizational Architecture

By organizational architecture, we mean the pattern of interconnectedness of all soft and hard systems that exist and function in an organization: its plants and subsidiaries, its information systems, its material distribution systems, its documentation and policy systems, its education and communication systems, and the managerial and leadership philosophy that guides the whole organization. The interconnectedness and interrelation of all systems and subsystems form an architecture in which the whole is always greater than the sum of its parts. These days, a new approach to designing an organizational architecture is required.

If you compared a number of organizations to find out which is the best, you would probably see that all of them have almost the same set of systems. All of them have planning in finance systems, material supply and control, information systems, operations systems, etc. If you could go inside those organizations, you would probably not see a big difference between them. If they are from the same industry, they have similar process flows, equipment, etc. But there is an invisible architecture that makes a difference. It is like the effect of a kaleidoscope: always the same parts, but they form different patterns when you rotate the kaleidoscope and form new interconnections of the parts, and the effect of the whole is always greater than the sum of its parts.

The changes in the internal and external environments strongly require organizations to change their way of thinking about organizational architecture. This is because of the tremendous change in technology, in the customers' perception of quality, in the way people think. Knowledge workers require different relationships, different forms of management and motivation. As Trudy and Peter Johnson-Lenz write, "We are now laying the foundation of a new social architecture made of fiber-optic cable, silicon chips, high-speed switches, display screens, and software through which digital information travels as pulses of light.... 'Reach out and touch someone,' 'the information superhighway,' and 'We bring the world to you' are literally becoming true. Every day millions of people worldwide use the Internet, an early version of the electronic global brain that connects our small planet."[15] As you see from this extract, the architecture of the world is rapidly changing. Should we also think of rapidly changing or shaping our organizational architecture? Many years ago, we started to change the old architecture that was based on linear thinking, pyramidal and hierarchical thinking, and creating boundaryless organizations. We have made some progress in changing the way we work and communicate. Cross-functional teamwork, information technology, and telecommunications have also influenced the way we think about an organization. Nevertheless, there is still a long way to go before we start thinking systemically and globally and learning how to see patterns.

Just as the architecture of buildings is different all over the world, our organizations will differ from each other. Just as the architecture of cities is influenced over time, organizations are also impacted by the requirements of time. However, there are always some major principles that underpin the architecture of a contemporary organization. To succeed and stay in business, organizations must continuously follow the requirements of time and keep their architecture adapted to contemporary requirements.

1.5.2 Three Major Principles that Underpin the Organization's Architecture

There is no one way or one set of principles that can be recommended as a guideline for creating a successful organizational architecture. Every organization has its own needs and specifics, and the architecture should be designed in such a way that it feeds those peculiarities. However, there are always some principles that seem to fit most organizations. In their book *Dynamic Manufacturing*, Robert H. Hayes et al. identified a few major principles that underpin the architecture of a world-class manufacturing organization.[16] Our studies and observations of a number of organizations from different countries allowed us to incorporate those major principles in the architectures of the Manufacturing Services Group (MSG) organizations, because they coincide with the MSG philosophy and objectives in creating extraordinary organizations. Below is a short description of these principles, which in our opinion deserve the reader's attention.

The first principle is that an organization's main purpose is to *increase value* for the ultimate customer. Focusing on process improvements, which finally results in an increase of value for the final customer, means continuously searching for ways to reduce non-value-added activities that will result in a reduction of scrap, rework, additional inspection, testing, etc. This principle reflects Taguchi's philosophy, which requires the continuous reduction of deviations from the customer's target that causes losses to society (see Section

13.2). In other words, continuous improvement efforts should be oriented on value creation through the elimination of all kinds of waste, independent of whether these measures are undertaken by the supplier or by the customer.

The second principle is *strict discipline*. Here the term "discipline" has a special meaning. It is related to the issue of keeping the promises made between people, departments, and other groups. This includes promises made by the supplier to the customer, promises made by one division to another in the same company, etc. This principle may sound trivial, but if we analyze the reasons for all the problems within an organization, we will probably find that a majority of them stem from not keeping promises: the product was not shipped on time because the supplier overestimated his capabilities; the engineering department failed to make the necessary changes in the design on time; etc. The strict discipline factor is concerned with relationships that are based on keeping the promises that are made to each other. The absence of such discipline introduces a lot of unjustified losses and creates unnecessary tension in the system.

The third principle is *simplicity*. It has become an unwritten rule in AMD that every year engineers from every part of AMD get together annually and conduct a special session to reduce the complexity of our processes. We usually cover the walls of the conference room with paper to generate a dialogue on this subject, and we always end up with plenty of ideas posted on these pages. Considering the tendency of processes to increase in complexity, it is easy to imagine how complicated the processes would be if activities to continuously simplify the processes were not in place. Process simplification is not only a topic for engineering or manufacturing departments. It is a topic for every area in the organization where any activities are going on: in planning, finance, supply management, etc. Process simplification is a source of reducing variation in cycle times, quality levels, yields, setup time, etc.—a source of decreasing uncertainty in the manufacturing system.

The above-mentioned principles—value, discipline, and simplicity—should be the foundation of organizational architecture. Just as an architect designing a building needs customer information to ensure that the functionality of the building matches the needs of the customer, so does management need to ensure that the architecture of the organization satisfies the needs of the three principles mentioned above.

References

1 Peter Drucker, "Introduction: Toward the New Organization," in *The Organization of the Future*, Frances Hesselbein, Marshall Goldsmith, Richard Beckhard, editors, The Peter F. Drucker Foundation, Jossey-Bass Inc., San Francisco, CA, 1997, p. 4

2 Gareth Morgan, *Image of Organization*, Sage Publications, Inc., Thousand Oaks, CA, 1997

3 George Lakoff and Mark Johnson, *Metaphors We Live By*, The University of Chicago Press, Chicago, IL, 1980, p. 3

4 Russell L. Ackoff, *Creating the Corporate Future*, John Wiley & Sons, Inc., New York, NY, 1981, p. 6
5 Arie De Geus, *The Living Company*, Nicholas Brealey Publishing Limited, London, UK, 1997, pp. 17–18
6 Ibid., p. 19
7 Ackoff, *Creating the Corporate Future*, p. 35
8 W. Edwards Deming, *The New Economics*, Massachusetts Institute of Technology, Center for Advanced Educational Services, Cambridge, MA, 1994, p. 95
9 Peter Drucker, *Post-Capitalist Society*, HarperCollins Publishers, Inc., New York, NY, 1993, p. 54
10 Daniel Kemmis, *Community and the Politics of Place*, University of Oklahoma Press, Norman, OK, 1990
11 Sally Helgesen, *The Web of Inclusion*, Doubleday, New York, NY, 1995, p. 13
12 Ibid.
13 Andrew S. Grove, *Only the Paranoid Survive*, Doubleday, New York, NY, 1996, p. 32
14 Edited by The Princeton Language Institute, *21st Century Dictionary of Quotations*, The Philip Leif Group, Inc., New York, NY, 1993, p. 29
15 Edited by Kazimierz Gozdz, *Community Building*, New Leaders Press, Sterling & Stone, Inc., San Francisco, CA, 1995, p. 246
16 Robert H. Hayes, Steven C. Wheelwright, and Kim B. Clark, *Dynamic Manufacturing*, The Free Press, New York, NY, 1988, p. 202

Chapter 2

Systems: A General Concept

Systems and systemic thinking has started receiving more and more attention from management. We have begun to recognize that an organization is not just a collection of departments with different functions that all try to do their best. Their interconnections and compatibility, the effect they have on one another, common values, shared vision, and the relationships with suppliers and customers and other stakeholders makes the difference. We started to understand that by doing our best separately, without considering the interaction of the parts, we would not receive the best results for the organization as a whole.

As Russell L. Ackoff wrote, "If each part of a system, considered separately, is made to operate as efficiently as possible, the system as a whole will *not* operate as effectively as possible."[1] Understanding the importance of relationships and the interaction of the parts has become even more important in a global environment, where we need to deal with different cultures and the organization is spread out all over the world.

Our organizational systems are becoming so complex that it is no longer sufficient to only have knowledge of the separate parts. In management, as in other areas of the organization, we can expect that the interconnections and interactions between the organization's components (departments, plants, divisions, subsidiaries, joint ventures, etc.) will often be more important than the separate components themselves. This is why it is paramount for managers to understand the concept of systems and systems thinking.

In this chapter, we will just touch the surface of this subject. The aim here is to introduce you to the principles of systems theory and to refer you to some sources for further study.

2.1 THE MEANING OF SYSTEM

We live and work within social systems such as family, education, organization, etc. We design and produce different physical systems such as the automobile, airplane, computer, television, telephone, and other systems in the form of products. Systems are everywhere around us. But what is paradoxical is that the principles governing the behavior of systems are not widely learned or understood.

The theory of systems and their applications still remain paramount. Before describing the concept of system, we must establish that we are in agreement with its definition. There are many different definitions of "system" in the literature.

Jay W. Forrester defines "system" as "a grouping of parts that operate together for a common purpose."[2] For example, a computer is a system of components that work together to perform mathematical or logical operations. Together, an autopilot and an airplane form a system for flying a specific route.

A system may include people as well as physical elements. For example, the engineer working on the computer and the computer itself form a system for computer-aided design (CAD). Management is another type of system of people for allocating resources and regulating business activities. Your family is still another type of system for living and perhaps raising children together. We could continue with a lengthy list of different types of systems surrounding us (which in most cases we are a part of), but these few examples may give you a feeling for the broad variety of systems with different functions, purposes, structure, and concepts.

In this variety of systems, there are a limited number of requirements that determine any type of a system. Russell Ackoff[3] gives a broader definition of "system" by describing three conditions that must be satisfied for a group of two or more elements to be considered a system. These conditions are:

1. *The behavior of each element has an effect on the behavior of the whole.* For example, in a computer each of its parts—integrated circuits, resistors, capacitors, disk drive, power supply, and so on—have an effect on the computer's performance as a whole. A malfunction in one device may impact the performance of the computer.
2. *The behavior of the elements and their effects on the whole are interdependent.* This implies that the way each element behaves and the way it affects the whole depends on how at least one other element behaves. For example, no component in the computer has an independent effect on the system as a whole. A component may affect the performance of the disk drive, which may affect the performance of the computer as a whole.
3. *The elements of a system are so connected that independent subgroups of them cannot be formed.* Therefore, Ackoff concludes, a system "is a whole

that cannot be divided into independent parts. From this, two of its most important properties derive: every part of a system has properties that it loses when separated from the system, and every system has some properties—its essential ones—that none of its parts do."[4] For example, if the monitor was disconnected from the computer system, it would not continue to function as it did before it was disconnected. If the printer was detached from the computer system, it would not print by itself. On the other hand, the computer system as a whole can receive signals from all over the world, translate and display or print the information into patterns, music, text, and calculations, and many other things that none of its parts can do by themselves.

Everything that a computer can do as a whole system comes from the interaction of its parts-computer monitor, printer, microprocessor, memory, mouse, etc.—not from their actions taken separately. Therefore, as Ackoff says, "...*when a system is taken apart it loses its essential properties*. Because of this—and this is the critical point—*a system is a whole that cannot be understood by analysis*."[5] This critical point has a very important practical implication: We need to learn how to think systemically.

From this, we can conclude that a set of elements (or parts) can be viewed as a system only if this set satisfies the three major conditions mentioned above. In the same way we used a computer system to see how it satisfies the three conditions, we could use any social, living, or other physical system to prove that only if they satisfy these three conditions can they be considered systems.

To get a feeling for how these three conditions work, we would suggest that you take any system you are familiar with, such as a car, airplane, organization, school, plant, any natural or physical system, and you will see that they will fulfill these three requirements because they are systems. By performing this experiment, you will conclude that the essential property of a system, taken as a whole, derives from the interrelations of its parts, not their actions taken separately. For example, all essential properties of a computer system—receiving and sending messages, performing computations, storing information, etc.—come from the interaction of the system's parts and not from the parts themselves. Therefore, we may conclude that a system is a whole that cannot be divided into independent parts.

2.2 THE CONTROL SYSTEM

What is the commonality between an operator, technician, engineer, designer, or manager and the president of the corporation? The operator works at the workplace to fulfill an operation; the designer works on a project to develop a new product; and the president manages people and directs them to achieve goals. All these people perform different functions on different levels, but they have one important commonality: Using words from cybernetics, we can say that all of them are "control devices," whose actions are directed toward achieving specific objectives. Essentially, they organize an object and bring it closer to perfection.

For example, the operator's aim is to keep the process in a state of statistical control and to produce the required amount of parts to the specification. For the president, the aim is to keep the corporation in business, creating jobs for people and making profits and a return on investment for the corporation and its shareholders. We would have no difficulty describing the objectives of control devices such as an engineer, designer, accountant, secretary, etc.

But to succeed in control it is not enough just to know the objective. We also need to know how it can be achieved. We must be able to influence the object under control in such a way that our plans can be fulfilled. For example, the operator needs to know how to influence the manufacturing process when the product is outside the specification limits; the president needs to know how to act when the customer order rates are falling, etc. And most of the time this is more difficult to do than defining the objectives.

This brings us to a consideration of one of the most fundamental concepts of modern cybernetics, namely the *control algorithm*. A control algorithm is a method of achieving the stated goal or objective—a rule for action. The control algorithm for the operator, for example, is to stop (or adjust) the process when a point falls outside the control limits. For some automated operations, the control operation is very simple. The process can be corrected just by pushing a button. Even for a pilot who controls a complicated aircraft, the control algorithm is sometimes very easy. The pilot just pushes one of the multiple buttons on the control panel. In the worst case, the pilot pushes the eject button to leave the aircraft in an emergency situation. But what about the control algorithm of an executive? What button should the executive push when he/she is losing the competitive edge?

So far we have described the commonality of different "control devices" and the differences in the complexity of the control algorithms. Despite those differences, we can consider a general control system independently of the specific peculiarities of controlling a particular object.

Figure 2-1 shows a model of a general control system. Here the interaction of the object with the control device is indicated by two arrows, A and B, which represent the communication channels between the object and the control device. The "object" can be a simple process operation or an overall manufacturing process; it can be a corporation or even an entire country. The control device can be an operator, a manager, a president, or any other device whose aim is to control. The control device receives information about the object via Channel B and then acts on the object via Channel A, and this *controls* it.

We would like to emphasize here the importance of Channel B. This is the feedback from the object under control. Without this element, the control system does not work. How can you control anything if you do not know what the

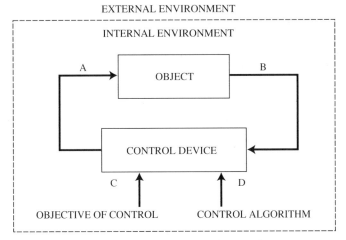

Figure 2-1 A Model of a General Control System

object needs? Can we say that we have constant feedback (Channel B) from all of our objects of control? The reader should not have difficulty in answering this question, having in mind a particular example. Speaking in general, we should say that statistical process control (SPC) is an example of how a process (object) can give feedback to the control device (operator, engineer, supervisor).

Channel A has the same level of importance. You can have constant feedback from the object, but if you do not know how or what to do, you cannot do anything to the object when it asks for it, and again control is impossible. Do we always react to the signals that are coming through Channel B?

The picture of a modern system of control will not be complete if we do not introduce two more elements to the system. To control effectively, we need to know *what* to do with the information received and *how* to use it to control the object. For this purpose, we introduce the following inputs to the control device: objective (aim) of the control process (see Fig. 2-1C) and the algorithm (or method) of control (see Fig. 2-1D). This information (data) has to be fed into the control device beforehand. So, if the control system is to impose the required order on the objective, it must contain two essential elements:

1. The objective of control
2. The control algorithm showing how the objective is to be achieved

In simpler terms, we need to know (1) where we want to go and (2) how to get there. We have described a control system that is varied for any controlled object. The purpose of describing this general system is to propose a way of taking a second look at our control activities from a cybernetic point of view. In addition, if we find that the control system we are involved with is short of just one of the four elements reflected in Figure 2-1, this system will not work.

The control system is influenced by its internal and external environments and is capable of adapting itself to changes in the environments. For example, if the object is a manufacturing process with the objective of control being to have a product output with a quality level of 5,000 parts per million (ppm), and the customer (environment) changes the requirement to 50 ppm, then the system must change its objective of control to survive. By doing this, the need arises to also change the control algorithm to fit the new objectives. In our example, the control algorithm (the method of control) may require the introduction of the six-sigma concept (see Section 14.6).

This control system can be applied to any controlled objects. The system will operate properly only in conjunction with a control program or algorithm (arrow D in Fig. 2-1), which must be incorporated in the control device. This is what enables the control device (operator, engineer, supervisor, etc.) to organize the object and to bring it into the desired condition. Understanding the major principles of control is a very important component of systems thinking.

2.3 THE PLANNING AND CONTROL SYSTEM

Although in organizational practice almost all functions and operations can be designed as systems of action, planning has a unique peculiarity in that it involves establishing the objectives necessary for the whole organization. In addition to this, a manager must plan ahead to have a full understanding of what kind of organizational relationship is needed and what form of control is to be applied. Planning, controlling, feedback, and corrective actions are essential managerial elements that form a system of planning and control (see Fig. 2-2).

As with any system, to be meaningful, the planning/control system should have an aim. In this case, the aim is to perform continuous control by comparing the planned (desirable) effects with the actual outcome. If there is a match between the plan and the actual results, Feedback 1 is used for new planning. As Figure 2-2 shows, any time a cycle is performed, learning occurs and knowledge is created. If a mismatch is observed after actualization of the plan, then Feedback 2 is used to introduce corrections in the implementation and in future planning. When there is no gap between the desired effect and actual results, learning also occurs and knowledge is created for future use. The system should be designed in such a way that it is capable of absorbing the necessary information from the external and internal environments, such as customer requirements, competitor's intents, supplier's capabilities, corporate goals, capacity availability, employee turnover, and other elements of information that make the plan realistic and competitive. It is important to note that in addition to an

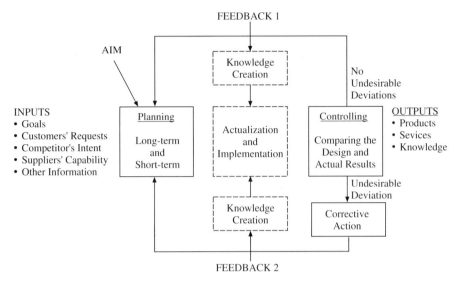

Figure 2-2 The Planning and Control System

output in the form of a product or service, there is always an output in the form of knowledge, which comes from fulfilling the plan and introducing corrections. The main point we want to make here is that planning and control are inseparable parts of the system and that any attempt of control without planning or planning without control guarantees failure. To expect an effective result from our employees, we need to make sure that they know the aim and that they know the results. This should be done continuously through the feedback loops—loops of learning.

As a final thought, while recognizing the important role of feedback, it is also important to plan special time for analyzing the feedback information. This should become a part of the culture while we learn how to plan and execute the plan. We also need to learn to understand and utilize the feedback. In real life, we have time for almost everything, but we usually do not spend enough time reflecting on the feedback gathered and learning to utilize this for future actions.

2.4 ANALYTICAL AND SYSTEMS THINKING

There are two approaches to thinking—analytical and systems thinking. Both of these are important in managerial work. However, managers and engineers are more used to analytical thinking than systems thinking. In this section, we provide a brief description of these two approaches to thinking, which may help improve your thinking process.

28 SYSTEMS: A GENERAL CONCEPT

2.4.1 Analytical Thinking

Before describing the systems thinking process, we want to briefly describe analytical thinking, so you will have the opportunity to compare them.

Usually when we want to understand how an organization, a machine, or any other complicated system works, we start trying to understand how the parts work and then try to see how the whole thing works. We usually use the three-step process of (1) taking the thing apart, (2) trying to understand how the parts work separately, and (3) putting the thing back together to understand how the parts interact to make the whole thing work. This process has become a basic form of inquiry when we want to understand how a company, corporation, or other system works. It is also used in reverse engineering or for other purposes when we want to know how things work. This form of inquiry is called analysis.

2.4.2 Systems Thinking

In contrast, systems thinking requires synthesis, which means putting things together, and is the reverse of the three stages of analytical thinking (see Table 2-1). This is holistic thinking, where we start with (1) identifying a containing whole (system) of which the thing to be explained is a part, then we (2) try to explain the behavior or properties of the containing whole, and only then do we (3) try to explain the behavior or properties of the thing to be explained in terms of its role(s) of function(s) within its containing whole.

It is very important here to recognize that in systems thinking, synthesis (or putting things together) precedes analysis (taking things apart). In an analytical way of thinking, the thing to be explained is treated as a *whole* to be taken

Table 2-1 A Comparison of the Three Stages of Analytical and Systems Thinking

Stage	Analytical Thinking	Systems Thinking (Synthesis)
1	Take apart the thing to be understood.	Identify a containing whole (system) of which the thing to be explained is a part.
2	Try to understand the behavior of the parts taken separately.	Explain the behavior or properties of the containing whole.
3	Try to assemble this understanding into an understanding of the whole.	Explain the behavior or properties of the thing to be explained in terms of its role(s) or function(s) within its containing whole.

Source: Adapted with permission from Russell L. Ackoff, *Ackoff's Best*, © John Wiley and Sons, 1999, p. 8 and 17[6]

apart. In systems thinking, the thing to be explained is treated as a *part* of a containing whole. An example may help illustrate the difference.

Suppose an auditor received a special assignment to investigate the reason(s) for high turnover in an organization. Also suppose that he is free to select the method he uses to get the job done. If he were an analytical thinker, he would proceed by disassembling the whole organization until he reached its elements, moving from organization to department, from department to team, until he reaches the smallest organizational area where the real work gets done. After receiving information from the parts, the auditor would then try to assemble an opinion of the organization as a whole.

A systems thinker who had the same assignment would begin by identifying a system containing the organization, which in this case is a corporation. Then he would study the situation in the whole electronics industry, which is still another level—a suprasystem containing the system in question. All this would allow the investigator to get an image of the whole and its environment. Finally, the systems thinker would assess the situation related to turnover in the corporation in relation to the higher level of the system—the industry.

2.4.3 Summary

This is not to say that either of the systems we just described is any better or worse than the other. These two approaches are not contradictory; they complement each other.

Analytical thinking allows us to focus on *structure*; it reveals how things work. Systems thinking allows us to focus on *function*; it reveals why things operate as they do.

To summarize, analytical thinking yields *knowledge*; systems thinking yields *understanding*. Analysis helps us *describe*; synthesis helps us *explain*. Analysis looks *into* things; synthesis looks *out of* things (see Table 2-2).

Table 2-2   A Comparison of the Attributes of Analytical Thinking and Systems Thinking

Analytical Thinking	Systems Thinking
Focuses on structure (how things work)	Focuses on function (why things operate as they do)
Yields knowledge	Yields understanding
Enables us to describe	Enables us to explain
Looks into things	Looks out of things

Source: Based on Russell L. Ackoff's description of analysis and synthesis in his book, *Ackoff's Best*, © John Wiley and Sons, 1999, p. 18[7]

2.5 DEMING'S VIEW OF A SYSTEM

"A system must have an aim. Without an aim, there is no system."[8]

Dr. W. Edward Deming

When describing his philosophy for organizational transformation, Dr. Deming pays great attention to the understanding and appreciation of systems. He places more emphasis on the human side of systems theory and application, and his view of a system is mainly focused on the process. When he uses the term "system," it is synonymous with the term "process," where the aim is customer satisfaction.

Dr. Deming defines a system as "a network of interdependent components that work together to try to accomplish the aim of the system."[9] This definition is oriented to living systems, such as organizations, where people are the major part of the whole system. There are other systems that do not have an aim. For example, a machine does not have an aim by itself and it can only serve as a means to achieve another's aim. From an organizational point of view, without an aim, there is no system.

Emphasizing the importance of having an aim, Dr. Deming also emphasizes the importance of making the aim clear to everyone in the system. Dr. Deming is not an advocate for mechanistic, bureaucratic thinking. He recognizes that people should have some room for self-expression. He says, "The components [of the system] need not all be clearly defined and documented: People may merely do what needs to be done."[10] This is a very important point, because if we treat people as parts of a machine that have a narrow function and can be easily replaced as the parts wear out, the organizational system will not work. In a machine when there is too much friction in the interaction of parts, some lubrication can help. We need a lot of knowledge and experience to find the "lubrication" for a human system. The more sophisticated and global organizations become, the more complicated the interactions among the elements of the organizational system become and the more knowledge and experience are required from the managers to manage the system. As Dr. Deming writes, "A system must be managed. It will not manage itself. Left to themselves...components become selfish, competitive, independent profit centres, and thus destroy the system."[11]

As the process of globalization accelerates, the speed of learning must also accelerate. While developing subsidiaries in different parts of the world and treating them as separate systems, management should, at the same time, make sure that they perform as components of a larger system—the corporation. Subsidiaries, while having their own purpose and aims, must at the same time work as subsystems that work under one larger corporate aim to serve the ultimate needs of the customer. Deming's systems philosophy recognizes the importance of cooperation between the system's components. He writes, "The secret is cooperation between components toward the aim of the organization. We cannot afford the destructive effect of competition."[12] In real life, we can often observe how two teams or manufacturing plants who are parts of one system compete with each other and, because of this, introduce difficulties in knowledge transfer, information sharing, and other forms of organizational collaboration, which may negatively impact the final results of the whole system.

To make a system work properly, Dr. Deming gives a major role to the management. To illustrate this point, he uses the remark from Mr. H. R. Carabelli of Michigan Bell Telephone Company that "a company could have the best product engineer, the best manufacturing engineer, the best man in the country in marketing, yet if these men do not work together as a system, the company could be swallowed whole by the competition with people far less qualified, but with good management."[13] He concludes that, "If the various components of an organization are all optimized (each for individual profit, each a prima donna), the organization will not be. If the whole is optimized, the components will not be."[14] This is a very important remark because in an organization where systems thinking is not a part of its culture, there is an assumption that if every team leader and every manager will optimize his part of the system, the whole system will be optimized. This wrong assumption creates many problems, and to fix these problems, the first step should be changing the wrong assumption and starting to think systemically.

Another responsibility of a manager who leads a system is managing innovation. As an example, Dr. Deming uses the vacuum tube, which went through a number of innovations and led to integrated circuits.[15] Dr. Deming says, "A system includes the future.... Preparation for the future includes lifelong learning for employees. It includes constant scanning of the environment (technical, social, economic) to perceive need for innovation, new product, new service, or innovation of method. A company can to some extent govern its own future."[16] As you can see, Dr. Deming's perception of a system is very broad. It includes all aspects of the organization's life. Learning how to manage a system is actually learning how to use a holistic approach to managing an enterprise. In conclusion, below are some remarks that came from Deming's philosophy on systems theory and application.

Some Concluding Remarks on Deming's Philosophy

1. A system must create value (results).
2. A system must have an aim. Without an aim, there is no system.
3. It is the obligation of the leaders to sponsor, energize, and direct the efforts of all components toward the aim of the system.
4. A system must be managed. It will not manage itself.
5. Managing a system requires imagination and innovation, as well as knowledge of the interrelationships between all its components.
6. A system cannot understand itself. It needs guidance from outside.
7. A system must be able to continuously scan its environment.

8. The secret of a system's success is in the cooperation between its components. Competition between the system's elements is destructive. Collaboration is a must.

Introducing systems that create value, set challenging aims, and mobilize people to achieve their goals, reacting to the external environment, and using a systemic approach to motivate creativity and innovation are some of the things Dr. Deming advises managers to do if their aim is success through organizational transformation.

References

1 Russell L. Ackoff, *Ackoff's Best*, John Wiley & Sons, Inc., New York, NY, 1999, p. 18
2 Jay W. Forrester, *Principles of Systems*, Productivity Press, Portland, OR, 1968, p. 1–1
3 Ackoff, *Ackoff's Best*, pp. 15–16
4 Ibid., p. 15
5 Russell L. Ackoff, *Creating the Corporate Future*, John Wiley & Sons, Inc., New York, NY, 1981, p. 16
6 Ackoff, *Ackoff's Best*, pp. 8, 17
7 Ibid., p. 18
8 W. Edwards Deming, *The New Economics*, Massachusetts Institute of Technology, Center for Advanced Educational Services, Cambridge, MA, 1994, p. 50
9 Ibid.
10 Ibid.
11 Ibid.
12 Ibid.
13 Ibid., pp. 71–72
14 Ibid., p. 72
15 Ibid., p. 54
16 Ibid.

Chapter 3

The Total Continuous Process of Improvement and Innovation (TCPI^2) Macro System

Advanced Micro Devices, like many other organizations around the world, benefited from the knowledge and management philosophy of Dr. W. Edwards Deming. The philosophy reflected in his 14 points for management[1] and later in his System of Profound Knowledge[2] has become a foundation for our Total Continuous Process Improvement and Innovation Macro System (TCPI^2-MS).*

The purpose of this chapter is to describe our macro system, which was gradually developed and implemented in the Manufacturing Services Group (MSG) and has become the main vehicle for our organizational transformation.

3.1 THE EVOLUTION OF THE MACRO SYSTEM

Influenced by Dr. Deming's seminars, in 1985 we developed our first system designed to improve the quality of our process (see Fig. 3-1). The implementation of this system allowed us to significantly reduce the variation of our processes and reduce manufacturing cost.

Although encouraged by the results of applying statistical principles, we quickly recognized that this was not the panacea for all aspects of improvement. Therefore, we decided to bring on board new tools and principle, such as Total Productive Maintenance (TPM), Error-Free Performance (EFP), and other techniques for process improvement. Later, we felt the need for further employee involvement through teamwork and other forms of empowerment.

---

*To simplify things, in further text we will use the term "macro system" to mean the Total Continuous Process Improvement and Innovation (TCPI^2) Macro System.

34   THE TOTAL CONTINUOUS PROCESS OF IMPROVEMENT AND INNOVATION (TCPI²)

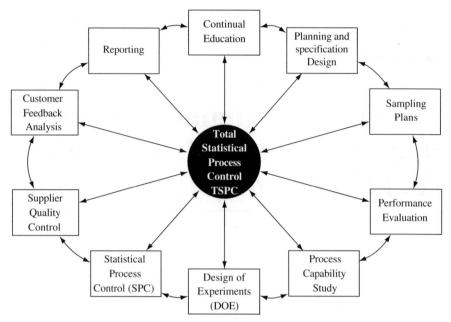

Figure 3-1   The Total Statistical Process Control (TSPC) System

At that time, we started to look for the best practices around the world and started to apply the concept of benchmarking. Gradually we came up with a broader system of process improvement (see Fig. 3-2).

After bringing the manufacturing processes into a state of statistical control, we initiated process improvement activities by applying process capability studies and design of experiments. This allowed us to achieve significant results. However, the trend of improvement gradually declined, and we recognized that more innovative actions were needed to make breakthrough improvements. This is why TCPI became TCPI², Total Continuous Process Improvement and Innovation.

The introduction of a broad variety of methods and techniques generated the requirement for a body of knowledge consisting of a large number of "hard" and "soft" disciplines. We were looking for a set of core disciplines to form a system of profound knowledge that would satisfy the need for a contemporary high-technology global organization. This is when we became interested in Dr. Deming's system of profound knowledge, which he described in his book *The New Economics*. We will elaborate on this system in Section 3.3. The philosophy of profound knowledge was incorporated into our new TCPI² macro system (see Fig. 3-3).

A DESCRIPTION OF THE MACRO SYSTEM 35

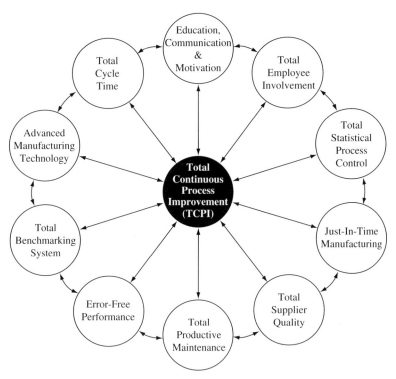

Figure 3-2 The System of Total Continuous Process Improvement (TCPI)

3.2 A DESCRIPTION OF THE MACRO SYSTEM

The purpose of the macro system was to introduce a systemic approach to improvement and innovation in all aspects of organizational life. Such a system was needed to catch up with the continuously changing external environment and, in particular, with customers' demands. The macro system was designed in such a way as to create and utilize the best practices and to continuously upgrade and deploy the knowledge required for organizational transformation.

As in any other system, all elements of the macro system are interconnected and complement each other. To achieve the goal—becoming an extraordinary form a system of profound knowledge that would satisfy the need for a organization—every element of the macro system was developed into a separate subsystem with its own elements and aim (see Table 3-1).

Table 3-1 shows the elements of every subsystem (see Fig. 3-3), which are further developed into lower level systems. To give you a feeling for the complexity of the macro system, Figure 3-4 shows the globalization subsystem. Figuratively speaking, the macro system is a network of interconnected elements

THE TOTAL CONTINUOUS PROCESS OF IMPROVEMENT AND INNOVATION (TCPI²)

Figure 3-3 The Total Continuous Process Improvement and Innovation Macro System (TCPI²-MS)

arranged in such a way that every subsystem on different levels can have an aim and an owner who will manage the system. As Dr. Deming constantly reminds us, "A system must have an aim.... A system must be managed."[3] The development of a system that embraces all components required for organizational transformation allows us not only to see the whole system but also not to lose sight of all its elements and their interactions. For example, when developing an educational program at AMD, we made sure that the programs for education covered the requirements of all elements of the macro system.

In the center of Figure 3-3 is the system of profound knowledge that, as the name suggests, was designed to provide the knowledge necessary to fulfill the main organizational goal—becoming an extraordinary organization. Considering that most of the techniques reflected in the periphery of Figure 3-3 are familiar concepts that are embedded in most organizations' cultures, here we will pay more attention to describing the central part of Figure 3-3—the system of profound knowledge.

Table 3-1 A Description of the Subsystems of the TCPI² Macrosystem

Subsystem	Description
Organizational Culture	• Purpose, Vision, Mission, and Goals • Shared Core Values and Principles • Behavior Standards and Ethics • Rituals, Symbols, and Stories
Total Employee Involvement	• Cross-Functional Teamwork • Dialogue • Process Control • Creativity and Innovation
Advanced Statistical Methodology	• Statistical Process Control • Design of Experiments • Precontrol • Capability Studies
Planning and Logistics	• Forecasting • Scheduling • Cycle Time Control • Production Control
Customer and Supplier Relations	• Quality Function Deployment • Supplier Quality • Customer Services • Customer and Supplier Education
Education and Training	• Conferences • Seminars • Symposia • On-the-Job Training
Globalization Management	• Global Thinking • Global Strategy • Multicultural Teamwork • Global Culture
Total Benchmarking System	• Current Status Assessment • Internal Benchmarking • External Benchmarking • Best Practice Identification
Advanced Manufacturing Technology	• Design for Manufacturing • Automated Manufacturing • World-Class Manufacturing • Computer-Aided Engineering and Design
Information System	• Electronic Documentation System • Computer-Based Training • Computer Business System • Computer-Based Modeling

38 THE TOTAL CONTINUOUS PROCESS OF IMPROVEMENT AND INNOVATION (TCPI²)

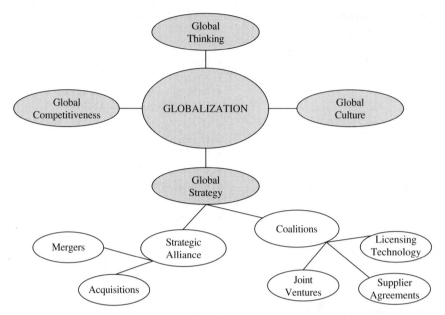

Figure 3-4 A Simplified Example of the Globalizations of a Subsystem and Its Further Development

3.3 THE SYSTEM OF PROFOUND KNOWLEDGE IN ACTION

"Without theory, experience has no meaning."

W. Edward Deming[4]

The further development and implementation of the macro system required a number of theoretical disciplines that would satisfy our needs in knowledge. In other words, we were looking for a body of knowledge that would serve us by filling the gap between what we knew and what we needed to know to accomplish our goal of transformation.

This need took us to Dr. Deming's system of profound knowledge. He asserts that it is this base of knowledge, or the lack of it, that is the primary determinant of organizational success. To apply Dr. Deming's concept of profound knowledge in our environment, we modified it slightly by adding to his four disciplines,[5] appreciation for a system, knowledge about variation, theory of knowledge, and psychology, one more discipline—globalization. We developed the whole system according to the specific needs of the electronics industry.

As you can see from the model of the macro system (Fig. 3-3), the system of profound knowledge in action is at the heart of the TCPI² macro system. Here, knowledge occurs because of the relationship of all elements of the macro system. The theory contained in the center is used as input to all peripheral subsystems. The output from the subsystems is not only value that is materialized in the product but also value in the form of the experience and knowledge that goes back to the center and enhances the body of profound knowledge. This cycle is continuous, and because of this, the $TCPI^2$ macro system is also a system of knowledge creation (see Fig. 3-5).

3.3.1 What is Profound Knowledge?

Profound knowledge may be interpreted as a clear perception of the truth, which comes from a deep understanding. Knowledge is considered to be profound if it is based on theory. This is in alignment with Dr. Deming's philosophy, which he expressed at different times by saying, "Knowledge is built on theory"[6] and "Without theory, experience has no meaning. Without theory, one has no questions to ask."[7]

As you can see from these short statements, Dr. Deming indicates a strong connection of knowledge and experience with theory. However, as you have probably seen in practice, some managers do not pay a lot of attention to the theoretical part of their knowledge or experience. Somehow we think that theory is not for practitioners. In the process of implementing our macro system, we observed a hunger for theory. We saw that it is impossible to receive deep knowledge without the appreciation of theory. Good questions came from the application of theory. For example, during the introduction of statistical methodology, we did not really understand the losses arising from tampering with a process until we learned the theory of variation.

Looking for the right name for a system that would embrace the major disciplines required for the implementation of our macro system, we decided to use Dr. Deming's term "profound knowledge," not only because our system

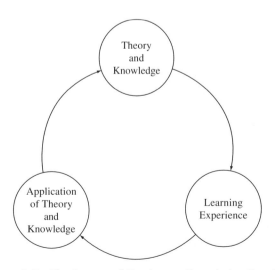

Figure 3-5 The Process of Continuous Knowledge Creation

was based on his philosophy but also because "profound knowledge" suited what we were looking for. Dr. Deming probably coined the term because it contains the essence of what contemporary organizations need to fill the gap of knowledge.

Profound knowledge means penetrating knowledge that is not trivial. It is knowledge that allows us to organize what we know about our process and improvement needs. It penetrates in the sense that it allows us to develop a process that can be understood and communicated to all levels of the organization. It allows us to develop a theory with which to measure and predict behavior, and this is what management is all about. As Dr. Deming said, "Management is prediction. The theory of knowledge helps us to understand that management in any form is prediction."[8] In these turbulent times, management has lost most of its belief in prediction. We predict one outcome at the end of the quarter, real life reveals another outcome, and we do not know the reason why. Even though prediction is difficult these days, we sometimes do not recognize that planning and prediction is mainly what management does. When hiring someone, we predict that he/she will make a difference. When we expand our capacity, we predict that there will be a growth in demand. When we acquire another organization, we predict that the money we pay for buying that organization will bring new "brains" into our organization. Prediction is everything, and this requires profound knowledge.

Profound knowledge is the most essential and necessary condition for continuous improvement and innovation. When opportunities for improvement rose to the surface and there was little external competition, we did not feel a very strong need for profound knowledge. Now that the lifetime of a product has become shorter and shorter and breakthroughs are necessary to stay in business, acquiring profound knowledge has become paramount. Those in the manufacturing industry usually feel more comfortable with concrete knowledge. Profound knowledge is a mix of "hard" and "soft" knowledge. It is often abstract, different, and enigmatic. This means that it is not often easily acquired. Concrete knowledge can be observed and recorded. Profound knowledge often requires challenging our assumptions and changing our mental models. This is because some areas of profound knowledge are quite different and contradict the way we usually think. Profound knowledge often requires us to learn how to unlearn, and this is not an easy thing to do. It requires bringing our cultural values and beliefs to the surface and changing our behavior. Once in awhile, it requires us to ask ourselves, "Is this right?"

As final thought, in recent years we have heard from management that knowledge is a core capability—the main capital of an organization. This is true because we live in a knowledge era. However, we need to emphasize here that it is very important to connect profound knowledge with organizational activities and with actions. As Zeleny (1989) said, "Separation of knowing from doing (knowledge from action) in the sense of some know and others act, like the separation of managers (coordinators) from the doers (workers) is a gaping and self inflicted wound of modern management."[9]

This is why we put profound knowledge at the center of our macro system (see Fig. 3-3). Only by connecting profound knowledge with the subsystems could we create a macro system in which theory could be materialized and enhanced and theory and action work together to achieve our aim for the system.

3.3.2 Deming's Philosophy on Profound Knowledge

In his book *The New Economics*, Dr. Deming describes his system of profound knowledge, which was developed gradually over many years. His famous 14 Points (see Insert 3-1) and Seven Deadly Diseases (see Insert 3-2) also form a system that interconnects with the system of profound knowledge. And, finally, his Plan-Do-Check-Act (PDCA) cycle also has a strong connection to his knowledge philosophy because knowledge occurs only when you act. This is why we put all these elements together into a model (see Fig. 3-6) that more fully reflects Dr. Deming's philosophy on knowledge and continuous improvement.

Insert 3-1

Dr. Deming's Fourteen Points

1. Create constancy of purpose toward improvement of product and service with a plan to become competitive—and to stay in business.
2. Adopt a new philosophy. We are in a new economic age. We can no longer live with commonly accepted levels of delays, mistakes, defective materials and defective workmanship.
3. Cease dependence on mass inspection. Require instead, statistical evidence that quality is built-in to eliminate need for inspection on a mass basis.
4. End the practice of awarding business on the basis of price tag alone, instead, depend on meaningful measures of quality along with price.
5. Find problems. It is management's job to work continually on the system.
6. Institute modern methods of training on the job.
7. Institute modern methods of supervision of production workers. The responsibility of foremen must be changed from sheer numbers to quality. Improvement of quality will automatically improve productivity.
8. Drive out fear so that everyone may work effectively for the company.
9. Break down barriers between departments. People in research, design, sales and production must work as a team to foresee problems of production that may be encountered with various materials and specifications.

(continued)

> **Insert 3.1** (*continued*)
>
> 10. Eliminate numerical goals, posters and slogans for the workforce asking for new levels of productivity without providing new methods.
> 11. Eliminate work standards that prescribe numerical quotas.
> 12. Remove barriers that stand between the hourly worker and his right to pride of workmanship.
> 13. Institute a vigorous program of education and retraining.
> 14. Create a structure in top management that will push every day on the above 13 points.
>
> *Source*: Reprinted with permission from *Quality, Productivity and Competitive Position* by W. Edwards Deming, © The MIT Press, 1982, pp. 16–17.[10]

> **Insert 3-2**
>
> **Enumeration of Dr. Deming's Seven Deadly Diseases**
>
> 1. *Lack of constancy of purpose.* A company that is without constancy of purpose has no long-range plans for staying in business. Management is insecure, and so are employees.
> 2. *Emphasis on short-term profits.* Looking to increase the quarterly dividend undermines quality and productivity.
> 3. *Evaluation of performance, merit rating, or annual review of performance.* The effects of these are devastating—teamwork is destroyed, rivalry is nurtured. Performance ratings build fear, and leave people bitter, despondent, and beaten. They also encourage mobility of management.
> 4. *Mobility of management.* Job-hopping managers never understand the companies that they work for and are never there long enough to follow through on long-term changes that are necessary for quality and productivity.
> 5. *Running a company on visible figures alone.* The most important figures are unknown and unknowable—the multiplier effect of a happy customer, for example.
>
> Diseases 6 and 7 are pertinent only to the United States:
>
> 6. *Excessive medical costs.*
> 7. *Excessive costs of warranty, fuelled by lawyers who work on contingency fees.*
>
> *Source*: Reprinted from *The Deming Management Method* by Mary Walton, Mercury Books, 1989, p. 36.[11]

In one of his presentations, Dr. Deming said, "Hard work and best efforts, put forth without guidance of profound knowledge, may be well at the root of our ruination. There is no substitute for knowledge."[12] This is just one of many statements that Dr. Deming had during his lectures and in his publications, which suggest that he was putting knowledge ahead of everything else when it came to continuous improvement. By the term, "a system of profound knowledge," Dr. Deming meant "a map of theory by which to understand the organizations that we work in."[13] He also called it a lens that allows us to get an outside view of the organizations that we work in.

As the center of Figure 3-6 suggests, the system of profound knowledge consists of four parts, which may be viewed as subsystems that interconnect and complement each other. These subsystems are: (1) appreciation for a system, (2) knowledge about variation, (3) theory of knowledge, and (4) psychology. We placed these four parts in the center of the model because, according to Dr. Deming, they constitute the core of his management philosophy. Even the theories of management outlined in his 14 points cannot be applied unless managers receive knowledge in at least these four disciplines. Emphasizing the importance of this system, Dr. Deming wrote, "The 14 points for management (*Out of the Crisis*, Ch. 2)...follow naturally as application of this outside knowledge, for transformation from the present style of Western management to one of optimization."[14] Profound knowledge is the basis for organizational success when applying the 14 points. Behind the philosophy of profound knowledge, there are the many years of Dr. Deming's work and thoughts before publication in his book, *The New Economics* in 1994. Following

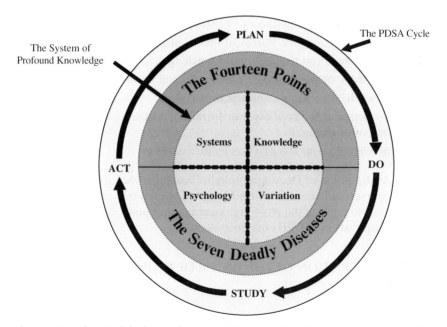

Figure 3-6 The Model of Knowledge Creation Based on Dr. Deming's Philosophy

is a synopsis of Dr. Deming's thoughts on this subject that we adapted from John C. Anderson, Kevin J. Dooley, and Susan D.A. Misterek's article, "The Role of Profound Knowledge in the Continual Improvement of Quality," published in *Human Systems Management* in 1991. This synopsis can help the reader better understand the magnitude of the body of the system of profound knowledge and direct him or her to a number of particular disciplines that need to be studied to become a profound knowledge manager.

A Synopsis of Dr. Deming's Thoughts on the Subject of Profound Knowledge[15]

1. Knowledge about the statistical concepts of variation
2. Knowledge of the losses resulting from tampering with a stable process and missed opportunities for improvement of an unstable process
3. Knowledge of procedures aimed at minimum economic loss from these mistakes (statistical process control)
4. Knowledge about interaction of forces (systems theory)
5. Knowledge about losses caused by demanding performance that lies beyond the capability of the system
6. Knowledge about loss functions and problem prioritization (Taguchi loss function and the pareto principle)
7. Knowledge about the instability and loss that result from successive application of random forces (butterfly effect—chaos theory).
8. Knowledge about the losses from competition for share of market (win-win versus win-lose)
9. Knowledge about the theory of extreme values
10. Knowledge about the statistical theory of failure
11. Knowledge about the theory of knowledge
12. Knowledge of psychology and intrinsic and extrinsic motivation
13. Knowledge of learning and teaching styles
14. Knowledge of the need for transformation to the new philosophy (management of change)
15. Knowledge about the psychology of change

For AMD, this list of Deming's thoughts became a source to create different types of courses and seminars for all levels of employees, and the philosophy described above became the basis in designing our Total Continuous Improvement and Innovation (TCPI^2) macro system.

3.4 THE DESCRIPTION OF THE SYSTEM OF PROFOUND KNOWLEDGE IN ACTION

Figure 3-7 is a model of MSG's system of profound knowledge, which was developed to serve our specific needs. The system consists of five major subsystems that interact with each other. The aim of this system is to develop and maintain the necessary level of theory and core knowledge for organizational transformation. The subsystem in the center of this system is the theory and practice of organizational psychology.

The main purpose of this knowledge is to influence and continuously improve the way the organization operates. It is mainly related to organizational culture and its explicit and implicit values and shared beliefs. This knowledge is especially important for organizations that operate in a global or international environment.

When an organization consists of a variety of people from different national backgrounds, management cannot assume that all of them will share "common" values and relate to "common" norms. Furthermore, in an operating environment in which managers are separated by distance and time, shared management understanding is often a much more powerful tool than formal structure and systems in coordinating diverse activities. This system works as a "pacemaker" to provide a rhythm of life for the whole system of profound knowledge.

The remaining four systems (see Fig. 3-7) contain all the elements needed to deal with variation, creating knowledge, building skills for continuous improvement and innovation, developing a global environment, and applying a systems approach to organizational transformation. As in any system, each of these parts has a separate function, and together they build, deploy, and maintain the necessary profound knowledge and influence the development of skills and knowledge for many other disciplines.

In this section we only provide a short overview of the system of profound knowledge in action because all the following chapters will be dedicated to the same subject. We will start here with an explanation of the meaning of profound knowledge and then describe every element of the system separately.

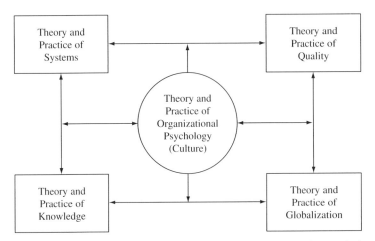

Figure 3-7 The Architecture of the MSG System of Profound Knowledge

This will prepare you for a better understanding and appreciation of the structure and context of the whole book.

3.4.1 The Theory and Application of Systems

In Chapter 2 you became familiar with some principles of systems. In this section, we provide some extra descriptions of the systems approach based on Dr. Deming's theory, which will complement the material described in Chapter 2. This will help you better understand why the theory and application of systems is a major element of profound knowledge.

According to Dr. Deming's definition, "A system is a series of functions or activities (subprocesses, stages—hereafter components) within an organization that work together for the aim of the organization."[16] In this definition and in his other publications, Dr. Deming uses the term "system" interchangeably with the term "process." The definition also suggests that Dr. Deming is making a strong connection between the system and the organization. He is placing the system "within the organization." The system works together with other systems "for the aim of the organization."

From an organizational point of view, Dr. Deming's definition deserves special attention because it suggests that an organization is a suprasystem consisting of many subsystems that work together for the aim of the whole organization. Dr. Deming gives the term "system" a very broad meaning. In his book *Out of the Crisis*, he answers the question, "What is the system?" by saying, "To people in management, the system consists of

Management, style of
Employees—management and everybody
The people in the country
   Their work experience
   Their education
   The unemployed
Government
   Taxes
   Reports
   Tariffs
   Impediments to trade and industry
   Requirements to fill positions by quota, not by competence
   Quotas for import and export
Foreign governments
   Quotas for import and export
   Manipulation of currency

Customers
   Shareholders
   Bank
   Environmental constraints"[17]

The description above shows the broad applicability of the systems approach. It also shows that the internal and external environment of an organization is a net of interconnected systems that work together as a whole. Dr. Deming considers the whole production process of an organization to be a complex system. In his famous 14 points, point five states, "Find problems. It is management's job to work continually on the system (design, incoming materials, composition of material, maintenance, improvement of machine, training, supervision, retraining)."[18] We emphasize Dr. Deming's broad application of the term "system" because this is not just a matter of terminology. By applying this term, Dr. Deming is drawing our attention to the fact that design, incoming material, maintenance, training, supervision, etc., are interconnected and interdependent parts of the manufacturing system. This means that by fixing one of these parts without considering it as a component of a larger system is like trying to fix a person's heart without considering the arteries that lead up to it. Sometimes we do not pay enough attention to the concept of seeing the whole object and to the importance of acting from a systems point of view. If a problem occurs (in the organization), and we fix it as if we were dealing with our automobile, the same problem often reoccurs. For example, often the situation occurs in which we hire a new accountant only to find that six months later we need to look for another accountant. This problem may repeat a number of times, and we do not know why. If the management would practice analyzing this situation, they would probably find that they were fixing the wrong part of the system.

Dr. Deming successfully convinced management that statistical principles and design of experiments are good tools to analyze the manufacturing process. Organizations benefited greatly by reducing the variation and optimizing the process. However, the signal that Dr. Deming was continually sending to management indicated that we needed to use a systemic approach in management. Somehow this did not get through. The 14 points we mentioned earlier are not just a list of recommendations. These points can also be considered to be a system of interconnected thoughts that give the most if applied as a whole. Today, with organizations becoming more global and competition getting tougher, an appreciation for systems has become even more important.

3.4.2 The Theory of Knowledge

Dr. Deming stressed that knowledge comes from theory. He wrote, "Without theory experience has no meaning."[19] In our working lives, we continuously accumulate experience. By supporting this experience with theory, we receive knowledge. "Theory leads to prediction," said Deming, "Without prediction, experience and examples teach nothing."[20] You could say that these days prediction is almost impossible. This is true, but prediction still remains a form of learning. It is much better to predict than to say that you don't know what your next step will be. You predict and then measure the deviation of the outcome from the prediction. You correct the error, and you learn. This is the only way to progress.

Argyris and Schön (1978) described a basic learning cycle to produce knowledge. It starts by moving from the discovery of problems, to the invention of solutions, onto the production of solutions in action, and to reflection on the impact of these actions, and then back to discovery (see Fig. 3-8). Chris Argyris calls this process "actionable knowledge." He wrote, "The word *action* conjures up images of individuals doing, executing, and implementing."[21]

As you can see, actionable knowledge comes not from the classroom but from the implementation of ideas and from reflecting on the results. In real life, we are so busy with inventing and producing that there is no time to reflect, but new discoveries only come from reflection (see Fig. 3-8). This is why the theory of knowledge is an important factor for success. We discuss this subject more in Part V.

3.4.3 The Theory and Practice of Globalization

In his book, *The Work of Nations*, Robert Reich, the U.S. Secretary of Labor in the Clinton administration, wrote that "We are living through a transformation that will rearrange the politics and economics of the coming century. There

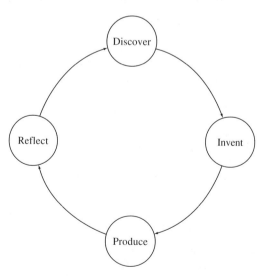

Figure 3-8 The Learning Cycle

Source: Reprinted with permission from *Organizational Learning: A Theory of Action Perspective* by Chris Argyris and Donald Schön, Pearson Education Inc., 1978.[22]

will be no *national* products or technologies, no *national* corporations, no *national* industries. There will no longer be national economies..."[23] These predictions are rapidly becoming a reality. The process of globalization is spreading, and transformation is on its way. This requires people who will possess knowledge about globalization, those with a global mind-set.

What do we mean by "global mind-set?" How does it differ from a traditional mind-set? As shown in Table 3-2, the global mind-set requires additional perspectives. It requires a lot of new learning, and a lot of things will need to be unlearned, which is sometimes even more difficult. Organizations will need to reshape their culture to motivate behavior based on the global mind-set. But how large is the contingent of people who need to have a global mind-set?

As George S. Yip wrote in his book *Total Global Strategy*, "Many managers are asking if they are in a global industry and whether their business should have a global strategy. The better questions to ask are: How global is their industry and how global should their business strategy be?"[25] This is to say that virtually every industry has (or will have) some aspects of globalization. The extent of globalization depends on the amount of intercountry connections. Today most of the organizations have some interconnection with other countries, and these interconnections will continue to grow. This means that a contemporary manager must possess knowledge about the peculiarities of managing in a global organization and knowledge about multinational culture, global strategies, etc. This is why we included the theory and practice of globalization as a major discipline of the system of profound knowledge. (You can find more on this subject in Part III.)

3.4.4 The Theory and Knowledge of Variation

The purpose of this section is to provide a general introduction to the concept of variation and familiarize the reader with Dr. Deming's view on this subject. This is to provide a functional explanation of why this particular discipline should be one of the elements of profound knowledge. You will learn more about it in Part IV of this book, where we describe various applications of the theory and knowledge of variation.

Table 3-2 Comparison of Traditional and Global Mind-sets

	Traditional Mind-set	Global Mind-set
Strategy/Structure	Specialize Prioritize	Drive for broader picture Balance contradictions
Corporate Culture	Manage job Control results	Engage process Flow with change
People	Manage self Learn domestically	Value diversity Learn globally

Source: Reprinted with permission from Stephen H. Rhinesmith, *A Manager's Guide to Globalization*, McGraw-Hill Companies, 1996, p. 28.[24]

As we all know well, nothing repeats itself precisely. Everything varies, that is, the quarterly results of an organization, stock market prices, the hourly production of a manufacturing process, the dimensions of the parts produced on a well-tuned machine, etc. It would not be difficult to continue listing examples of variation indefinitely because our lives are filled with variation. This natural "fluctuation" is a result of the interplay of numerous small variables and is called random or chance fluctuation, which is not traceable to any specific cause. Dr. Shewhart referred to these as "nonassignable causes." These small random—nonassignable—variations affect any manufacturing process and every manufactured product. It affects anything human (or within nature). To illustrate that nothing can be produced exactly the same, in his book *Economic Control of Quality Manufactured Product*,[26] Dr. Shewhart uses an example of writing a letter *a* and then trying to rapidly write the same letter several times in succession, while making each letter the same as the original. Try it, and you will see that you can achieve some consistency (similarity), but when you try to be perfect, you will introduce even more variation. The same is true of any manufacturing process.

In semiconductor manufacturing, for example, the output will vary slightly from device to device, from lot to lot, because of thousands of variables that we may or may not be aware of. This includes variations in the type of equipment, in electrical power, raw materials, the weather, the time of day, machine settings, measurement equipment, equipment maintenance, and so on. These are random variations in the manufacturing system. It is difficult (if not impossible) and uneconomic to trace the random variation to a particular source of variation. The manager has only two options here: (1) to live with this amount of natural variation, or (2) to invest money and efforts into improving or changing the process.

So far, we have described briefly the nature of random (or natural) variation. But there is another type of variation that occurs when the process starts behaving unnaturally and introduces extra variation that can be traced to a particular class of causes and eliminated. This can be done by using special statistical tools.

If the theory of variation and the methodology of separating assignable causes from nonassignable causes is unknown, we may react to issues and spend our time fixing them, but they are not traceable to the source of the variation. In other words, we must learn to recognize the two types of causes we mentioned above, and only then can we take the necessary corrective actions.

What we learned is that in any process there exist two types of variation. One type is the random (or natural) variation that we need to accept as a natural phenomenon or work on changing the system. The other type is the unnatural variation that must be recognized as soon as it appears and eliminated continuously. Not knowing how to recognize these two types of variation can be very costly.

Dr. Shewhart, who worked at Bell Laboratories, is the person who, in May 1924, described a new device that he thought might help control manufacturing quality. Shewhart's boss at that time, George D. Edwards, recalled, "Dr. Shewhart prepared a little memorandum only about a page in length. About a third of that page was given over to a simple diagram that we would all recognize today as a schematic control chart. That diagram, and the short text which preceded and followed it, set forth all of the essential principles and considerations which are involved in what we know today as process quality control."[27] How valued is Shewhart's concept today? As you will see in Part IV of this book, organizations that work in a low parts per million (ppm) environment use the Shewhart control chart and other techniques that are more appropriate for a rare manufacturing defect. Dr. Shewhart's principles of controlling variation are the foundation of the theory of variation. In our opinion, the literature that is referenced in Part IV and this section, plus other books on the theory of variation, must be part of the knowledge manager's bookcase. Knowing the theory of variation and having knowledge about statistical principles can help you not only control the processes but also improve their capability. Figure 3-9, which is reprinted from Western Electric's book, describes the basis of a process capability study. As this figure shows, the first thing that is needed is to recognize and divide the total variation of the process into "variations that we must live with" and "variations that we don't have to live with." Only then can you determine what you, as a manger, can do to improve the capabilities of your process.

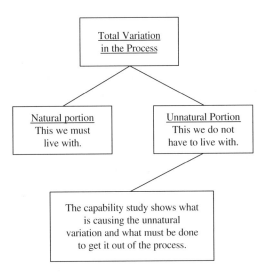

Figure 3-9 Theoretical Basis of a Process Capability Study

Source: Statistical Quality Control Handbook, AT&T, 1956, p. 35. © AT&T Corp. All rights reserved. Reprinted with permission.[28]

In our book, *Process Improvement in the Electronics Industry*, we shared our experience of a process capability study and also the application of other statistical techniques that we used to achieve six-sigma quality. The Western Electric book we just mentioned is an important reference, especially when introducing statistical principles.

Understanding the theory and knowledge of variation will help you see a number of things differently. It will help you to unlearn a lot of things and replace them with things that will make your life easier.

Deming On Variation

"Life is variation."

Dr. Deming[29]

Dr. Deming's view on variation is unique for a number of reasons. In particular, we refer to the broadness of his view on this subject. When he describes applying the theory of variation to the manufacturing process, he emphasizes the human side. Exploring the concept of variation, he used sensitive examples taken not from the manufacturing process but from real life. This was probably to make sure that people would pay greater attention to the point he wanted to make. He used a story about a six-year old girl who came home from school with a negative note from her teacher. She had been given two tests, and the results showed that she was below average in both tests. The news affected her adversely. She was humiliated to be considered inferior. The little girl's parents then put her into a school that nourished confidence, and she recovered her self-esteem. Dr. Deming concluded, "What if she had not recovered? A life lost."[30] How many times do we, as managers, make the same mistake as the first teacher of this little girl? A manager wants all the people who work for him to be above average. Those who are below average will feel insecure because the manager is not happy with them. This way of thinking exists because of the lack of understanding in the law of variation. After a performance evaluation, roughly half of the employees tested will be above the average and the other half will be below the average. Using a cholesterol test as an example, Dr. Deming says, "Half of the people in any area will be above average for that area in test of cholesterol.... There is not much that anyone can do about it."[31] The same thing can be said about measuring anything: the output of a product, the cycle time, the manufacturing process, etc.

How many times have you sat through a presentation in which the manager tries to explain the slight ups and downs of a particular output over a period of time? In reality, the process performs the same, with a variation that is natural to this process. There is actually nothing to explain here. Once the process has been brought into a state of statistical control, it has a definite capability, and if the manager is not satisfied with this average, he needs to take care of the system. Everything else is just a random variation. The irony here is that when the manager explains the low yield in May by bringing up some actual causes, some superiors may take this explanation seriously. In fact, we are just dealing with natural fluctuation here. On the other hand, when a process is not under statistical control, it does not have a definite capability. Its performance is not predictable. This holds true for any type of process. How many times have we tried to predict the outcome of a process, ignoring the fact that the process is not in control? How many times is our prediction in error, and we wonder why? These days, it is usually difficult to plan and predict the outcome, but if the process is out of control, prediction is certainly impossible.

Dr. Deming describes two kinds of mistakes that management frequently makes when attempting to improve the results of the system's output. These mistakes are related to the confusion of special causes and common causes of variation. They are:

"1. Ascribe a variation or a mistake to a special cause when in fact the cause belongs to the system (common causes).
2. Ascribe a variation or a mistake to the system (common causes) when in fact the cause was special.

   Over adjustment is a common example of mistake No. 1. Never doing anything to try to find the special cause is a common example of mistake No. 2."[32]

In describing the system of profound knowledge, Dr. Deming strongly emphasizes the relationship between process control and prediction of output and results. When a process is in a state of statistical control, results such as cycle time, cost, quality, and quantity can be predicted. If the process is not in control (not in a stable state), then its performance results are not predictable. Emphasizing these two different states of a process, stable and instable, Dr. Deming brings to our attention the fact that managing people in these two different environments is also different. As he said, "Confusion between the two states leads to calamity."[33] When a process is in a state of statistical control and its output is satisfactory, the management should be oriented on monitoring and controlling the process to make sure that corrective action is taken when it begins to get out of control. When the process is in a stable state but the results need to be better, management should orient their people to use engineering and statistical techniques to improve the capability of the process.

Dr. Deming's effort to promote statistical principles helped many managers apply these principles to improve their manufacturing processes. However, some managers have difficulty understanding that the same statistical principles are applicable to the management process, as for any process. In this regard, it is interesting to review Figure 3-10, which was adapted from Dr. Deming's work.

This simple picture suggests that when people work in a system, we cannot expect everybody to work in the same manner. The distribution of their results is represented by (A) in Figure 3-10. These are people from the system who together will produce average results that will reflect the capability of the process. All these people (in Zone A) do not need special help. The only thing they need is

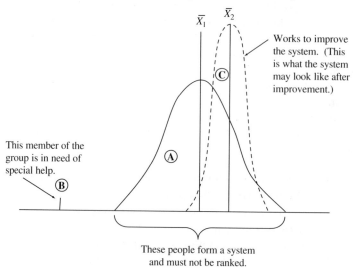

Figure 3-10 A Distribution that Reflects a System's Performance
Source: Adapted from W. Edwards Deming, *The New Economics,* © The MIT Press, 1994, p. 127.[34]

not to be disturbed. As the figure suggests, the only members who need special help are the people outside the distribution (see B). If the manager is not satisfied with the average outcome ($\bar{X}_1$), he or she needs to develop activities to improve the system. The dashed distribution (C) represents the process after the improvement ($\bar{X}_2$). Understanding the theory of variation will make it easier for a manager to understand that people are different from each other and ranking them will not help to make them the same or make them all perform "above the average." What is needed to improve the capability of a human process is a motivational system that will include continuous education, skill development, a creative environment, and a great culture in which they will enjoy the work. Recognizing the variation in people's capabilities will encourage management to pay more attention to people development.

Deming's Chain Reaction   Dr. Deming showed the graph in Figure 3-10 to the participants of a four-day seminar hosted by the Growth Opportunity Alliance of Greater Lawrence (GOAL) in Springfield, Massachusetts. He said, "Let's have a look at page three of the additional notes—chain reaction. You can't argue with a chain reaction. It will work."[35] Further describing the flow reflected in the chart, Dr. Deming said, "Chain reaction. Improve quality, what happens? Your costs go down. Half of the people here will understand that. The other half will not." His voice rose in indignation, "On the third day here, people will ask, 'Where do we stop improving quality? How do we know where to stop?' That is, where will further improvement not pay for the cost of improvement?"[36] Then, humorously, he said, "He that asks me that question will have his certificate recalled."[37] Continuing, he said, "But go ahead and ask. I want to make it clear that as you improve quality, your costs go down."[38] The reason we bring this extract from the seminar to your attention is to emphasize the importance of the continuous reduction of process variation.

If we were conducting the same seminar today, what percentage of people would understand and agree with the "chain reaction?" What percentage of participants would ask when to stop putting efforts into improving quality by reducing process variability?

Motorola's initiative of six-sigma quality, which was later adopted by many other organizations, supports Dr. Deming's philosophy and responds to the questions we raised above. Reducing the variation of any process should be continuous and will bring your organization the results reflected in Dr. Deming's Chain Reaction (see Fig. 3-11). Table 3-3 supports the idea of the "chain reaction" and shows the relationship between reducing the process variation and reducing the cost of quality. Less variation, fewer losses.

Later, in 1982, when Dr. Deming published his book *Out of Crisis*, he included the Chain Reaction Chart in it. Describing the results from the collaboration between specialists from Bell Laboratories and, in particular, the influence of Dr. Shewhart's book, *Economic Control of Quality of Manufactured Product*, Dr. Deming wrote, "The results were exciting showing that productivity does indeed improve as variation is reduced, just as prophesied by the methods and logic of Shewhart's book."[41]

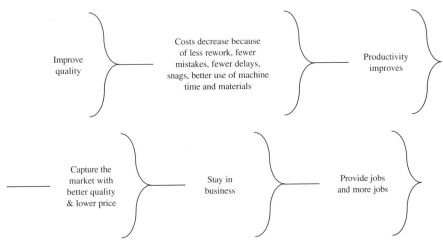

Figure 3-11 The Deming Chain Reaction

Source: Reprinted with permission from W. Edwards Deming, *Out of the Crisis*, copyright © The MIT Press, 1986, p. 3.[39]

Table 3-3 The Cost of Quality

Sigma Level	Defects Per Million Opportunities	Cost of Quality
2	308,537 (noncompetitive companies)	Not applicable
3	66,807	25–40% of sales
4	6,210 (industry average)	15–25% of sales
5	233	5–15% of sales
6	3.4 (world class)	<1% of sales

Each sigma shift provides a 10% net income improvement.

Source: From *Six Sigma: The Breakthrough Management Strategy Revolutionizing the World's Top Corporations* by Mikel Harry and Richard Schroeder, copyright © 1999 by Mikel Harry and Richard Schroeder. Used by permission of Doubleday, a division of Random House, Inc.[40]

In recent years, the quality of products in general has improved because of the requirements of the external environment. It is just impossible to stay in business very long with poor quality. Today, we can hear from some leaders of organizations that "Quality is considered a given factor to stay in business, an entry level for global business." But what we can also observe is that some organizations that achieved an acceptable level of quality stopped paying enough attention to further reducing the variation of their processes. At meetings we talk about the need for raising the productivity or reducing the cost of products without connecting all this to the major source of improvement—reducing the variability. This is why the theory and knowledge about variation remains one of the most important elements of profound knowledge today.

3.4.5 Organizational Psychology

In Webster's dictionary you can read that psychology is "the science dealing with the mind and mental and emotional processes"[42] According to Dr. Deming, "Psychology helps us to understand people, interaction between people and circumstances, interaction between customer and supplier,... interaction between a manager and his people and any system of management."[43] On basis of these two descriptions, almost all the chapters of this book include different aspects of psychology, because we wrote about relationships and about different mental processes. This is especially true for Part V of this book, which is dedicated to the subject of organizational culture.

In this section, we will provide only an overview of some aspects of psychology, mainly those related to Dr. Deming's view on this topic.

As we mentioned above when describing the theory of variation, people are different in their needs, abilities to work and learn, ways of thinking, behavior, values, cultures, mind-sets, and many other aspects. It is very important for a manager to recognize these differences and to use them to optimize everyone's abilities and inclinations. This is why Dr. Deming advised management not to rank people. He wrote, "Management of industry, education and government operate today under the supposition that all people are alike."[44]

How much psychological damage can a manager do by operating under the assumption that all people are alike? Under this assumption, we expect everybody to produce the same amount of work, to attend the same learning programs, and to use the same "best" style of leadership (see the Situational Leadership System). Under this assumption, we use the same incentives and the same forms of punishment. We want everyone to fit the organization's policy, which is also developed under the assumption that people are alike. We want all subsidiaries located in different countries to have the same set of values.

Take, for example, education. When we develop an educational program in the organization, we make attendance mandatory and we ask participants to sign in when entering the classroom. Can you imagine the reaction of a manager if an employee said, "I will learn about this myself from the books. Can I skip the class?" The manager's reasoning would be, "How do I know that he or she will really learn? How do I know that the person can learn by correspondence?" Mandatory participation in an educational program is like the saying, "You can lead a horse to water, but you can't make him drink." Our experience has shown that if the educational programs are good, and have the proper selection of "hot" topics, then we do not need to control people's attendance. They will consider this program an opportunity and a privilege to attend. They will stand in line for it voluntarily. They will "come to the water" voluntarily, and when they do, real learning will occur.

Dr. Deming says, "People learn in different ways, and at different speeds. Some learn a skill by reading, some by listening, some by watching pictures, still or moving, some by watching someone do it."[45] The manager must learn how to recognize and differentiate people's individual capabilities and talents and to create an environment of learning that will fit everybody's abilities and needs, one where all this learning will be materialized in organizational values.

Utilizing All Sources of Motivation Motivation is an important component of psychology. Through proper motivation, we can influence people's behavior. There are two major forms of motivation, *intrinsic* motivation and *extrinsic* motivation. For many years, organizations mainly recognized the extrinsic sources of motivation by using monetary rewards to improve employee performance. This motivator—money—works very well when people receive less than necessary to maintain their quality of life. However, as soon as monetary compensation reaches a level that is above the amount required to maintain the quality of life, money loses the power of motivation. People may enjoy receiving more money as a measure of their contribution, but this will not improve their performance. What money can do is crush their intrinsic motivation. Dr. Deming uses a simple but psychologically powerful story that he heard from Dr. Joyce Orsini.

A little boy took it into his head for reasons unknown to wash the dishes after supper every evening. His mother was pleased with such a fine boy. One evening, to show her appreciation, she handed to him a quarter. He never washed another dish. Her payment to him changed their relationship. It hurt his dignity. He had washed the dishes for the sheer pleasure of doing something for his mother.[46]

By using only extrinsic motivation, we may develop an organizational behavior where money is expected for any extra move an employee makes, and later it will become difficult to expect extra effort even for money. For example, in the former Soviet Union, when productivity was low, a monetary system was developed to motivate people in all kinds of activities to improve productivity. In the beginning, the new system worked very well; the productivity improved to some extent. But later, people would not make any extra effort without asking for an extra ruble. Still later, in some areas, even the rubles stopped working. Coal miners, for example, were already being paid relatively more than others because of their difficult and dangerous work. The country needed a larger output of coal, so to motivate the miners to work overtime 7 days a week and on holidays, they were offered almost double pay. However, because their income exceeded the demands of their quality of life, most of the miners refused to work the extra time.

At AMD, we introduced a motivation system that includes intrinsic and extrinsic motivation. For example, people who graduate from the Sigma College, an internal institution where people learn about the theory and practice of variation, receive a book on statistical principles and a letter of appreciation from the Group Vice President. It is interesting to note that even though attendance in the Sigma College is voluntary, we have had full enrollment and high attendance. In 2001, we will celebrate its tenth anniversary. The employees take the Sigma college courses because they need the knowledge, because management recognizes it, and because the managers at all levels are also students of the Sigma College.

Another example is motivating for innovative projects. Teams who participate in the development and implementation of innovative projects receive a monetary award, which is by itself an example of extrinsic motivation. However, this money is delivered with a special plaque that has the team members' names engraved on it. The awards are presented to the participants in front of a large audience at an international conference or symposium. The formal award presentations are followed by a nice dinner, music, and dancing. All this, plus applause and collective congratulations from their peers, has a significant impact on the people who receive the awards, more than just a monetary award. All together, this has become a strong motivation, which is a mix of intrinsic and extrinsic motivational factors. But what is more important is the fact that when we asked the employees what they think about the this form of awarding innovative projects, most of them responded that while they like it; the strongest motivator is the opportunity to participate in the company's development.

In a dialogue with a large group of technicians in AMD Thailand, we asked the question, "What motivates you to work?" The first reaction to this question was, of course, money. They came here to make money. This answer was on the surface of their minds, but as the dialogue continued, one of the participants spoke up and said, "To me, just coming to work every day is by itself a great motivator." Then another person said, "I just got married to a lady who works here. I probably would not have met such a nice woman without having the opportunity to work in this company." The technicians started talking about all kinds of social motivators, such as celebrations, motivational events, participation in cultural rituals, and others.

We do not need special awards, but they are stronger than just money. Is money important? This is almost the same as asking if oxygen is important. Without oxygen, we would all die, but does this mean that breathing is a motivator for us to live? Money is a necessary element for survival, but there are many intrinsic motivators to create, learn, and work.

All these elements of intrinsic motivations naturally became a part of the organization's system. Usually, they are not formally reported and formally managed. They are just a part of the organizational life. In AMD, every plant is a part of the corporation, and at the same time, a part of its own country, with its own cultural system of intrinsic motivation. The two systems—intrinsic and extrinsic—interact with each other and form an effective motivational system that holds people together, makes them happy, and has a positive impact on the organization's results.

References

1. W. Edwards Deming, *Out of Crisis*, Ch. 2, Massachusetts Institute of Technology, Center for Advanced Engineering Study, Cambridge, MA, 1986
2. W. Edwards Deming, *The New Economics*, Massachusetts Institute of Technology, Center for Advanced Educational Services, Cambridge, MA, 1994
3. Ibid., p. 50
4. Ibid., p. 103
5. Ibid., p. 93
6. Ibid., p. 102
7. Ibid., p. 103
8. Ibid., p. 101
9. Zeleny, in John C. Anderson, Kevin J. Dooley, and Susan D.A. Misterek's article, "The Role of Profound Knowledge in the Continual Improvement of Quality," Human Systems Management, IOS Press, Netherlands, 1991, p. 256
10. W. Edwards Deming, *Quality, Productivity and Competitive Position*, Massachusetts Institute of Technology, Center for Advanced Engineering Study, Cambridge, MA, 1982, pp. 16–17
11. Mary Walton, *The Deming Management Method*, Mercury Books, London, a division of W.H. Allen & Co. Plc, UK, 1989, p. 36

12 J.C. Anderson, K.J. Dooley and S. Misterek, "The Role of Profound Knowledge in the Continual Improvement of Quality," Human Systems Management, IOS Press, Netherlands, 1991, p. 244
13 W. Edwards Deming, *The New Economics*, p. 92
14 Ibid., p. 93
15 Anderson et al., *The Role of Profound Knowledge in the Continual Improvement of Quality*, p. 244
16 W. Edwards Deming, "Foundation for Management of Quality in the Western World", p. 13, in William J. Latzko and David M. Saunders, *Four Days with Dr. Deming*, Addison-Wesley Publishing Company, Boston, MA, 1995, p. 35. Reprinted by permission of Pearson Education, Inc.
17 Deming, *Out of the Crisis*, pp. 317–318
18 W. Edwards Deming, *Quality, Productivity, and Competitive Position*, Massachusetts Institute of Technology, Center for Advanced Engineering Study, Cambridge, MA, 1982, p. 17
19 Deming, *The New Economics*, p. 103
20 Ibid., p. 103
21 Chris Argyris, *Knowledge for Action*, Jossey-Bass Publishers, San Francisco, CA, 1993, p. 1
22 Chris Argyris and Donald Schön, *Organizational Learning: A Theory of Action Perspective*, Addison-Wesley, Reading, MA, 1978
23 Robert B. Reich, *The Work of Nations: Preparing Ourselves for 21st Century Capitalism*, Alfred A. Knopf, New York, NY, 1991, p. 3
24 Stephen H. Rhinesmith, *A Manager's Guide to Globalization*, McGraw-Hill Companies, New York, NY, 1996, p. 28
25 George S. Yip, *Total Global Strategy*, Pearson Education Inc., Upper Saddle River, NJ, 1992, p. 1
26 Walter A. Shewhart, *Economic Control of Quality Manufactured Product*, Van Nostrand, 1931
27 John Butman, *Juran—A Lifetime of Influence*, John Wiley & Sons, Inc., New York, NY, 1997, p. 34
28 *Statistical Quality Control Handbook*, AT&T, Basking Ridge, NJ, 1956, p. 35
29 Deming, *The New Economics*, p. 98
30 Ibid., p. 98
31 Ibid., p. 99
32 Deming, *Out of the Crisis*, p. 318
33 Deming, *The New Economics*, p. 100
34 Ibid., p. 127
35 Mary Walton, *The Deming Management Method*, Mercury Books, a division of W.H. Allen & Co. Plc., London, England, 1989, p. 25
36 Ibid., pp. 25–26
37 Ibid., p. 26
38 Ibid., p. 26
39 Deming, *Out of the Crisis*, p. 3

40 Mikel Harry and Richard Schroeder, *Six Sigma*, Doubleday, New York, NY, 2000, p. 17
41 Deming, *Out of the Crisis*, p. 3
42 *Webster's Twentieth Century Dictionary*, 1979, p. 1454
43 Deming, *The New Economics*, pp. 107–108
44 Ibid., p. 108
45 Ibid., p. 108
46 Ibid., p. 110

PART II

Managing a Knowledge-Based Organization

Knowledge has always been important to organizations. However, only in the last 5–10 years have we seen an increase of interest in learning and creating knowledge. Today we can hear many executives say that the competitive advantages of their corporations depend largely on the knowledge they possess and their capability to create new knowledge, in other words, on their capability to learn. The question you may ask is, Why knowledge and why now? Isn't knowledge just one of the important components of the organization's success? Why should we consider knowledge as the most important thing an organization needs to stay in business and succeed?

These questions apply to any industry or any organization, small or large, newly formed or well established. Take the electronics industry as an example, in particular, the semiconductor manufacturing industry. Who makes microprocessors? What makes up the cost of a microprocessor? The major part of it is the knowledge required to make this "black box." There is a little bit of silicon inside, a little bit of metal, and plastic. These elements cost just cents. What the microprocessor makers are selling is their knowledge. To package this knowledge into the form of a useful product that can be plugged into a computer and sold, the producer of the microprocessor must have knowledgeable people who cost more than manual workers and must continuously invest heavily in research and development, sophisticated robots, automation, and highly technological processes that are all based on people's knowledge. Sure, we need land, labor, and capital, but take away knowledge and these three things are just not enough. At the end of the competition, the winner will be the organization with the smartest ideas that is capable of transforming ideas into marketable products faster than its competitors. This again brings up the need for having more knowledgeable and smarter workers.

But where do these knowledgeable and smarter people come from—the best colleges? Yes, but most organizations recruit people from these colleges. So, do these people come from our competitors? Sometimes. But if we hire someone from our competitor, we can only achieve what our competitor has already achieved. So, how can we take it at least one step further? There are certainly no rules on this matter. Every organization has its own way of building a pool of talented and knowledgeable people. To nurture these people, you need to recognize that knowledge is one of the major core capabilities. What is needed is a core group of innovative, talented, and knowledgeable people who, as a magnet, will pull in other knowledgeable people, and the effect will snowball. An old saying is that "money comes from money." Well, we can say that "knowledge comes from knowledge." In a knowledge-based organization, there is a culture, a language, and a set of values, which create a human magnetic field that holds those talented people together.

In 1992, Bill Gates wrote, "Take our 20 best people away and I tell you that Microsoft would become an unimportant company."[1] This figure would probably be larger today, but we interpret this statement as saying that an organization needs to have a core group of people who have extraordinary knowledge in a particular field, people who speak "the same language." And this small number of people will "roll the snowball" of knowledge. Without studying Microsoft's data, it is still easy to say that this is an example of a knowledge-based organization, where the knowledge workers and the knowledge managers work and learn together to better serve the customers' needs.

Creating a knowledge-based organization is a very broad topic. The entire contents of this book cover some aspects of creating and applying knowledge in an organization, but Chapter 4 is more specifically about the methodology of knowledge creation.

Chapter 4

Organizational Learning

4.1 WHAT IS LEARNING AND WHEN DOES IT OCCUR?

Sometimes managers describe and measure the learning activities in their organizations by referring to the amount of money they spend on educational programs and information systems, the number of classes or courses conducted, or the number of inventions they have registered, etc. These things obviously support the development of a learning infrastructure, but it is not actual learning. Learning has a deeper meaning, and it is very important for knowledge managers to understand this if they are interested in creating a learning environment in their organization. Peter M. Senge, the author of *The Fifth Discipline*, wrote, "...Learning...involves a fundamental shift of movement in mind.... Most people's eyes glaze over if you talk to them about 'learning' or 'learning organizations.' Little wonder—for, in everyday use, learning has come to be synonymous with 'taking in information.'"[2] This means that learning does not occur in the classroom and it is not measured by the amount of information we are capable of storing in our computers, but it can only occur in action, in creating our reality, in changing our world. Learning also means enhancing our capability for producing new results that involve a fundamental shift of mind-set.

When does learning actually occur? Chris Argyris defines learning as occurring under two conditions. "First, learning occurs when an organization achieves what it intended; that is, there is a match between its design for action and the actuality or outcome. Second, learning occurs when a mismatch between intentions and outcome is identified and it is corrected; that is, a mismatch is turned into a match."[3] Figure 4-1 describes the first condition for learning to occur. The organization (or individual) creates a

66 ORGANIZATIONAL LEARNING

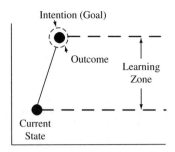

The first time learning occurs is when an organization achieves what it intends to achieve.

Figure 4-1 A Match Between Intention and Outcome

vision or sets a desired goal that is significantly greater than the current reality.

By doing this, the organization builds up a creative tension that motivates it to design and introduce the actions necessary to make the vision a reality. To produce substantial learning, the creative tension must be strong enough to build up a challenge. It is important to make sure that your intentions for the future improvement of change are far away enough from the current reality to produce enough creative tension to motivate people toward new achievements.

Figure 4-2 describes the second condition for learning to occur. In this case, we can see that there was a mismatch between the intended design for action and the actual outcome. The results were analyzed, and additional or new actions were taken to correct the undesirable results. After the corrections were undertaken, the outcome demonstrated a match between the intended

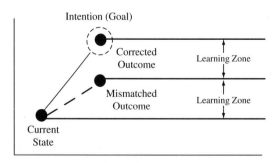

The second time learning occurs is when a mismatch between intention and outcome is identified and corrected

Figure 4-2 A Mismatch Turned Into a Match

design for action and the actual outcome. Learning occurred here when the mismatched outcome was identified and corrected.

We will explore deeper into these two concepts of learning in the following sections of this chapter.

4.2 LEARNING HOW TO LEARN

While visiting one of our new overseas plants, we were surprised at how fast the organization managed to introduce a new technology with sophisticated equipment. The training department had a very limited amount of time to train a large group of local employees in the repair and maintenance of the equipment. When we entered the facility, we saw that almost all the signal lights that indicate equipment readiness were green, which indicated that this manufacturing floor was functioning well. There was only one machine with a red light. When we approached the stalled machine, we saw a young technician with a smile on his face. "Ready!" he said proudly, looking at the managing director who was escorting our team of visitors. "What was it?" asked the managing director. "You want to become a technician?" humorously replied the young man, as he rapidly continued, "Honestly speaking, I don't know exactly, but it works now, so I feel I fixed the problem. "Good," said the managing director, "keep it up. By the way, how much trouble do those machines give you? They probably keep you busy all the time. "Not really, the testers are very reliable," replied the technician. Continuing the conversation, the technician told us that he was very happy with his new job and shared some of his work "secrets" with us. He said that he keeps a technical diary in which he records the problematic symptoms, and when a problem arises, he refers to his notes. "I already memorized all this, but sometimes when I forget, I look it up in my notes and this helps a lot," he said. The managing director interrupted by saying, "Based on this principle, we are now in the process of computerizing all these notes. This will allow us to have a troubleshooting system that every technician can use. This will help consolidate the organization's experience into one organizational memory."

What we just described is an effective form of learning, and it is appropriate for taking care of everyday problems and other repetitive issues. The same pattern of work can be observed at all levels of the organization, particularly in professionals and managers. We spend the majority of our time solving problems. It makes us feel more secure and valuable. As our experience increases, we develop mental models, record them in our minds, and act accordingly. We tend to feel safer because our previous way of doing things has proved to be the "right way" many times. This kind of behavior develops a barrier to new and deeper learning, and we gradually become poor learners. On the surface, this kind of reasoning may sound paradoxical, so we will try to elaborate on this issue later in this section.

4.2.1 How We Usually Work and Learn

We all have goals, desires, and expectations. We expect a certain level of product quality. We want all our machines to work properly. We want to stay within the budget, etc. If we had all this, there would not be any problems, but real life is different. There are always some deviations from our desired results, and as soon as a deviation occurs, we have a problem. So, how do we resolve it? We usually act on the problem according to our previous experiences. We have fixed a problem previously and achieved the desired results, so we use the same reasoning when the same kind of problem occurs. If we succeed again, the way of solving this type of problem becomes a mind-set. Should the same type of problem reoccur, we would react the same manner. Why? Because that solution worked well before. Hence, there is no need to look for better ways to resolve the problem. Very rarely do we ask ourselves why the way we acted worked well. And this frees up our mind to think differently. It frees us from learning. So, if a person works in such a way in a company, in the same position for 20 years, then he learns effectively and accumulates experiences for the first 2–3 years, and then repeats it over and over again. Because organizational learning depends on the way individuals within the organization learn, this would not be just the individual's problem but a problem for the whole organization. When we are successful, we do not think much of learning. Why should we? And when things are going badly, there is no time and resource for learning. However, if we want to stay in business, learning should happen anytime and continuously.

4.2.2 How We React to Problems

Organizations usually react to sudden dramatic events. For example, a large customer refused to continue buying our product. We called a meeting and assigned a committee to investigate, and we sought for the root causes of the problem. But how many times have we participated in a special meeting to discuss the results of success? Probably not often, because most of our time is spent discussing the strategy of how to resolve problems. Figure 4-3 shows that we may continue to apply our experiences again and again to resolve almost the same problems in certain specific situations. However, it is much more difficult and challenging to be able to apply our past experiences when the situation and the environment are significantly different. We take on a greater risk when we act only from our past experiences in a situation where a totally new experience is required.

The model described in Figure 4-3 may be familiar to you, because this is the way we usually learn to react to problems. As we mentioned above, a problem is a discrepancy between the desired and actual results. When a problem occurs, we usually react to it on the basis of our prior experiences. We store our experiences in our brain, from where we can recall them when analogous problems occur. Having had a similar experience, we are reasonably assured that this is the best way to resolve such problem, and the more times we repeat

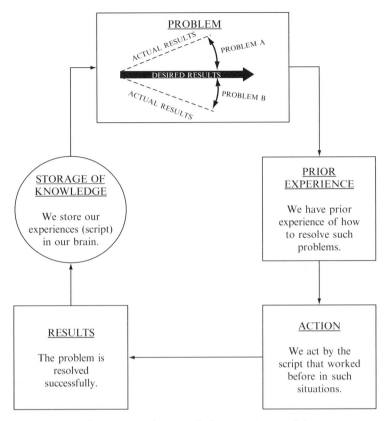

Figure 4-3 The Knowledge Freezing Model

the cycle, the less we even question whether there is a better way to resolve such problems. We gradually develop such a stereotype in thinking and reacting to particular problems that we don't find the time to analyze them and see what impact they are going to have on our future. We freeze our existing knowledge and stop learning.

4.2.3 Two Scenarios of Learning

In the past, when the speed of change was significantly slower, the same technology was maintained for years. When the external environment was relatively stable, we could allow people in the organization to learn from an established procedure—learning from a supervisor who would share the experiences that he had accumulated over the years. But the world has changed, and the lifetime of sophisticated equipment or products is significantly shorter. This requires us to seek new forms of learning. How many times have you heard someone say, "Let's not reinvent the wheel," meaning that if there is an old script for how to think, why bother to look for something more progressive? It is safer and cheaper. We have already tested "our way." As soon as an organization gets trapped into a mental model that does not fit today's requirements, learning will not occur, and that organization will begin to lose business. Some managers use the phrase "Let's keep it simple" in the wrong way. Simplification is a very important component of progress, but making our life simple by using an older, already-proven script can be the beginning of an organizational disease called Learning Deficiency.

What distinguishes a knowledge manager from an ordinary manager is that the knowledge manager recognizes that as the world changes so does our ability to learn from our experiences. The knowledge manager will need to learn how to act in an environment in which the speed and complexity of change are growing exponentially and it is no longer possible to rely on certainty and predictability. Risk has become a modern word for a knowledge manager. We know many ordinary managers who have worked successfully all their lives under relatively constant conditions, who have made very few mistakes in their lives; but when a manager does not fail, he also does not learn. This does not mean that we must fail in order to learn. What it means is that we must build creative tension in our work and we must take a calculated risk. By doing this, we have a greater chance of failure but at the same time a greater opportunity to learn. This type of "failure" is a way toward ultimate success. We will elaborate on this thought by describing two different scenarios.

Scenario 1: Suppose the process for your product has a high failure rate, which has caused a low yield and noncompetitive cost. As the manager, you decided to introduce a process improvement program. A goal was established for an incremental reduction of the process variation. Every time an incremental improvement occurs, your team celebrates the success. This model of improvement guarantees no failure. You can observe a continuous trend of nonconformity reduction, and the yield is gradually going up. Most of your processes are brought into a state of statistical control and are approaching a, let's say, four-sigma quality level. You are happy, your boss is happy, and there is no risk of failure.

Scenario 2: Like Motorola, you decided to take a long-term goal of achieving a six-sigma quality level in your organization, which is an improvement over a particular period of time. Your team may work hard, and as time passes, you may find that your processes are running at a five-sigma quality level. You failed because you did not accomplish your goal. As you reached a quality level of five-sigma, you felt that you had hit a concrete wall. To overcome this barrier, you needed new knowledge—knowledge on how to achieve, maintain, and work in a low ppm environment. You learned that achieving a six-sigma quality level requires more time, so you revised the goal, and your organization was criticized for taking unreasonable goals. You delved deeper into understanding your processes. You spent more time, money, and human energy, and finally you achieved your goal. Your organization is running at close to a zero ppm quality level, and the cost of your product is competitive. Your people obtained new knowledge while tackling new barriers. This was a failure that finally brought you to great success.

This is what we mean when we say that managers who do not experience failure often enough will also not have the opportunity to develop good learning skills and refresh their knowledge. This type of manager will eventually fail. This is why learning to learn has become a discipline that should be included in the curriculum of preparing knowledge managers.

In addition, the whole learning architecture should be improved to facilitate fast learning. For example, most motivational systems do not encourage the managerial behavior described in Scenario 2. So these motivational policies have encouraged a behavior in which we do not want to take great risks that could result in a temporary failure. Under these conditions, we prefer to take goals that have very high confidence of success. At the moment, organizations do not have instruments to measure the structural tension of the goal we set. The motivational system we adopt does not consider the risk we are taking. What actually counts in a good performance review is the number of goals achieved. This is why it is very important to include the performance appraisal system in the learning architecture.

Establishing new goals that contain a rational (or even irrational) risk creates a tension between the current reality and the selected goal. This will lead the organization to success. It also prepares the employees for new and more difficult problems.

4.3 SINGLE-LOOP AND DOUBLE-LOOP LEARNING

We begin our process of learning at the time we are born. We learn to crawl, walk, and talk; we learn how to read, write, and work. We learn from successes and failures. Our whole life is a learning process. Through individuals' learning, organizations also learn how to succeed in the turbulent marketplace. So, do we know how to learn? This may sound naive, but in fact it has been proven that as we become smarter and better educated we also become worse learners. This is a dilemma that must be resolved for individuals and organizations to succeed in the future.

Chris Argyris wrote, "... Success in the marketplace increasingly depends on learning, yet most people don't know how to learn. What's more, those members of the organization that many assume to be the best at learning are, in fact, not very good at it. I am talking about the well-educated, high-powered, high-commitment professionals who occupy key leadership positions in the modern corporation."[4] So, if people who hold key leadership positions are not the best at learning, how can we expect the whole organization and its employees to be good learners? This is an obvious dilemma that needs to be addressed. As Chris Argyris suggests, "Most companies not only have tremendous difficulty addressing this learning dilemma; they aren't even aware that it exists."[5] To address this dilemma, we need a deeper and broader understanding of what learning is all about and what we can do, as individuals and as an organization, to become better learners.

As we mentioned above, sometimes people define learning as going to school, as reading a book or a manual. Even though this is a very important component in the learning process, this alone will not result in learning. You cannot say, "I read a book on how to become a pilot, so now I am ready to fly." Learning requires action, risk taking, and the accumulation of experience. This experience should not come from problem-solving activities alone. It should come from analyzing the mental models you hold about the environment and your resulting behavior. Chris Argyris coined the terms *single-loop* and *double-loop* learning to capture the way we learn. As an analogy, he uses a thermostat (a simple device for regulating the temperature of a room) (see Fig. 4-4).

As the readings on the thermostat go above a chosen setting, the fuel supply to the furnace is progressively reduced, and conversely, as its readings fall below that setting, the fuel flow is increased. This example of a familiar control system is used here to describe the concept of *single-loop* learning (see Fig. 4-5). The thermostat is capable of maintaining a selected level of temperature by regulating the fuel, but the thermostat cannot ask, "Why am I set at this particular temperature?" and then explore whether or not some other temperature might more effectively achieve the goal of maintaining a necessary room temperature. If the thermostat was capable of asking and exploring such a question and acting accordingly, this would be analogous to *double-loop* learning.

If we would analyze the work we usually do in our organization, we would probably find many examples of single-loop learning. For example, when the yield of a particular manufacturing process goes down, we apply the experience we have and the yield is brought back to its normal level. Later, when for some reason the yield drops again, we seek a solution to eliminate the cause of the problem. This activity is analogous to the work of a thermostat and is

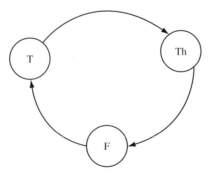

where:
T is the variation of the temperature
Th is the variation of the thermostat
F is the fuel control for the furnace

Figure 4-4 A Closed-Loop System

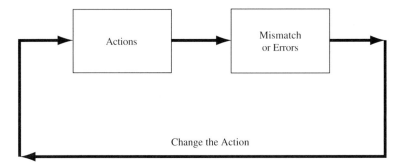

Figure 4-5 Single-Loop Learning
Source: Overcoming Organizational Defenses by Argyris, Chris, © 1990, p. 96. Reprinted by permission of Pearson Education, Inc., Upper Saddle River, NJ.[6]

considered single-loop learning. Another example could be the activity of budget setting, which is frequently based on past experience. When an expenditure exceeds the budget, the budget is adjusted by cutting other expenses and reducing the cost to bring the organization's budget back into balance. Later, when the problem arises again, the process of budget normalization is repeated. Budgeting is the "thermostat" that keeps the budget in control and is mainly a single-loop learning activity. In a single-loop learning mode, we may even introduce innovative actions and achieve significant improvements, but if these actions are based on the existing organizational policies, mental models, and governing values, it is still considered single-loop learning.

Sometimes the external environment forces organizations to move from single-loop to double-loop learning. For example, several years ago almost all integrated circuit (IC) packaging technology consisted of a leadframe with wires connecting the leads to the die. For years, the electronics industry utilized single-loop learning to incrementally improve the package technology. This included the development of finer-pitch wire bonding processes, smaller package outlines, finer pitch leads, and improved lead quality such as coplanarity. Obviously, improvement and learning occurred from such activities. This type of single-loop activity was acceptable up to the time at which no significant changes were made to the package requirements in relation to the number of leads needed. However, while the number of transistors in a die dramatically increased, the die itself was shrinking, so the wire bonding process became more complicated. A provocative question arose, "Why do we need leads, and why do we need wire bonding in the first place?" This question challenged the assumption that had existed for years that an electronic device must have leads to be connected to the circuit board. When the assumption was changed, the mental model of packaging design also changed and a solution to make packages without leads was found. This is an example of double-loop learning because it required an alteration of the organization's governing assumptions and values (see Fig. 4-6).

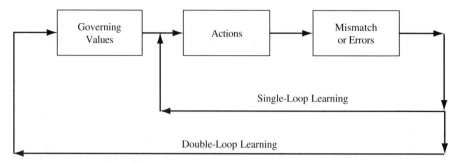

Figure 4-6 Double-Loop Learning

Source: Overcoming Organizational Defenses by Argyris, Chris, © 1990, p. 96. Reprinted by permission of Pearson Education, Inc., Upper Saddle River, NJ.[7]

4.4 OTHER CLASSIFICATIONS OF LEARNING

To further develop the concept of organizational learning, in addition to single-loop and double-loop learning, researchers and practitioners come up with other concepts and classifications. In his article, "A Model for Changing the Way Organizations Learn," Robert M. Fulmer[8] describes a group of learning classification that deserves attention. Familiarization with these learning classifications will help you better understand the single-loop and double-loop learning concepts and will also provide a feeling for what type of learning is appropriate during different periods of organizational development. Based on Fulmer's classifications, we will provide a short description of three different models of learning: maintenance learning, shock learning, and anticipatory learning.

4.4.1 Maintenance Learning

Maintenance Learning is related to the activities of the continuous improvement of the already-existing processes. In other words, it is related to doing things right without being conscious of the issue of doing the right things. Maintenance Learning, as single-loop learning, is not sensitive to the external environment and quite often misses important signals from the environment that require change. Maintenance Learning requires less risk because it is based on past experience and does not require significant changes in the organization's strategy or processes. Because of this, the organization that applies only Maintenance Learning can miss emerging new business opportunities. Maintenance Learning is usually applied in organizations that want to maintain the status quo and have short-term strategies. Therefore, such organizations are not ready to react to the environmental changes and sooner or later (if it is not too late) must introduce other more progressive forms of learning.

4.4.2 Shock Learning

Shock learning is a reactive form of learning and usually occurs in a period of crisis. This form of learning, which occurs during a lot of stress and tension, may also aggravate the problems. During the period of crisis, we usually observe the development of a sense of urgency in the organization, which may accelerate learning. However, at the same time, we may also observe that all learning applies only to short-term goals. It is very difficult to address long-term issues, let alone introduce innovation, in a high-stress environment. People in this kind of environment will strive to do business using their past experience, sure that what they do will work because it has worked in the past. Hence, Shock Learning is also considered single-loop learning in special situations.

4.4.3 Anticipatory Learning

This form of learning consists of two parts: Participatory Learning and Future-Oriented Learning. Participatory Learning is where everyone in the organization participates in the learning process and can freely explore alternative solutions and personal opinions. Future-Oriented Learning occurs where the focus is on the future. It requires research to better understand and foresee tomorrow's opportunities. At the same time, Future-Oriented Learning is involved in learning about the past to better understand the future (see Fig. 4-7). These two elements of Anticipatory Learning are closer to the requirements of double-loop learning because they include modification of the organization's underlying mental models (norms, policies, and objectives).

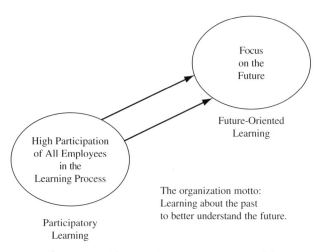

Figure 4-7 The Anticipatory Learning Model

4.5 EMOTIONAL LEARNING

You may recall some unpleasant circumstance that had a permanent or long-term effect on your behavior. In a conversation with one of our friends, we heard a story that will be a good example of the topic we want to introduce here.

Our friend recalled something that happened to him many years ago, when he was just starting his managerial career. He said, "During a visit with the CEO, I expressed my view on a very sensitive problem that, in my opinion at that time, could lead to disaster if it was not reacted to immediately. I had probably expressed my thoughts too strongly as I quickly realized that my view was not aligned with the CEO's. His reaction was bad, and my visit almost cost me my job (and possibly even more since this was in Russia). It was my good luck that the CEO was promoted to a higher position in a different industry. This single emotional incident affected my future behavior significantly, and took me years to get over. I became a different person. I was worried about expressing my opinion on things I felt strongly about. I found that this *learning* experience had a deep effect on me, and it was difficult to *unlearn*."

Psychologists would call this kind of behavior *escape conditioning*. Our friend said, "The new CEO was a different type of person, but I never had the nerve to go to him with a new proposal, because I was worried that the previous bad experience would be repeated."

Even though our organizational culture may protect us from such emotional learning, we certainly have come across many people who suffer from this kind of emotional stress. If something bad happens once, a person can learn not to repeat the experience if he or she is trying to avoid the problem. Examples of this can be a failure in a major project, making the wrong decision, failing in a new position, or any other organizational behavior that has caused the person to suffer. These experiences can introduce fear into the personality that will force the person not to do things that may in fact ultimately bring him/her pleasure and emotional satisfaction. The problem is that such stress is not on the surface where management can see it and "fix" it. This kind of conditioning lies deep within a person, and may be totally unknown to the person harboring these feelings. Employees who have been traumatized by this kind of emotional learning will usually miss a lot of great opportunities and stop progressing in their careers. As Mark Twain once observed, "A cat that steps on a hot stove once will never step on a hot stove again... but neither he will step on a cold one." The same behavior can be observed in people, who have the capability of overlearning, overgeneralizing, and overdramatizing experiences they have had.

The manager's role is to create an environment in which people can take risks and learn from their own and others' mistakes. If, for some reason, a person has been traumatized by a bad experience, the manager should become the facilitator and help the person to recover and regain confidence in himself and others. Helping an employee to overcome the powerlessness created by a painful experience is a managerial responsibility because the manager is also responsible for his/her people's emotional well-being.

One of the authors remembers an episode that can illustrate the way we can create an environment that may help overcome emotional stress. "In a military school for fighter pilots, two students were killed when they-crashed their plane. This was my first experience with seeing how a simple mistake could take a human life. Needless to say, this tragedy demoralized the students. However, what surprised me was that while marching back to our quarters from the cemetery, the commander ordered the orchestra at the front of the troop to start playing an enthusiastic march, and then ordered the troops to sing a song. We weren't in the mood, but we followed the order. Early the next morning we were told to start the flying training program, even though we were originally scheduled to take it later. Only later did I realize how important this arrangement was to prevent the students from sinking into grief and fear."

It is your own responsibility to pull yourself out of a strong emotional situation—you must do it yourself. You can probably remember situations in which someone made the wrong decision and was labeled as a failure, and the only way out was for the employee to leave the company and start anew elsewhere. Although emotional events can damage your spirit, courage, and willingness to take risks, it also provides a positive learning experience that can not be learned any other way. In this case, management needs to help by creating the right environment.

4.6 LEARNING FROM PARTNERS

Let us assume that an organization assessed its existing level of knowledge and then compared that with the estimated needs to achieve its vision (see Fig. 4-8) and determined the gap to be filled in. There are different ways to close the gap, but from a global perspective this can be done by adopting some knowledge from other organizations.

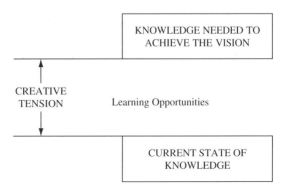

Figure 4-8 Learning Opportunities

Gary Hamel and C.K. Prahalad called this form of learning *borrowing*. They wrote, "Through alliances, joint ventures, inward licensing, and the use of subcontractors, a firm can avail itself of skills and resources residing outside the firm. At the extreme, borrowing involves not only gaining access to the skills of a partner, but actually internalizing those skills by learning from the partner."[9] In this sense, globalization is a process of learning; a process of borrowing knowledge and then internalizing it and making it a part of your own assets. What is interesting here is that nobody loses; rather, both parties win by raising their level of knowledge and becoming more competitive. This, for example, happened when AMD formed a joint venture with Fujitsu, or when AMD licensed IBM's packaging technology (see Section 10.1.2).

It is important here to note that internalization is usually a more efficient approach to acquiring new knowledge than acquiring an entire company. For example, when AMD bought Monolithic Memories Inc., (MMI), it paid for some new knowledge and also for the knowledge and skills that AMD already had. In a global environment, organizations expand their boundaries and activities by building new relationships with global organizations. This may be done in the form of joint ventures, subcontracting, mergers, and other forms, which if organized properly may become a real source of knowledge. By itself, this is an important source of core competency.

4.7 WHAT FORM OF LEARNING IS PREFERABLE?

Now that we are familiar with the distinction between single-loop and double-loop learning and other classifications of learning, what form of learning would you prefer for your organization? You would probably prefer all forms, because they are not contradictory. Which form of learning you select depends on whether the organization is focusing on the improvement of the already-existing processes, methods, and tools without changing the governing assumptions, the organization is in a crisis and needs fast, radical solutions, or the organization is willing to challenge the existing assumptions of current practices. We used an example above that showed how the mind-set of semiconductor packaging technology changed. Below is another example to illustrate that, besides the need to move from single-loop to double-loop learning, both modes of learning work together to complement each other.

For many years, the molding process of packaging for electronic devices was designed and produced in such a way that any time the size of the device was changed a new mold was required. This was acceptable at that time, when there were few different die sizes and they changed infrequently. A process of continuous incremental improvement was in place, and people learned from this experience. However, when a requirement for packaging very small devices arose and the rate of changing the device configuration increased, a problem was created. This generated an absolutely different approach to molded packaging. We began to mold devices in strings, and some molds could be easily applied to different sizes of devices.

This example demonstrates how the external environment forces us to move from one mode of learning to another. In other words, there was a need for double-loop learning. In this mode, new learning occurs and more knowledge is provided. However, after the new approach to molding is in place, a need for continuous improvement of the new methodology arises again. So, single-loop learning "returns."

In most organizations, there is a place for both approaches of learning. However, in recent years, a large group of scientists and writers have focused on the transformative side of learning. They argue that incremental learning does not totally satisfy the needs of today's organizational environment.[10] While understanding the concerns of those authors and recognizing the power and benefits of transformative learning, we also observe a great need for continuous improvement. Incremental improvement, which leads to single-loop learning, is a way to polish the process, where people who work in the process improve their skills and enhance their existing experiences. This is in contrast with transformative learning, which generates double-loop learning and leads to radical improvement that often results in a breakthrough effect.

Organizations need to find a balance in the application of different forms of learning, and this balance will depend on the specific needs of the industry and organization. For the electronics industry, in our opinion, learning investments should occur concurrently with improving the existing conditions (single-loop learning), and with creating the organization's future (double-loop learning). Figure 4-9 shows that the incremental and transformative approaches toward progress work very well together and produce better products and deeper learning.

Conclusion

It is impossible to give an organization a prescription for how it should learn. It depends on the organization's peculiarities and environmental conditions.

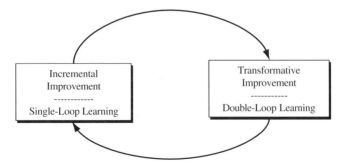

Figure 4-9 The Complementary Effect of Single-Loop and Double-Loop Learning

However, it is obvious that a modern organization cannot be satisfied with only Maintenance Learning or Shock Learning, which is a model of single-loop learning. As Robert M. Fulmer suggests, "The best learning approach is one that helps individuals and organizations assume responsibility for creating the future they want long before a crisis limits the organization's choices."[11] This means that out of all the forms of learning mentioned above, the future-oriented form of learning, which facilitates double-loop learning, is the most appropriate form in today's turbulent environment.

References

1. Richard Karlgaard, "Bill Gates," *Forbes Magazine*, New York, NY, December 7, 1992, p. 71
2. Peter M. Senge, *The Fifth Discipline: The Art and Practice of the Learning Organization*, Doubleday, New York, NY, 1990, p. 13
3. Chris Argyris, *On Organizational Learning*, Blackwell Publishers, Cambridge, MA, 1992, p. 8
4. Ibid., "Teaching Smart People How to Learn," *Harvard Business Review*, Boston, MA, May-June 1991, p. 99
5. Ibid.
6. Ibid., *Overcoming Organizational Defenses: Facilitating Organizational Learning*, Pearson Education, Inc., Upper Saddle River, NJ, 1990, p. 92
7. Ibid., p. 94
8. Robert M. Fulmer, "A Model For Changing The Way Organizations Learn," *Planning Review*, volume 1, 1994, pp. 20–24
9. Gary Hamel and C.K. Prahalad, *Competing for the Future*, Harvard Business School Press, Boston, MA, 1994, p. 166
10. Anthony DiBella, *How Organizations Learn*, Jossey-Bass Inc., Publishers, San Francisco, New York, NY, 1998, p. 50
11. Robert M. Fulmer, "A Model For Changing The Way Organizations Learn," *Planning Review*, volume 1, 1994, p. 24

Chapter 5

Systemic Problem Solving (SPS) as an Effective Way of Learning

In the meetings you attend, pay attention to the number of times the word "problem" is repeated. You will probably find that there are not any meetings where someone doesn't say, "The problem is...." This is not surprising. After all, what are meetings for?

But what *is* a problem? Where do they come from? How do we reduce them? Can we live without problems? These and other questions were brought up in a seminar we conducted called "The Problem With Problems." This chapter is a result of that seminar, which suggests that the knowledge manager should know more about this subject. We are not talking about the six or eight steps of problem solving or special techniques for effective problem resolution (like you will find in a lot of textbooks). Rather, we want to describe the anatomy of a problem, define different types of problem resolution, demonstrate the importance of applying a systemic approach to problem solving, and talk about other aspects related to this subject. All this will enlarge the meaning of problem solving, and we will start to see it not only as a vehicle of process improvement but also as a vehicle for continuous learning and knowledge creation.

5.1 THE PROBLEM OF PROBLEMS

If you asked a firefighter what he prefers more: playing cards while waiting for a fire alarm or actually fighting a fire, what do you think his response would be? Whatever the answer, he probably would not say that he hates fire fighting because that is his profession. He joined the fire department because he likes fighting fire. He hates to see people suffering from a fire. He does not want to be wounded or see his colleagues wounded while fighting the fire, but he likes the job. It creates a tension, and when the fire is in control, the tension is released and he feels good about his contribution to saving people's lives and assets. He feels important and needed. He is a hero.

We used this example just to emphasize that in the manufacturing environment we also have "firefighters." In the general working environment, fire fighting is not as dangerous—in the worst case, you could lose your job, but not your life. However, it is a rare case when a manufacturing worker loses his job from not "putting out a fire." What often happens is that you get promoted faster if you are a good "firefighter." People go from meeting to meeting making notes about problems and come back to their subordinates and share the "alarms" they heard about. Teams attack the problems; everybody is busy; and everyone feels good because there is structural tension and an opportunity to demonstrate their talent, knowledge, and expertise. This gives us a feeling of security; a feeling that we are needed. In the process of solving the problem, the tension is gradually released, and some positive results are achieved. Because of this, management's interest in the problem is reduced, and if a new "fire" occurs, you may be transferred to take care of the new, more important problem, even though there may still be work to do on the previous problem. Because of this, the problem you just attempted to resolve may come back; although you may not recognize it because it may have taken on a new pattern and may have more causes added to it. All this keeps you busy and makes you feel important. You hate problems, but in a way you like them because they have become as much as part of your working life as fire is for a firefighter.

We may have exaggerated a little here because we want to bring your attention to the problem of problems that need to be elaborated on and then eliminated.

5.2 RAPID CHANGE REQUIRES FAST LEARNING

From the moment of its conception and throughout its lifetime, an organization is busy taking care of problems. Working on problems remains an important part of the organization's activities, independent of the progress and changes made. But how do time and change influence the structure of a problem? Is there any difference between a problem today and one ten or twenty years in the past? Donald A. Schön[1] first brought our attention to the relationship between change and problems. In its simplest form, his argument can be described as follows.

As the rate of change continues to increase, the complexity of the problems we face also continues to increase. The more complex these problems become, the more time is needed to solve them. The more the rate of change increases, the more the problems that we face change and the shorter the life of the solutions we find to them (see Fig. 5-1). Because of this, by the time an organization finds the solutions to the problems, the problems have so changed that our

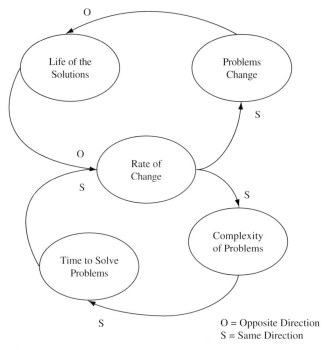

Figure 5-1 The Relationship Between Change and Problems

solutions are not exactly valid or effective. In other words, many of our solutions fit problems that no longer exist in the form in which they were solved. This can take the organization further and further behind the times.

Many experts suggest that the problem created by accelerating change can be solved by improving the organization's ability to adapt and by fast learning. In his book *Creating the Corporate Future*, Russell L. Ackoff gives considerable attention to the subject of learning and adaptation. He recommends, "It is better to develop greater immunity to changes that we cannot control, and greater control over the others."[2] So, if we cannot influence the speed of change, we should learn how to react to problems faster and more effectively. This is why understanding the anatomy of a problem and knowing the forms and methods of reacting to problems will help us avoid situations in which the solution to the problem becomes obsolete before we implement it, in other words, to avoid the introduction of the right solution to the wrong problem.

5.3 A HOLISTIC VIEW OF PROBLEM SOLVING

One way to increase the effectiveness of problem-solving activities is to learn how to see them as wholes, or as they are related to larger wholes. You can probably remember a time when people would say, "That's not MY problem." Every plant and department had its own problems (or part of a larger problem) to take care of. An organization was divided into segments, and if a problem occurred in one of these segments, it was solved there. When a larger problem was identified, it was divided into parts and each part was given to a different person or department to solve. There was no thought about how all the solutions would fit together. Departments and individuals all had a tendency to ping-pong problems back and forth and find ways to reason that "this is not our problem." This situation reminds us of an old Russian movie in which a man's body found lying on the street had been moved back and forth by the policemen from different parts of the city to make sure that the body was not found in their territory.

Segmentalism of the problem-solving process restricts the ability of the organization to utilize its existing knowledge and experience and to see the problem from a holistic perspective. Organizations that learn faster and continuously create knowledge should introduce a holistic approach to problem solving. This will allow the organization to see any problem as a part of a larger whole. Organizations that practice the application of the cross-functional problem-solving approach have the opportunity to introduce integrative thinking and to create an environment in which new ideas, experiences, and knowledge can be shared across organizational boundaries. This will keep organizations from seeing problems from too narrow a perspective or independent from the context. To solve problems properly, they need to be viewed as connected to and independent from any other problem, as we see problems within the human body as independent yet connected to the whole. A holistic and systemic view in problem solving is a way toward knowledge creation and deployment, a way to improve the organizational effectiveness.

5.4 THE ANATOMY OF A PROBLEM

A problem can be viewed as a gap between the current state and a desired future state, in other words, the gap between the state where we are in a particular moment and the state of where we want to be (see Fig. 5-2).

The lower part of Figure 5-2 represents the current state (where we are). This can be the quality level of a technology process, the market share of an organization, or any other level of the initial state. The upper part of the figure represents the desired future state, the goal or vision for improvement and change that reflects our desires. Between these two parts is a gap that represents the magnitude of a problem. So, as you can see from this figure, if there is no desire, then there is no problem.

Using the terminology of Robert Fritz,[3] this gap creates a *structural tension* that requires resolution—the greater the gap, the greater the tension. When there is no desire for change or improvement, the structural tension is zero. By viewing a problem as a gap between the current and desired states, we may

THE ANATOMY OF A PROBLEM 85

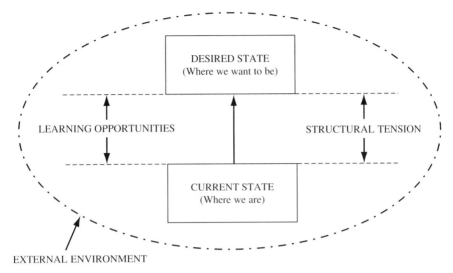

Figure 5-2 The Structure of a Problem

conclude that only by setting a goal or creating a vision can we design a problem. This is not exactly correct. A problem occurs any time there is a deviation from our desired state. You desire, for example, to drive safely and never have a malfunction of your car. So when you have a flat tire, you have a problem; which is a deviation from the desired state. It creates a structural tension that can only be resolved by changing the tire. Such terms as goal, vision, and mission do not fit here. You just need to resolve the problem by changing the tire; and that is considered a reactive action.

However, in our description of the anatomy of a problem, we would prefer to bring to your attention the proactive part of taking care of problems because proactive actions are the only way to progress. Using reactive problem resolution is important to maintain the status quo but insufficient to move forward. From this perspective, without challenging goals or creative visions, there will not be innovative problems. There will only be problems that relate to everyday routines.

There is one more important element in the anatomy of a problem. This is the environment in which the problem is imbedded. Any problem happens in space and time. The time and the circumstances can create a new problem, can challenge our perception of problems, and can accelerate or dissolve problems. In short, to understand the problem, we need to understand the environment in which the problem appears or is created. For an illustration, let's go back to 1969, when the United States first sent a person to the moon. Was it a problem not to have the capability of sending a person to the moon? Probably not. Russia was the only country in the world that appeared to be striving for this capability. Was it a problem for the other countries? It only appeared to become a problem for the U.S. when President John F. Kennedy declared a vision to "send a person to the moon and to return him safely no later than 1970." Only after a challenging vision was created did a lot of problems occur. And these problems arose from the need for U.S. superiority in space. At that time, Russia had achieved temporary superiority, and the only way for the U.S. to regain superiority was to do something that no country had done before: send a man to the moon and return him to Earth safely. This example demonstrates how a problem can be imbedded in the context and influenced by the environment. We can find many examples in organizations where competition, customer demands, and the struggle to stay in business created problems, which created structural tension, and this tension, in turn, created continuous progress.

The gap between the current state and the future vision state is not only a gap for action, it can also be seen as a learning zone filled with opportunities to learn. Acting on problems is one of the most effective ways to create continuous learning and accumulate new knowledge. Our experiences have shown that adults are not inspired to learn in the classroom alone, but, rather, show them the problem and when they feel that their experience and existing knowledge is not enough, they will seek sources of knowledge. People tend to be self-directed to look for learning opportunities. So, having introduced a system of problem solving, we also introduce a system of continuous learning.

5.5 WHY PROBLEMS REPEAT THEMSELVES

In many management meetings, we frequently hear, "We must get rid of the problem." If the intent of problem solving is just to get rid of something we do not like, then we will always be overloaded with problem solving and not move forward. As Robert Fritz suggests,[4] problem solving has a built-in structural tension that oscillates. And, as we already know from the previous section, oscillation is like sitting in a rocking chair: it gives you feeling of movement, but you are not actually moving anywhere. This is why, when describing problem-solving activities and seeking improvement, it is very important that the knowledge manager understand the structural tendency of oscillation in problem solving.

For purposes of illustration, let us first observe the traditional approach of problem solving. As we mentioned earlier, we usually start by defining the problem. Then the undesired problematic situation provokes actions designed to reduce or eliminate the gap between the actual and desired states. Our intention in this case is to get rid of the problem. The larger the gap, the more attention is required by the problem. However, by taking action the tension is reduced, and this reduces the attention required by the problem. We feel better about the situation. And, even if the problem is not yet solved, this improvement of the situation reduces the problem intensity and probably the interest given to it. All this creates oscillation. In his book, *Corporate Tides*,

Robert Fritz describes in detail how the traditional process of problem solving forms an oscillation pattern (see Fig. 5-3).

As Figure 5-3 suggests, high intensity of the problem leads to actions taken to solve it, which in turn lead to lower intensity of the problem, which further leads to less action. This leads to reintensification of the problem if it was left unsolved. What we have just described is a predicable pattern of oscillation. If we adopt this strategy, at best we can expect a temporary release and some satisfaction from the illusion that the problem is gone. It is like putting a Band-Aid on a serious wound. It will not totally heal the wound, but it will give you some temporary relief.

In summary, if an organization focuses its activities mainly on reacting to problems, sooner or later it will have a real problem staying in business. Conversely, if an organization focuses its activities mainly on achieving desired results, it will move successfully into the future. To demonstrate the difference between reacting to a problem and reacting to a desired vision, Robert Fritz uses an analogy that emphasizes the difference between building demolition and architecture. He writes, "One is taking action to have something go away [building demolition], the other is taking action to have something come into being [architecture]. This is the difference between problem solving and driving the organization by a vision of what we want to accomplish."[5]

The solution to the problem about problem solving is to reduce the oscillation of the structure in such a way that will allow us to have a constant structural tension, which will generate continuous progress. Robert Fritz advises organizations that introduce Total Quality Management (TQM) or other change systems that they will enhance performance, provided they are not used as problem-solving devices. He writes, "When TQM or other change systems are used as problem-solving devices, they do not work. This is because the

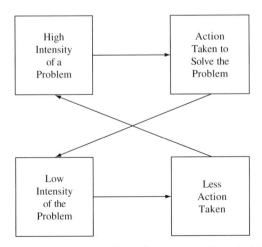

Figure 5-3 The Tendency Toward Oscillation In Problem-Solving Activities

generative force within the condition-analysis-improvement cycle moves toward oscillation as improvements are made. However, when this cycle is placed into the context of structural tension, it moves toward advancement."[6] This is why, in AMD, the Total Continuous Process Improvement and Innovation (TCPI^2) macro system was designed so that it moves the organization forward while making oscillation difficult.

5.6 THREE FORMS OF PROBLEMS

According to Russell L. Ackoff, problems can be treated in three different ways: they can be *resolved*, *solved*, or *dissolved* (see Table 5-1). To *resolve* a problem, we need to find the right means that will result in an outcome that is good enough and satisfies a particular design. Ackoff calls this approach "clinical" because it relies mainly on past experience and it is rooted deeply in common sense. Managers, because they are always short of lead time, are usually problem resolvers. Like most clinicians, managers almost always argue that there is no time to seek alternative solutions and claim that their selected methodology is preferable because it minimizes risk and saves time.

To *solve* a problem, we need to select a means that is expected to yield the best possible outcome, one that optimizes alternative actions. Ackoff calls this "the research approach" because it is based on scientific methods and techniques. The research approach to solving problems is usually used in engineering and R&D departments whose main obligation is to search for future improvement rather than solve day-to-day problems that are mainly related to the subject of meeting short-term goals. Managers who use the research approach often resort to clinical treatment when a problem cannot be treated quantitatively.

To *dissolve* a problem, we need to either change the nature of the entity that has it or alter its environment to remove the problem. Problem dissolvers are

Table 5-1   Three Ways of Treating Problems

Ways of Treating Problems	Approach	Lifespan	Outcome
Resolving	Clinical	Shorter than "solving"	Satisfy (good enough)
Solving	Research	Shorter than "dissolving"	Optimize (the best possible result)
Dissolving	Design	The longest life span of the three forms	Idealize (results in a system or environment change)

Note: Few if any problems are ever permanently resolved, solved, or dissolved.

Source: Based on Russell L. Ackoff's *Creating the Corporate Future*, John Wiley & Sons, Inc., 1981, pp. 170–171[7]

usually involved in changing the system and bringing it closer to an ultimately desired state, one in which the problem cannot or does not arise. Ackoff calls this the design approach.

Understanding the difference between resolving, solving, and dissolving problems allows us to better understand the term "problem" and recognize that taking care of problems has a much broader definition than just maintaining the status quo. In the next section the reader will find an example that will provide better understanding of the difference between the three terms mentioned above.

In this section, we describe three ways to react to problems. Depending on the intention, we resolve, solve, or dissolve problems. On a routine basis, when we react to a deviation from the desired result, we take actions to *resolve* the problem. For example, we perform activities to bring the process back into a state of statistical control.

When our intention is to achieve greater results, we *solve* problems. For example, we may introduce a "design of experiment" to achieve a higher process capability. However, if we find a completely new solution that allows us to eliminate the occurrence of a chronic problem, we *dissolve* the problem. For example, we introduce a new software package to totally eliminate the possibility of shipping the wrong product to a customer.

The differentiation of the activities related to taking care of problems allows the knowledge manager to better understand the meaning of a problem and apply different actions to achieve greater results. The application of all three forms in the process of continuous improvement and innovation will not only increase the effectiveness of work but also enhance the learning process.

Summary

In this section, we used three terms—resolving, solving, and dissolving—to distinguish between the different types of activities that usually take place in closing the gap between the current state and a future state. In doing this, we enlarged on the term "solving," which people used to think of only as a general term for problem solving. Therefore, we will use the term Systemic Problem Solving (SPS) as a comprehensive definition that includes all the terms related to closing the gap between the current and desired states (i.e., using all means to achieve the goal).

5.6.1 An Example to Demonstrate the Three Types of Problem Resolution

For more than fifty years, the semiconductor industry has used wire-bonding technology as the main concept for performing the connections between die and lead frames. To maintain and hopefully improve the established yields and resolve clinical problems, organizations were resolving problems by using their past experience and good sound judgment. However, as the years passed, the functionality and performance of the integrated circuits and the number of input/output (I/O) connections required increased. This forced the semiconductor industry to spend more on research and development and to design more complicated wire-bonding machines that are capable of bonding to smaller bond pads. This allows a greater number of pads to be squeezed onto a die while keeping it as small as possible. *Solving* problems in this way was possible for a relatively long time, but when the I/O connections became very large, the wire-bonding methodology could no longer be considered optimal. Further designs to improve the product, machines, and technology in the existing frame of thinking were very difficult and not economical. A new solution based on a different mind-set was needed. When IBM first came up with flip-chip technology, it allowed them to *dissolve* a large number of problems that were prevalent in wire-bonding technology. This became possible because wires were no longer used to connect the die to the substrate. Flip-chip technology is a process that uses solder instead of wires to provide the metallurgical means for forming interconnections. In other words, solder is applied to the bond pads, and the die is flipped over and interconnected to the package base.

To prevent damage to the underlying die circuitry, wire bond technology restricted us to using only the periphery of the die for bonding. Because of this, a larger die was needed to have the space for a larger number of wire bond connections. Because the flip-chip process does not damage the underlying die circuitry, bond pads can be placed in an array on the active area of the die instead of just on the die's perimeter. By using an area array design, this resulted in a reduction of the die size by 30–50%. Replacing wire-bonding technology with the flip-chip technology is an example of problem *dissolution*. All the problems with die size, distance between the wire bond pads, etc., were dissolved. Certainly, flip-chip technology is not free of problems, but these problems are at a higher level of development, which provides us with greater opportunities for further progress in the semiconductor industry. The AMD-K6® and AMD Athlon™ and other high-speed microprocessors would have been very difficult, if not impossible, to design without changing the mental model of forming interconnections. As you can see, the formulation of three different levels of dealing with problems sheds a broader light on the difficulty of working on problems.

5.7 SYSTEMIC PROBLEM SOLVING (SPS)

SPS is an important block of the $TCPI^2$ macro system. It includes all the elements of the activities related to closing the gap between the current and desired states. The SPS block is interconnected with other $TCPI^2$ subsystems such as motivation, innovation, information, education, etc. SPS allows an organization to continuously resolve, solve, and dissolve problems and to create new knowledge for future activities.

Figure 5-4 is a model of SPS, which includes two major modes of closing the gap between current and desired states. The first mode—reactive—has only one

SYSTEMIC PROBLEM SOLVING (SPS)

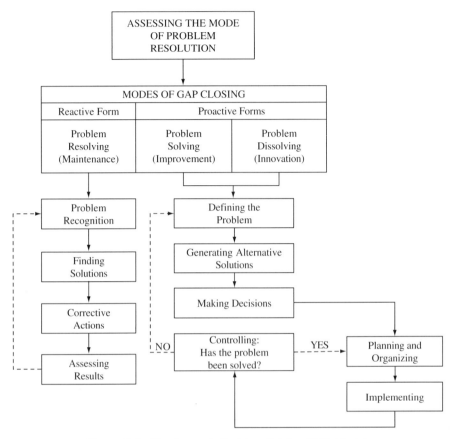

Figure 5-4 The Systemic Problem Solving (SPS) Model

form of overcoming a problem. The second mode—proactive—has two forms of overcoming a problem: problem solving and problem dissolving. These three forms can be used separately or together, depending on our objectives. For instance, to maintain an established level of process quality in manufacturing, we usually use a reactive mode to *resolve* problems whenever a deviation from the target is observed, for example, if machines stop working properly, or the yield drops, or the customer returns products, etc. As Figure 5-4 suggests, to close the gap in this mode, we resolve the problem by going through a process of problem recognition, finding solutions, introducing corrective actions, and assessing results. The feedback (shown with a dotted line) emphasizes that this is a maintenance activity, or "keeping your hand on the pulse" of the process. As a deviation occurs again, the cycle of resolving the problem is repeated. What we have described so far may remind you of a thermostat that goes into action when the temperature drops below the desired level. Resolving problems takes up a lot of time and is a very important part of maintaining the operation and manufacturing results at a desired level. However, to stay in business, an organization also needs to perform activities for continuous improvement, which is another form of closing the gap.

Activities related to problem *solving* come into play here. These activities are related to goal setting and finding the means to continuously raise the bar of desired results. What was good yesterday may not be satisfactory today, and what is good today will need to be further improved for the company to stay in business tomorrow. As Figure 5-4 suggests, problem *solving* is a proactive activity. As soon as the mode of problem resolution is determined, the process flows through the steps of defining the problem, seeking alternative solutions, making decisions on the best alternative, planning and organization, implementing the selected solution(s), and controlling the outcome. If the outcome matches the desired results, we plan for the future, and the process continuously loops from planning to implementation to controlling. If the outcome is not satisfactory, this is a signal that we probably worked on the wrong causes or we defined the problem incorrectly. In this case, we restart the cycle from the definition of the problem. Solving problems is an effective way to improve the process, and problem solving usually involves teamwork.

While recognizing the importance of problem solving as a form of process improvement, we should also say that this activity has its limitations. For example, sometimes you can improve the yield of a process to some extent, but afterwards you face a barrier where continuous improvement ends or gets very expensive. Just like the wire-bonding technology we mentioned above, even after many improvements, very often manufacturers still need to find a new approach to achieve significant improvement [flip-chip technology (read more about this in Section 5.6.1)]. Sometimes a total change in the organizational mind-set is needed to progress further. The development of flip-chip technology was a move toward *dissolving* a problem.

This activity is called problem dissolving because by moving to a new concept, we dissolve the problems that blocked future progress. Problem dissolving is based on activities that require innovation, creativity, and new knowledge. It usually results in a breakthrough that creates the opportunity for tremendous improvement and double-loop learning. Problem dissolving goes through the same cycle as we described earlier for problem solving (see Fig. 5-4). However, the means and the approach in the process of problem dissolution are different. They take more time, more effort, and perhaps more capital investment. As in any system, the three forms of SPS mentioned above work as interconnected subsystems that complement each other. For example, when Advanced Micro Devices implemented flip-chip technology (problem dissolution), this immediately created a need for continuous improvement activities (problem solving) to enhance the new process. It was also necessary to continuously advance the processes toward a new target, which required problem resolution.

Therefore, these three forms do not contradict each other but work together and complement each other. When the form of problem dissolving is applied, a breakthrough is usually achieved. At the same time, old assumptions and mental models are also dissolved. Because of this, problem resolving and problem solving are applied at a higher level of performance, and all of this is called progress—a means for organizational survival.

5.8 THE CURRENT REALITY TREE

How many times have we taken care of a problem and were absolutely sure that we had fixed it, and then later wondered why the problem reoccurred? This is mainly because we worked on the symptoms of a problem while thinking we worked on the real cause. How do we differentiate the symptoms from the real causes? The methodology that follows is one of the right tools to get at the root causes of a problem. It is called the "Current Reality Tree" (CRT), which is part of the Theory of Constraints (TOC) that was developed by Dr. Goldratt in 1986.[8]

Below is a brief introduction to the TOC; we then use a step-by-step approach to introduce CRT, followed by an actual example that demonstrates how to use the CRT to locate the root cause(s). The main purpose of this section is to give the reader a feeling for how the concept of CRT works.

5.8.1 A Glimpse of the Theory

Dr. Goldratt uses the metaphor of a chain to illustrate an organization. He notes that the various departments function as a complete chain altogether. A strain in one of the links weakens the ability of the entire chain to act as a force. The strength of the chain is solely dependent on this weakest link. If we want to improve the strength of the chain, it makes sense to improve this weakest link first before focusing on any other links. The same applies in an organization. No matter how strong the organization is, there is a relatively weaker link. This constraint in one department limits the organization from performing at its optimum, hence obstructing it from reaching its goal. A constraint is identified as "whatever impedes progress toward an objective or a goal."[9] It can be assumptions, conflicts, or internal policy. In fact, it was noted that "most constraints in organizations today are policy constraints rather than physical constraints"[10] because "many times, when we finally break a constraint, we do not go back and review and change the rules and policies that caused the constraint initially."[11] Schein also noted that "perhaps the most limiting factor to a company's ability to make money is its internal policy."[12]

Therefore, if the application of TOC has successfully helped break a constraint, then we should not rest on our laurels and allow inertia to rule. This would eventually become a constraint in itself. Instead, we should repeat the process of TOC to resolve the next weakest constraint, and so on. It is a continuous process of improvement.

In an interview with *Management Roundtable*, Dr. Goldratt said, "When I talk to managers, I find that one of the biggest problems in most companies is that most of their people don't see the company as a whole. They see fragments.

Because of this, you get localized optimums, many wrong decisions, and much miscommunication."[13] Instead of focusing on the trees (processes), TOC looks at the forest (entire system). It helps us look at things from a wider angle and understand the interdependence of separate parts. This is important because by understanding the interdependence of existing subsystems, we can then make improvements as a whole unit. The improvement of a stand-alone unit may not be desirable if it is deemed to defeat the progress of the unit as an entirety. "Thus, every action taken by... any part of the organization should be judged by its impact on the overall purpose."[14]

TOC consists of five major tools: 1) the Current Reality Tree, 2) the Evaporating Cloud, 3) the Future Reality Tree, 4) the Prerequisite Tree, and 5) the Transition Tree.

1) Current Reality Tree: The Current Reality Tree shows the cause-and-effect relationships between the present undesirable effects (UDE) and their causes. The purpose is to look for the primary problem. Once the problem is resolved, all the undesirable effects should be eliminated as well.

2) Evaporating Cloud: The major role of the Evaporating Cloud is to help resolve conflicts by means of uncovering the assumptions or actual causes of the conflicts. This is achieved by verbalizing the assumptions. This can be done through brainstorming or any idea-generating process. Once verbalized, the assumptions can be understood by the involved parties, thus resolving the conflict. It is important to note that the objective here is not to find a compromise, but rather to establish a common goal and invalidate the problem.

3) Future Reality Tree: The Future Reality Tree projects the desired reality. This is expressed through drafting the cause-and-effect relationships between the proposed solutions that we want to implement and the resulting desirable effects. While drafting the desirable effects, we will also identify any potential negative effects resulting from the proposed changes. In this way, this what-if exercise enables us to think through and check for deficiencies in our solutions before we invest resources.

4) Prerequisite Tree: The Prerequisite Tree identifies the obstacles that stand in the way when we want to implement the changes that lead us to our target. It spells out the minimum conditions necessary to overcome obstacles, so that the target can be realized. Because there may be more than one obstacle and condition in planning to move toward this major target, this tree also determines the intermediate objectives to be achieved. Each may be the target of different obstacles and conditions.

5) Transition Tree: The fifth and last, but by no means, least important tool of the Thinking Process, is the Transition Tree. As the name suggests, the Transition Tree takes us from the current state to the identified, desired state in the future. It differs from the other Trees in that it is a systematic, step-by-step action plan that focuses on implementing the

how-to actions to reaching the identified goal. Besides the actions, it also identifies and communicates the reasons for those actions.

5.8.2 Current Reality Tree

As the intent of this section is to provide a taste of what TOC is, and in particular, the Current Reality Tree (CRT), we do not intend to go into great detail. Moreover, the concept of TOC is too broad to be covered and understood within a few pages. Therefore, we will restrict the scope of discussion to the CRT. We will broadly outline the major steps in constructing the tree, after which we will illustrate the steps with a supportive example. Readers who are interested in having a deeper understanding of the TOC concept are strongly encouraged to refer to the books referenced in this chapter.

Steps to Building a Current Reality Tree (CRT)

Step 1. Identify Area of Constriction This essentially refers to how far you can stretch your authority, directly or indirectly. It is important to realistically identify this area as it affects the root problem that we single out from the tree ultimately. We would not want to go through the problem-finding process only to realize that we are not in control of the situation.

Step 2. Identify Undesirable Effects These are the negative effects that spring from the problem you are facing. Hence, start by writing down the problem in simple terms, for example, why is the turnover rate so high? Next, list as many undesirable effects as possible. However, if there are too many to deal with, you will need to prioritize them and take only the worst few. Keep the rest, though; they may be needed later.

Step 3. Connect the Undesirable Effects Determine whether there are any links among the undesirable effects and which effects are linked.
Start with two undesirable effects. Put one vertically below the other. Does the lower undesirable effect cause the upper undesirable effect? If yes, well done. You have established a cause-and-effect relationship. If no, there could be other intervening steps between the two. Evaluate this again, or refer to those you sieved out in Step 2, and repeat the process of connection until all possible connections are made. It is all right if there are a few undesirable effects left unconnected.

Step 4. Build the Tree Every connection of undesirable effects forms a branch of the CRT. This step traces every branch downward, with the objective being to reach a common root cause of the problem.
Start with a branch. Determine what contributes to the cause. Write the reason below it. Then determine what contributes to this reason, and write the answer below it. Keep determining the "whys" of every cause, and

write each reason below the cause, until you have reached the lowest reason. On reaching the bottom of the branch, look for horizontal connections with other undesirable effects.

When all undesirable effects are connected, either vertically or horizontally, look for those causes that have many arrows branching out from them, without any arrows going in. These are the various root causes (RC). The core problem (CP) will be the one from which the majority of the undesirable effects originate. A good guideline to determine a "majority" would be that it accounts for 70% or more of the undesirable effects.[15]

Step 5. Decide on the Core Problem   Ask yourself whether the core problem identified in Step 4 is within your area of authority to correct. If not, you will have to drop it and recognize that it is not within your means to solve. Take a look at the other root causes, starting from the next worst, and stop at the one that is within your area of authority to do something about. Remember that the objective of the CRT is to find out the source of the problem and be able to solve it.

Step 5a. Review the Tree   It is always advisable to let a third party vet and comment on the CRT you have constructed before finalizing it. The rationale is simple. We tend to think our masterpiece is perfect because we have taken so much pain to create and fine-tune it. However, more often than not, feedback from an objective party may highlight some very critical, but overlooked, areas. Where the problem affects more than yourself, it is even more crucial to have another view.

5.8.3   A Manufacturing Example

The example we selected to demonstrate the methodology is a relatively simple one. In many cases, a problem consists of many undesirable effects (UDE), and the CRT allows us to cut them down to a limited number of real causes. This illustrative example is provided as an introductory illustration to using the Theory of Constraint, which really must be tried to be understood. This concept is not as simple as it looks. If it is used properly, the results are tremendous.

We will illustrate the construction of a CRT to locate the core problem. This example is taken from the manufacturing environment in the semiconductor industry. The department concerned was responsible for the testing of various devices.

A Brief Summary of the Background   In 1999, a manufacturing plant was able to maintain excellent on-time delivery because of excess capacity and its ability to react to market dynamics. Beginning in January 2000, the plant management noticed that there was a steady increase in the volume of devices that were being shipped in for testing. This was good news, because it meant that there was a surge in market demand. However, even though the assembly run rate was growing to 900K per week, the current volume to be tested remained at 630K per week.

The plant's Industrial Engineering (IE) department forecasted that there would be a capacity shortage throughout the year 2000, especially with LTX testers. While waiting for a long-term solution from Corporate Engineering to sourcing for testers (these testers are difficult to source in the market), this offshore plant had to take some immediate action. The concept of TOC was employed to look into the problem. Below is the step-by-step approach taken to construct the CRT and locate the core problem (see Fig. 5-5).

Step 1. Identify Area of Constriction The situation was first analyzed by mentally identifying and differentiating the areas in which action could be taken by the local plant (span of control). Then those areas that were not under the local plant's control (span of influence) were set aside for consideration by the upper management.

Step 2. Identify Undesirable Effects A problem statement was generated: The quantity of pack-outs (finished product) was so low. Answering the "why" threw up a total of 16 UDEs, of which the worst 5 were selected (listed below). All the UDEs were independently negative, even though they were connected.

A. The number of units tested is below the standard set by the plant.
B. The untested inventory is too high.
C. The release rate of parts is slow.
D. The testers were not fully utilized.
E. The loading plan is frequently revised.

Step 3. Connect the Undesirable Effects All the UDEs were written on Post-It notes, which made it easier to shuffle them around and connect them in a manner that would show meaningful cause-and-effect relationships.

UDEs B and C were arranged into a vertical cause-and-effect relationship; UDEs A and D were also found to be the effect and cause, respectively, and UDE E was connected to C through intermediate entities and to D through an ellipse.

Step 4. Build the Tree The UDEs were connected starting from the top of the tree. At each connection, the team asked themselves the reason(s) for the cause of the stated UDE(s). They gradually reached the bottom of the tree, where the root causes were identified.

Appropriate places were found in the tree to connect additional UDEs. Arrows and ellipses were drawn to complete the tree. Intermediate causes and effects were introduced onto the tree as required.

Step 5. Decide on the Core Problem Root causes were identified. To qualify as a core problem, a root cause must account for 70% or more of the UDEs. To determine the core problem, the following root causes were analyzed (see Table 5-2).

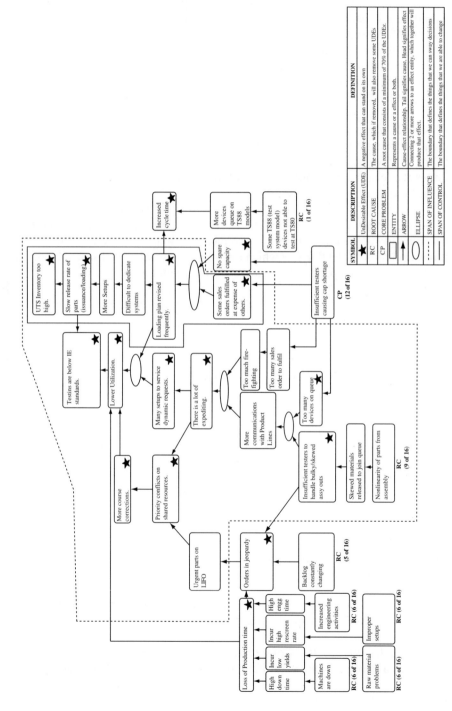

Figure 5-5 An Example of a Current Reality Tree

Table 5-2 Root Causes

Root Cause	Total No. of UDEs	No. of UDEs Eliminated	% UDE Eliminated	Selected Core Problem
Machines are down	16	6	38%	
Increased engineering activities	16	6	38%	
Raw material problems	16	6	38%	
Improper setups	16	6	38%	
Backlog constantly changing	16	5	31%	
Nonlinear assembly parts	16	9	56%	
Insufficient testers	16	12	75%	**
Different configurations of various testers	16	1	6%	

**Identified as the core problem.

The core problem "insufficient testers" was identified based on this rationale:

- Izt accounted for 12 out of 16 UDEs, which was about 75%;
- It would negate more than half of the UDEs;
- It matched the team's initial intuition that it was the core problem.

Step 5a. Review the Tree Even though the problem affected only the Operations department, a third party (a scheduler with knowledge of CRT and the problem at hand) was asked to review the tree.

Conclusion The application of TOC is not limited to manufacturing, management, or quality in organizations but applies to virtually all aspects of life; for example, in career path planning, family planning, direction in studies, or even interpersonal relationships, TOC can help us to analyze and solve problems. The important point to note here is that we cannot solve the symptoms of a problem, but rather, its root, as demonstrated by the real-life example above. Having solved the problem, we do not stop there but proceed to look at the next most problematic area. Problem solving is a continuous process. We are not proposing that the concept of TOC is the only methodology to solve problems. It is, however, a very effective approach to tracking and eliminating the root cause of a problem. It is consistent with the SPS problem solving mentioned above.

It is good to keep in mind that although TOC is an excellent improvement methodology, we should not expect it to resolve every problem. Rather, we should adopt the methodology to look at problems from another paradigm while we make our improvement efforts.

References

1. Donald Schön, *Theory In Practice: Increasing Professional Effectiveness*, Jossey-Bass Publishers, San Francisco, a division of John Wiley & Sons, Inc., New York, NY, 1992
2. Russell L. Ackoff, *Creating the Corporate Future*, John Wiley & Sons, Inc., New York, NY, 1981, p. 5
3. Robert Fritz, *The Path of Least Resistance*, Fawcett, Columbine, NY, 1989, p. 115
4. Ibid., *Corporate Tides*, Berrett-Koehler Publishers, Inc., San Francisco, CA, 1996, p. 47
5. Ibid., p. 50
6. Ibid., p. 54
7. Ackoff, *Creating the Corporate Future*, pp. 170–171
8. Eliyahu M. Goldratt, and J. Cox, *The Goal*, North River Press, Great Barrington, MA, 1986
9. Thomas B. McMullen, *Introduction to the Theory of Constraints (TOC) Management System*, CRC Press LLC, Boca Raton, FL, 1998, p. 20
10. Sid Sytsma, *The Theory of Constraints: Making Process Decisions Under Conditions of Limited Resources, Capacities, or Demand*, Marcel Dekker, Inc., New York, NY, 1997, p. 1. Available at: http://www.sytsma.com/cism700/toc.html
11. Ibid.
12. Robert E. Stein, *The Theory of Constraints—Applications in Quality and Manufacturing*, Marcel Dekker, New York, NY, 1997, p. 43
13. Eliyahu M. Goldratt, "On Saddam Hussein, milestones, and how the theory of constraints applies to project management," *The Management Roundtable: Product Development Best Practices Report*, Waltham, MA, August 1998. Available at: http://www.roundtable.com/PDBPR/goldratt.html
14. Eliyahu M. Goldratt, *The Theory of Constraints*, North River Press, Great Barrington, MA, 1990, p. 4
15. H. William Dettmer, *Goldratt's Theory of Constraints*," ASQ Quality Press, Milwaukee, WI, 1997, p. 75

Chapter 6

Knowledge-Based Innovation

"Learning is a source of innovation."

Dr. W. Edwards Deming[1]

In 1985, Peter Drucker emphasized the importance of knowledge-based innovation by writing, "Knowledge-based innovation is the 'super-star' of entrepreneurship. It gets the publicity. It gets the money. It is what people normally mean when they talk of innovation."[2] Today, knowledge-based innovation still remains the superstar. Because in these turbulent times, for an organization to survive, there is a great need for its continuous renewal by introduction of new, innovative products and processes that are based on new knowledge. In today's information-rich environment, there is an ongoing circle of information being exchanged in organizations, which is continuously transformed with new knowledge. This knowledge is the base for innovation.

6.1 CREATING AN INNOVATIVE ENVIRONMENT

We often hear in presentations given by managers that we need to build an innovative organization and that it is important to create an environment in which people can feel free to create and innovate. But what does this actually mean? How different is an innovative organization from an ordinary organization? How does an organization become innovative?

Figure 6-1 is a simplified model of an innovative organization. As you can see, at the center of the model are some specific elements that are embedded in the organization's overall culture. This makes the organizational culture capable of operating in an innovative mode. Figure 6-1 also shows that innovation is

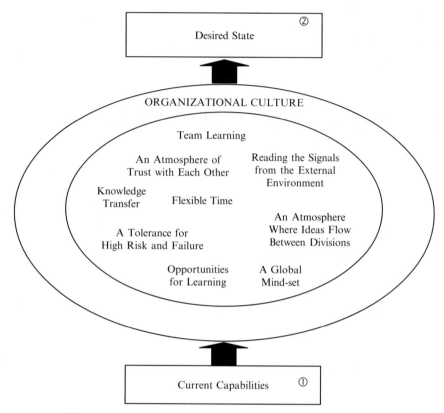

Figure 6-1 A Simplified Model of an Innovative Organization

not a separate phenomenon. It is an essential part of the whole organization, which works as one of the core capabilities to achieve the organization's vision, mission, and goals.

Before we explain the peculiarities of an innovative organization, it is important to note that to innovate, an organization needs to be capable of making a long-term investment. Innovation is not cheap, and there is a long wait for the return on this investment. Highly innovative companies such as DuPont, General Electric, Pfizer, and Rubbermaid, spend in the neighborhood of one billion dollars or more a year on research and development.[3] The time it takes to receive a return on this investment can sometimes add up to many years. Because of this, an organization should learn how to make money with its existing products and processes and find the right balance between short- and long-term investment.

To get into an innovative mode, an organization needs to develop a sense of urgency and to create a stretch that will build the tension toward new innovative achievements. To do this, it is very important to clearly understand the two major elements that create the innovative tension: (1) the current capabilities of the organization and (2) the desired state (see Fig. 6-1). When we talked about process improvement in Chapter 4, we mentioned that these two elements allow an organization to create the necessary stretch, but when we move into an innovative mode, the desired state—the purpose, vision, mission, and goals— should be more risky and challenging. It is also very important to look at the current state from an innovation point of view. This means to assess how innovative the existing products and processes actually are. How competitive are they?

The concept of the S-curve that we described earlier comes into play here. At this stage it is important to decide which products and processes should stay and which of them should be removed. This is a tough decision, in which two kinds of risk are involved. One is the risk of pulling a product off the market too early (it still may produce substantial revenue and profit for a number of years with little or no capital investment required). The other risk is not putting a new product on the market at the right time. Both of these risks can cause significant losses and undesirable effects.

Once the right creative tension has been developed, a lot of work must be done to build the right internal environment for innovation. First, it is necessary to ensure that the organization's culture fits the requirements for innovation. In particular, it is necessary to ensure that the employees and the management trust each other. This sounds very simple, but from our own experience we know that this is not always an easy task. We still want employees to work normal hours, we still want to know when the work will be done, we still cannot tolerate failure, and we still want everything to be under management's control. In short, we prefer short-term success with the odds at, say, 90% in our favor, to long-term success with the odds at, say, less than 50% in our favor.

This will not create a highly innovative environment. From the employees' point of view, trust is built when they are not penalized for negative results on a high-risk project or when they are rewarded for great results on a high-risk project, even when the results are somewhat below the set goal, etc.

An innovative environment also means having the opportunity for learning and knowledge sharing, accessing the necessary information, participating in the decision-making process, knowing what is going on not only in the employee's own department but across boundaries. Innovative people also need their own time when they can experiment and let their thoughts fly. For example, at 3M, managers have adapted the 15 Percent Rule, which allows 3M's employees to devote up to 15% of their working time to projects of their own choosing. This gives the employees the opportunity to work on ideas without prior approval from management, or without even having to tell their superiors what they are doing during this dedicated time. An innovator should have the choice of working in a team versus making an individual contribution to the company's success. We can incur a loss from forcing someone to be a team player. Individual contributors are still team players in the sense that their applied knowledge links to the organization's bottom line.

Some of these elements may sound chaotic, but they are very important components of building an innovative environment. This, however, does not suggest that people in an innovative organization are not concerned with the quarterly results or achieving the company's vision and goals. It only means that the managers in an innovative organization deal with knowledge workers, highly creative people that need management in the form of facilitating them to create. In short, management that is more concerned with creating the environment we just described (see Fig. 6-1) rather than management that is concerned with controlling the employees' actions. Innovative workers know how to do their job. What they need is the right direction and great support.

The First Step in Creating an Innovative Environment

What are the first steps toward creating an innovative environment? Without doing anything special, an organization always has innovative and creative people who use their creativity at work. But if we want to accelerate the process of innovation, it is good to start by forming a special team (see Fig. 6-2) with selected talented people. This team can be formed from a group of individuals who have knowledge in a particular field, such as specialists on test equipment for electronic components, or a team of people from different departments who

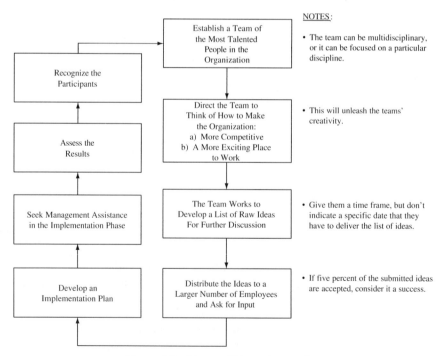

Figure 6-2 How to Trigger Innovation

have different knowledge and expertise—in other words, a multidisciplinary team. This depends on what the objectives are: Do we have a problem that requires deep knowledge in one discipline or a problem that needs to be looked at from different perspectives? For example, if the company wants to concentrate on all kinds of savings to reduce product cost, a multidisciplinary team is required. If the organization is looking for a breakthrough solution in a particular technological process, a team with knowledge in that particular field is required. We should note, however, that experience has shown that even when we need a homogenous team made up of specialists, it is sometimes beneficial to involve people who know little or even nothing in the field of interest. These "strangers" may find a solution to a problem that no specialist would think of.

When the team is formed, the manager should give it a direction to think of how to make the organization more competitive and, at the same time, an exciting place to work. Encourage the team, but do not tell them the details of what to do, and they will bring you a list of ideas, some of which may be brilliant. You, as a manager, should not try to judge what ideas will work and what ideas will not. Distribute the list of ideas to a much larger group, or even to the employees of the whole organization. Let them work on these ideas. Soon, you will get back from the people their own thoughts on what ideas are more important. After boiling down the list of ideas to the most important ones, the management team can develop an implementation plan. What is left is to assist the employees in the implementation phase, assessing the results on a periodic basis.

The last part in the flow of "how to trigger innovation" is motivating the participants. This should be done according to the organization's motivational system, which might include a combination of monetary and nonmonetary rewards. What is important here is to pay attention not only to the dollar value of the implemented idea but also to its originality. The motivational system should encourage risk taking. It is sometimes reasonable to award a project that actually failed but gave rise to a new idea or from which we learned a new way of thinking.

The flowchart (Fig. 6-2) is just an example to demonstrate the issue that innovation cannot be forced. It needs flexibility and a special environment. At the same time, innovation can be triggered, motivated, and challenged.

6.2 CHARACTERISTICS OF KNOWLEDGE-BASED INNOVATION

There are two major characteristics that differentiate knowledge-based innovation from other types of innovation. The first characteristic is the lead time for knowledge to become applicable technology and begin to be accepted in the market. The second characteristic is the convergences, which are related to the fact that knowledge-based innovation is almost never based on one factor but on the convergence of several different kinds of knowledge. Below we provide a short description of these two characteristics, which are described in depth in Peter E. Drucker's book *Innovation and Entrepreneurship*.[4]

6.2.1 The Lead Time of Knowledge-Based Innovation

According to Drucker, knowledge-based innovation has the longest lead time of all innovations. The lead time consists of two major components. The first is the time between the emergence of new knowledge and the time when this new knowledge becomes applicable to technology. The second is the time between when the new technology was created and the time it turns into products (see Fig. 6-3). Together, they make up the total lead time that is needed for an idea to become a reality. For example, by 1918 all the knowledge needed to develop the computer was available, but the first computer did not become operational until 1946, which is a lead time of 28 years.[5] The long lead time of knowledge-based innovations not only holds true for science and technology but applies equally to knowledge-based innovations that are nontechnological. For instance, after World War II, two Americans—B.F. Skinner and Jerome Bruhner, both at Harvard—developed and tested basic theories of learning. Yet only in recent years have learning theories gained in importance. The theory of learning became a hot subject in organizations, especially after the publication of Peter Senge's book, *The Fifth Discipline*, in 1990.

In 1985 Peter Drucker wrote that "... the lead-time for knowledge to become applicable technology and begin to be accepted on the market is between twenty-five and thirty-five years."[6] This still remains true for large discoveries. However, in new technologies, we can observe a trend toward shorter lead times, from the inception of the idea to the time of its actual implementation. For example, between 1962 and 1965, IBM developed a new technology called flip-chip, which could replace integrated circuit wire bond technology. The first actual implementation of this was in 1969, which was 4 years later, with a total lead time span from conception to implementation of 7 years.* The lead

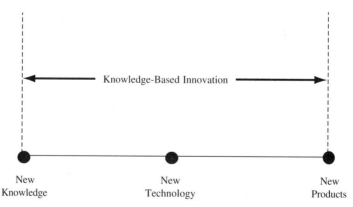

Figure 6-3 The total Lead-Time needed for an idea to become a Reality

*Based on information received from Raj Master, a former IBM employee, who participated in the implementation of flip-chip processing.

time of knowledge-based innovation should be seriously considered by organizations and their knowledge managers when planning change.

Even though we have seen some tendency for shortening the lead time, we have also seen a shortening of the lifetime of products that need to be replaced with a more innovative product. Considering these two tendencies together, the issue of accelerating the speed of innovation remains critical. Because of this, organizations should have a "container" of new ideas and innovations prepared, so that the "green light" can be given as the need arises.

6.2.2 The Convergences of Knowledge-Based Innovation

Another unique peculiarity of knowledge-based innovation that deserves special attention is *convergences*, because knowledge-based innovation is almost never based on one factor but on the convergence of several different kinds of knowledge. For example, a contemporary PC requires convergence with at least five different forms of knowledge:

1. A scientific innovation,
2. The audio tube,
3. A major mechanical discovery,

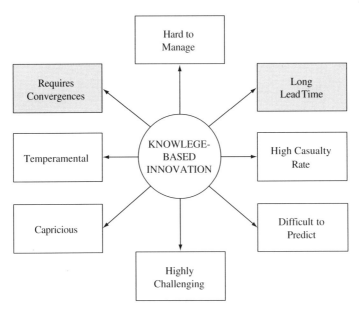

Figure 6-4 The Major Characteristics of Knowledge-Based Innovation

Source: Based on *Innovation and Entrepreneurship: Practice and Principles* by Peter F. Drucker, pp. 107, 111. Copyright © 1985 by Peter F. Drucker. Reprinted by permission of HarperCollins Publishers, Inc.[7]

4. Microprocessor technology and other electronic components, and
5. The concepts of programming and feedback.

Without any one of these factors, no computers could have been built. This demonstrates the point that until all forms of necessary knowledge converge, the lead time of a knowledge-based innovation will not even begin. If an organization does not consider this peculiarity of knowledge-based innovation, and starts to develop a new product without being certain of having the necessary knowledge, there is a very high risk of failure. Innovation only succeeds when all the necessary knowledge exists and is in action. All the knowledge needed may not be available to a particular organization, but it must already be in use somewhere for the organization to gain access to it.

There are other characteristics that differentiate knowledge-based innovation from all other types of innovation (see Fig. 6-4). Here we have limited ourselves to a description of two characteristics: lead time and convergences.

6.3 REQUIREMENTS FOR CREATING KNOWLEDGE-BASED INNOVATION

The requirements reflected in Figure 6-5 indicate that we need a special approach to generate and manage knowledge-based innovations. There are three specific requirements we would like to point out that are very important in

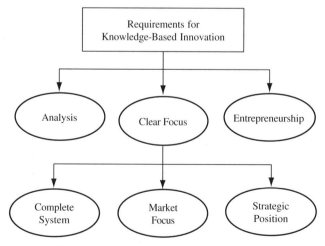

Figure 6-5 Three Major Requirements for Knowledge-Based Innovation

Source: Based on *Innovation and Entrepreneurship: Practice and Principles* by Peter F. Drucker, pp. 115–119. Copyright © 1985 by Peter F. Drucker. Used by permission of HarperCollins Publishers, Inc.[8]

management of knowledge-based innovative activities. They are careful analysis, clear focus, and entrepreneurial management.

6.3.1 Conduct Careful Analysis

Knowledge-based innovation requires careful and detailed analysis of the availability of all necessary factors: knowledge, technology, economy, etc. The main purpose for such an analysis is to determine which factors are available and which are not. On the basis of this determination, a decision is made on which factor can be produced. All these analyses will allow us to make the decision to go ahead with or postpone the introduction of the desired innovation. For example, to introduce a competitive high-speed microprocessor, the following factors must be in place: (1) the designers of the microprocessor should be knowledgeable and skillful enough to produce a new design; (2) the fabrication process should be capable of producing the new design with a reasonable yield; (3) there should be a packaging design for the new device; (4) there should be test equipment in place that can accurately and precisely measure the new parameters; (5) there should be computer software available to utilize the capabilities of the new design; (6) there should be computer manufacturers and other manufacturers that are willing to utilize the new device in their products; (7) there should be consumers ready to apply the capabilities offered by the new microprocessor; (8) there should be benchmarking information on the availability of the microprocessor or what is expected to come out. These and many other factors must be considered when planning to invest in a higher-level microprocessor. If any of these factors are not available and cannot be produced on time with high certainty, it would be better to postpone the implementation than to invest in it without having a clear idea of how it can be produced. The importance of this type of analysis is obvious; however, the "Let's do it and we will see" approach is still very often applied in some organizations.

As Peter Drucker warns, "Failure to make such an analysis is an almost surefire prescription for disaster."[9] Two things can happen if you do not take the time for such analysis: either the knowledge-based innovation will not be achieved, or the organization that invests in this innovation will lose the time and create an opportunity for a competitor to enjoy the fruits of its innovation. A classic example is the development of penicillin. The British discovered it, but because they failed to identify the ability to manufacture the stuff and did not have the knowledge of fermentation technology, they gave the opportunity to a small American company, Pfizer, to introduce this innovation. This small company, which took the time to develop the knowledge of fermentation, became the world's foremost manufacturer of penicillin. The need for an analysis to determine whether all the necessary knowledge exists or can be gathered before the cycle of innovation begins should be obvious. However, in real life we have many examples of organizations that started to invest in a serious innovation and then, while on the road to success, found a barrier or were missing some necessary component of knowledge, and all their efforts went down the drain.

6.3.2 Develop a Clear Focus on the Strategic Position

When a new innovation is introduced, it usually creates a lot of excitement and attracts many people and organizations, including competitors. Readers with children will probably remember when Beanie Babies were first produced. They were quite expensive, and you needed to stand in line to buy one. They created a lot of excitement in the market, and other companies copied them and produced their copies as quickly as they could. In a short while, the marketplace was flooded with Beanie Babies and interest in this toy fell. This was just a toy, and the producers expected children to get bored with it quickly. But then again, this same phenomenon occurs with more complex products. How do we take full advantage of an innovation? How do we keep our competitors from copying it? A clear focus on the strategic position can help, but a clear focus cannot be introduced tentatively. We must be right the first time. We will not have a chance to fix the problem if we are wrong. When we are dealing with knowledge-based innovation, the competition is very high. Almost immediately after the innovation is introduced, others will try to "help" us saturate the market. There are basically three major focuses for knowledge-based innovations: complete system, market focus, and strategic position (see Fig. 6-5). Below is a short description of these three major focuses, which together may help to build a clear focus of the strategic position.

Developing a Complete System The first focus is the focus on the development of a *complete system* that will work to dominate the marketplace. A classic example is IBM's approach in their earlier years of experience. Instead of selling computers to customers, IBM developed a leasing system and supplied their customers with the necessary software and instructions. Focusing on leasing instead of selling computers allowed IBM to preserve their monopoly in the computer business for a long time.

Developing a Clear Market Focus Here the aim of the organization is to create the market for the products that result from knowledge-based innovation. One example is optical fiber. When optical fiber was first invented, no one, including the inventors, realized that this optical technology would revolutionize the world. Now we know that optical fiber can handle more than 400 billion bits of voice, video, or Internet traffic each second. This technology created the market that, in turn, discouraged or even destroyed potential competitors (see Insert 6-1).

Occupying a Strategic Position Occupying a strategic position actually means concentrating on the major function of the organization. The semiconductor industry is a good example. Despite tough competition, a few leading manufacturers of microprocessors (such as Intel and AMD) can maintain their leadership position almost irrespective of individual computer manufacturers. Maintaining a continuous market focus is essential for an innovative organization.

The knowledge manager needs to pay great attention to the three focuses described above. Even though each of these focuses involves high organizational risk, not focusing on them is even more risky. "Edison's Success"(see Insert 6-2),

---

Insert 6-1

The Development of Optical Fiber

In 1970, Maurer and co-inventors Drs. Donald Keck and Peter Schultz designed and produced the first commercially feasible optical fiber, the building block of the information superhighway. It was little realized that this invention could revolutionize the world 30 years later. Optical fiber can handle more than 400 billion bits of voice, video, or Internet traffic each second. In just 15 years, well over 200 million kilometers of fiber-optic cable has been installed worldwide, according to industry analysts. That much alone can handle more information than all the billions of kilometers of copper wiring installed over the past century. This "optics" technology was applied to the communications field to reduce cost and improve cost and quality.

Source: Based on http://www.corning.com/news/company/press_releases/1999/991102_draper.asp[10]

Insert 6-2

Edison's Success

The power of a clear focus is demonstrated by Edison's success. Edison was not the only one who identified the inventions that had to be made to produce a light bulb. An English physicist, Joseph Swan, did so too. Swan developed his light bulb at exactly the same time as Edison. Technically, Swan's bulb was superior, to the point where Edison bought up the Swan patents and used them in his own light bulb factories. But Edison not only thought through the technical requirements; he thought through his focus. Before he even began the technical work on the glass envelope, the vacuum, the closure, and the glowing fiber, he had already decided on a "system": his light bulb was designed to fit an electric power company for which he had lined up the financing, the rights to string wires to get the power to his light bulb customers, and the distribution system. Swan, the scientist, invented a product; Edison produced an industry. So Edison could sell and install electric power while Swan was still trying to figure out who might be interested in his technical achievement.

Source: Based on *Innovation and Entrepreneurship: Practice and Principles* by Peter F. Drucker, pp. 118–119. Copyright © 1985 by Peter F. Drucker. Reprinted by permission of HarperCollins Publishers, Inc.[11]

adapted from Peter Drucker's book *Innovation and Entrepreneurship*, is a good example of what can happen with a knowledge-based innovation when clear focus is ignored.

6.3.3 Practice Entrepreneurial Management

Knowledge managers, especially those who work in high-technology industries, must apply entrepreneurial management to be successful. This requires a culture with elements such as practicing taking high risks, putting a premium on foresight, both financially and in managerial skills, being market- and customer oriented, etc. Entrepreneurial knowledge and skills should receive great attention and be included in the learning curriculum for knowledge managers.

The brief description provided in Section 3.3 may give you a feeling for what the requirements are for creating knowledge-based innovation.

6.4 THE LIFE SPAN OF AN INNOVATION

Technical innovations can work in two ways, either competence enhancing or competence destroying. A lot of attempts have been made by researchers and practitioners to predict competence-destroying cycles by describing technology life cycles as progressing along an S-shaped curve (also known as the Ultimate Curve; see Fig. 6-6).

At the bottom of the curve (1), the technology is new and untried. With its application, the technology improves its performance at an increasing rate (2). This is a period of innovation and creativity in which organizations are able to gain a lot of savings and improve product quality. With time, as the technology

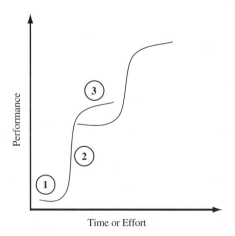

Figure 6-6 The S-Shaped Curve

matures, growth slows and levels off (3) as the curve reaches the performance limitation of a particular technology.

The art of a knowledge manager is to understand at any particular time where his organization is on the shape of this curve: at its lower part (1), on the vertical part (2), or as it approaches the asymptote at the top (3). This will allow management to make timely decisions on investment to obtain new technology.

Sometimes we may think that if a technology has served its purpose for so many years it will continue to serve this purpose forever, and we lose the momentum when we need to be ready for a new breakthrough. We may not be prepared for it and this is dangerous. So how do we know when it is the right moment for the existing technology to be replaced with a more innovative technology? We have to rely mainly on our experience and intuition. We have to learn to listen to the external environmental signals and know how to decode them. We cannot wait until we see strong evidence that the major technology that has kept our organization alive is on top of the curve. It will be too late then.

In 1965, one of Intel's founders, then the head of Fairchild Semiconductor's Research and Development Laboratory, Gordon Moore, came up with a principle that is valid even now for forecasting the pace of semiconductor advances. This principle is widely known in the semiconductor industry as Moore's law,[10] which states that the number of transistors on a semiconductor chip doubles approximately every 18–24 months. For microprocessors, that doubling of transistors is 24 months. The strength of this law is proven with time (see Fig. 6-7). Using AMD as an example, Figure 6-7 demonstrates that the number of transistors in semiconductor chips has increased from 29 thousand in the 8086 microprocessor in 1978 to 107 million in the AMD Hammer microprocessor in 2002. This is a 369-fold increase in microprocessor complexity, if transistors per chip are used as a measure. The producers of microprocessors forecast that the tendency toward increasing the number of transistors per chip will continue in the future. For example, in 2001 Intel expects to deliver chips that have one billion transistors.[13]

By using this indicator, organizations can have some forecast of the speed of change and plan accordingly. Because of the fast acceleration of the number of transistors per chip, it is very important for organizations to have stored ideas that can be taken from the "knowledge container" and developed before the organization reaches the upper curvature on the S-curve. While introducing a needed change before it is too late, there is also another side to it. This is to put a technology, which is still serving its purpose effectively, in the grave. For example, Clayton M. Christensen observed that by the time Hitachi and Fujitsu finally switched to thin-film technology for the production of computer disks, the two companies had ridden the ferrite oxide "S-curve far longer and had achieved ... over eight times the performance IBM seemed to have identified as the limit of the ferrite-oxide approach...."[14] So there can be great losses from switching to a new technology too early when it is still possible to extract more revenue from existing technology. However, it is suicidal to delay the ability to

114 KNOWLEDGE-BASED INNOVATION

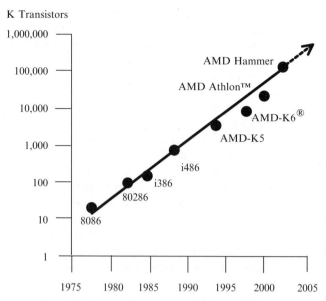

Figure 6-7 The Exponential Growth of Microprocessor Complexity
Source: Based on AMD's Internal Publications

move to a new technology. In both cases, it depends on the management's judgment, which certainly involves risk.

Conclusion

The S-curve is a very important factor in determining the organization's capability gap. Two errors can occur in determining when to retire a technology. The first is to err on the side of complacency, which is probably more likely to hurt the organization. The second is the possibility of overreacting to new technology and underestimating its compatibility with existing capabilities. This is less dangerous but can bring unnecessary losses to the company, in the same way as missing the approach of the asymptote at the top. However, it is also very important, because it allows us to more accurately identify the real capability gaps. Misreading or miscalculating the S-curve and moving too quickly to a new technology are just as detrimental. The failure to find a balance and seek synergy between new technological knowledge and current capability is not only dangerous but also very costly.

6.5 DR. DEMING ON INNOVATION

"He that innovates and is lucky will take the market."

Dr. Deming, 1994[15]

Dr. Deming's philosophy of quality management that is rooted in understanding the theory of variation is known worldwide. His seminars and books have helped many organizations achieve significant improvements in process quality. While recognizing the importance of incremental improvement, Dr. Deming also brings our attention to the importance of developing new innovative products and processes. He wrote, "No advance is made by the customer.... None asked for electricity, or the automobile, the camera, pneumatic tires, or the copying machine. The consumer can think only in terms of what you and the competition offer.... Improvement is important, but it's not enough."[16]

This is to say that incrementalism is important, but it must be continuously complemented with innovation to develop new competitive products and processes. And you should offer this to the consumer before your competitor does. For instance, if you are a computer manufacturer, you must offer a computer with higher speed, greater memory, and other attractive features before your competitor can take your market share. The consumer will appreciate all this and will pay for it, but as an innovative organization you must offer it first. Dr. Deming links innovation with the principles of statistical control, which in his words "open the way to innovation." Referring to a speech given by Mr. Conway, President of Nashua Corporation, Dr. Deming wrote, "Statistical control opened the way to engineering innovation. Without statistical control, the process was in unstable chaos, the noise of which would mask the effect of any attempt to bring improvement. With statistical control achieved, engineers and chemists became innovative, creative...."[17] These are symptoms that show innovation is connected with continuous incremental improvement. Considering innovation an important issue, Dr. Deming included innovation in his famous 14 points (see Section 3.3.2, Insert 3-1).

While emphasizing Dr. Deming's concerns about innovation, we should point out that the major attention to this subject occurred earlier in his career. In his later work, once he felt that management in the United States understood the value of innovation, he tended to put more emphasis on continuous process improvement. His aim was to make continuous improvement a part of the organizational culture in American corporations. This was during the period when organizations were running processes with high defective rates, and improving the process and product quality was very important to maintain a competitive position.

6.6 INCREMENTALISM AND INNOVATION: CAN THEY COEXIST?

In the last few years, organizations have tended to pay more attention to the importance of breakthroughs and the increasing the frequency of new innovations. This is understandable because the organizations that survive in today's tough competitive age are those that innovate and bring new products to the market. Some organizations are too attached to their old products and think that continuous improvement alone will guarantee their survival. This is wrong.

There are also people who see improvement as an enemy to innovation because it may divert the organization's efforts away from innovation. The question is, Can these two activities—innovation and improvement—coexist?

We recognize the fact that incrementalism may block management's attention to innovation; nevertheless, we do not think of incrementalism as a major, or even minor, enemy of innovation. If we have the right leadership and proper management, these two activities can coexist in peace and complement each other.

Incrementalism, as well as innovation, also requires profound knowledge, great creativity, and innovative thinking. Sometimes it is difficult to even draw a definite line between the two activities. Incremental improvement coupled with innovation can bring significant results. We will see this in an example from Motorola described below.

6.6.1 Motorola's "Obsession" with Quality

In 1987, Motorola introduced its renowned six-sigma quality initiative. During this period of time, the media would occasionally criticize Motorola's "obsession" with the reduction of process variation. Some criticisms were, "Why do we need almost perfect quality (3.4 ppm)?" and "Would it bring a severe cost penalty?" Motorola's results proved that the reduction of process variation not only improves quality but also delivers significant savings. In their book, *World Class Quality*, Keki R Bhote and Adi K. Bhote wrote, "Since 1979, Motorola had been tracking its cost of poor quality.... In 10 years, it saved more than $9 billion by reducing its cost of poor quality!"[18]

In 1981, Bob Galvin, the chairman of the board at Motorola, established a goal to ultimately bring a quality improvement of 1,000:1. The goal was not completely achieved throughout Motorola's operations, but the average improvement was an incredible 800:1.[19]

Was all this a result of incrementalism or innovation? The answer is both. It could not occur without people's creativity and innovation, without the incremental improvements that came from the application of simple statistical tools and complicated design of experiments. It was a change in Motorola's mind-set. Their example later influenced the way of thinking in many other organizations, including AMD. Motorola proved that it is reasonable to drive toward perfection. They proved that quality is not a cost issue—it is an investment that brings high returns. Motorola's story shows that if we want to send a great product to customers, we need to think about the product and the process in the same way. To reach the market, a new innovative product needs a process that is capable of consistently producing this new product. By working on the existing process, we prepare for the introduction of new innovations. For example, by achieving a 3.4 ppm quality level in an existing process we created a culture in which the upcoming innovative product or process would also be designed and produced at a 3.4 ppm quality level or better. So, in this sense, continuous improvement efforts are strongly interconnected with innovative efforts. This is why we think that continuous incremental improvement and innovation are two sides of the same coin and should work under one umbrella.

6.6.2 Tom Peters' View of Innovation and Incrementalism

"The only sustainable competitive advantage comes from out-innovating the competition."

Tom Peters[20]

In almost every book written by Tom Peters you can find something related to innovation. He has always considered innovation to be a major activity of a business. In his book *Thriving on Chaos* he made a table that he called "A world turned upside down."[21] He wrote, "I listed the major activities of a business. Then I listed the way it 'was/is' and the way it 'must become' as column headings. To my dismay, in all ten basic areas, almost a 180 degree flip-flop was required."[22] Tom Peter's list deserves great attention, but here we will limit ourselves by focusing only on one of the points—innovation (see Table 6-1).

As the table suggests, Tom Peters brings our attention to the need for a 180-degree "flip-flop." He argues that innovation not only should be driven by central R&D and big projects but should also take place in decentralized units, and it should be everyone's business. He also suggested that innovation must be driven by a desire to make small and customer-noticeable improvements. In real life, innovation sometimes occurs in unexpected places, where it is less influenced by the bureaucracy and politics of the organization. This is especially important today, when a lot of organizations grow in size, and because of this their core becomes more powerful.

Tom Peters, while emphasizing the importance of innovation, connects it with an organization's capability to change. He argues that innovation should come from everyone, and in this regard he writes, "No skill is more important than the corporate capability to change per se. The company's most urgent task, then, is to learn to welcome—beg for, demand—innovation from everyone. This is the prerequisite for basic capability-building of any sort, and for subsequent continuous improvement."[24]

Table 6-1 Extract from Tom Peters' List "A World Turned Upside Down"

	Was/Is	Must Become
Innovation	Driven by central R&D, big projects the norm, science-rather than customer-driven, cleverness of design more important than fits and finishes, limited to new products	Small starts in autonomous and decentralized units the key, everyone's business, driven by desire to make small and customer-noticeable improvements

Source: From *Thriving on Chaos: You Can't Shrink Your Way to Greatness*, by Tom Peters, p. 52. Copyright © 1987 by Excel, a California Limited Partnership. Used by permission of Alfred A. Knopf, a division of Random House, Inc.[23]

This quotation demonstrates that Tom Peters sees innovation as a base for "subsequent continuous improvement." In other words, he sees innovation and incrementalism as two links from the same chain of manufacturing success. To deliver the message to American corporations that incrementalism, which proved to work very well in Japan, can also work successfully in the United States, Peters used an extract of an appendix from Richard J. Schonberger's book *World Class Manufacturing* as an illustration (see Insert 6-3).

Schonberger's list contains 84 plants he has discovered in North America, which have been following the constant improvement strategy and have achieved a fivefold, tenfold, or twentyfold improvement in manufacturing lead time. As Schonberger noted, "The list is representative but by no means complete."[26] If you analyzed the names of the companies in Schonberger's list that he called the "Honor Roll: the 5-10-20s," you would easily conclude that most of them (if not all) represent highly innovative companies. 3M is a classic example. Its main core capability is innovation. All this is to demonstrate the importance of how incrementalism and innovation work together as a whole.

6.6.3 Friend or Enemy?

Challenging what we have said so far: Is incrementalism a friend to innovation or an enemy? You may think this question is strange. After all the proofs we made that incrementalism and innovation MUST work together and complement each other, there is no place for such a question. This question was

---

Insert 6-3

Phenomenal Results From Incrementalism in the United States

3M, Weatherford, Okla. (floppy discs): WIP [work in progress inventory] cut from six hundred to six hours, space per unit cut sixfold, productivity tripled

Omark, Guelph, Ontario (saw chain): Lead time cut from twenty-one days to one, flow distance cut from 2,620 to 173 feet

Omark, Onalaska, Wis. (gun cleaning kits): Lead time cut from two weeks to one day, inventory cut 94 percent

Omark, Woodburn, Ore. (circular saw blades): Order turnaround time cut from ten to fourteen days with 75 percent fill rate to one or two days with 97 percent fill rate, WIP cut 85 percent, flow distance cut 58 percent, cost cut 35 percent

Hewlett-Packard, Greeley, Colorado (flexible disc drives, tape storage units): WIP cut from twenty-two days to one day, whole plant on JIT [just-in-time]

Source: *World Class Manufacturing: The Lessons of Simplicity Applied* by Richard J. Schonberger, p. 174. Copyright © 1986 by Schonberger & Associates, Inc. Reprinted with permission of The Free Press, a Division of Simon & Schuster.[25]

brought to our attention while reading Tom Peter's best-selling book, *The Circle of Innovation*. In this marvelous book, Tom Peters supports Nicholas Negroponte's idea that "Incrementalism is innovation's worst enemy."[27] Tom Peters wrote, "Negroponte, who heads the highly innovative, highly respected Media Lab at the Massachusetts Institute of Technology, has allowed us no wiggle room at all. He could easily have said, 'Incrementalism is *an* enemy of innovation.' The idea would still be profound, but with a lower-case p. But he said that incrementalism is innovation's *worst* enemy. That's profound... with a capital P."[28] This profound idea took us by surprise. Negroponte and Peters perhaps used this strong statement to draw the reader's attention to the importance of innovation. Dr. Deming used the same strategy to emphasize incrementalism and suppress the importance of innovation when he felt that American companies needed to pay more attention to incrementalism. Most of his lectures were oriented on continuous process improvement, even though he also recognized the importance of innovation.

In our opinion, incrementalism is the greatest friend to innovation, and the greatest follower. Incrementalism is the wellspring of innovation. As we already mentioned, for many years the semiconductor industry used wire-bonding technology and continued to incrementally improve it up to the time when a "wall" was hit and no more improvement could serve the purpose. Some time before this occurred, the scientists from IBM, foreseeing that this would happen, were working on an innovation that would allow the semiconductor industry to move to an absolutely new technology that made wire-bonding technology obsolete for a large number of products. The innovation was flip-chip technology. Then when this revolutionary technology was introduced activities related to incremental improvement came into play. But only after all the bugs were eliminated did we learn how to apply flip-chip technology. The new technology demonstrated its power and effectiveness as compared to the old technology. What will come after flip-chip technology? Time will tell. What we do know is that incrementalism follows innovation and vice versa. They work in a circle and complement each other.

Figure 6-8 depicts the relationship between incremental improvement and episodic innovation. There are two things we want to reflect in this pictorial representation. (1) In recent years there has been a strong tendency for more radical, more frequent innovation. This process is presented here by making, with time, the vertical solid lines longer and the horizontal solid lines shorter. (2) Incremental improvement (represented by dashed lines) occurs in the time between innovations and enhances the results of the innovation. It is understood that the incremental results are smaller than breakthrough results, but the cost of a breakthrough is also much higher than that of continuous process improvement.

Together, these two activities involve a larger group of people, which makes the job more interesting for everyone. It is especially important to give serious consideration to incremental activities when considering the process of globalization that involves the Asian culture. In these countries, employee

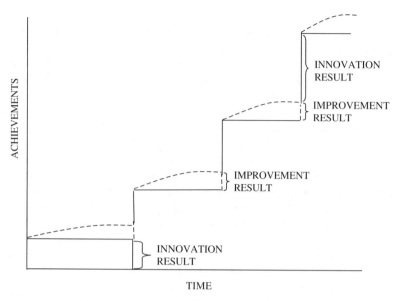

Figure 6-8 A Graphical Representation of Improvement Results and Innovation Results

involvement in the process by forming quality circles or other small group activities is still considered important to the organization's success.

6.7 CREATIVE THINKING: A WAY TOWARD INNOVATION

"Genius is ninety-nine percent perspiration and one percent inspiration."
Thomas A. Edison

Knowledge managers should be creative thinkers and should be able to create an environment that will allow people to think creatively. In one of our conversations with managers, someone said, "People's creativity comes from their DNA." This person was challenging the possibility that a person could develop qualities of creative thinking.

Sometimes when we see the results of creativity we think that it comes from people who were born with a special talent, which in a way is true. Most of the time, we do not notice that someone is going through the process of becoming creative. We only see the results. How many years does a gifted person work to become a great artist? How much time and effort is needed for an artist to have an art gallery display his work?

Visit a museum where you can see collections of work that reflect the life of a famous artist. You will see the gradual transformation of an artist who started with ordinary pictures; only through hard work was he able to create a masterpiece. This is not to say that if you work hard, you will become another

Picasso. The point we want to make is that if you find your place in life and work hard to obtain the required knowledge, apply this knowledge and develop new skills, you can develop a creative mind. Creativity is not something you learn by reading a book or sitting around waiting for a flash of inspiration to come. Creativity comes through learning, persistence, and hard work.

Organizations that know how to plan and manage creativity succeed in innovation, have a lot of inventors, and make the workplace interesting. However, creativity is not recognized as a leadership issue in many organizations. We forget that by raising the bar of creativity, we will also improve the level of innovation.

While writing this section, we had a conversation with Saragavani Pakerisamy, who has been an AMD employee for 26 years. He is the engineer whose responsibility is to make sure that the finished electronic devices preserve their quality while they are being handled during each process and on their way to the customer. He takes care of the methods and materials of packing the finished semiconductors. This is an area where one might think that there is no place for creativity. After all, what can you do with a tray or a box? However, when you walk into Saragavani's cubicle, you will see that the partitions are well decorated with patents, all related to one thing—packing.

From our conversation with Saragavani, we found that since he was a child he has been very curious and has frequently asked himself and the people around him, "Why?" This habit of questioning everything made him the way he is. When he started working here as an engineer and was assigned to resolve packing issues, he thought that this was a pretty boring place to work. But later, in the process of resolving problems, he learned what packing is all about and gradually became a specialist in this field. He became a creative engineer-innovator. We asked, "What can management do for you to help you in your work?" He humorously replied, "The main problem is that I don't have any more space on the walls to hang my patent awards," as he showed us a pile of new patents he had recently received.

This example shows that almost everything we do can become a source of creativity and innovation. It also shows that creativity is a process of questioning and of double-loop learning (see Section 4.3). Continuing his thoughts, Saragavani said, "A creative mind is like any muscle in your body—the more you use and develop it, the more exercise it will need. To survive and grow, we need some kind of exercise continually." We questioned whether his creative ideas came from resolving the problems or proactive thinking. He replied, "Both," and he started to explain the way he thinks about resolving an existing problem and about finding alternative solutions.

We never planned to have a formal interview with Saragavani, but it came out as an interesting dialogue. In conclusion, we asked him what he would suggest to a just-graduated engineer who came to work for AMD. In response, he pulled down a page from his cubicle wall with a caption that said, "Always think out of the box." We think that this is good advice, not only for engineers but also for everyone, including the CEO. Usually, with time each of us carefully constructs a "box" that contains our thoughts and assumptions, which we use for reference to make our decisions and act accordingly. When the new information challenges or contracts what is accumulated in the box, we resist; we are unable to take it in. Breaking the boundaries that we created and getting out of the box is a way to achieve creativity, innovation, and success.

6.8 KNOWLEDGE AND MENTAL MODELS

From a business perspective, the term "mental model" means the conceptual model that each member of the management team carries in his or her head to explain the way the organization operates. If managers have a good understanding of the organization's structure or budget, or how the manufacturing process works, then we can say that they have a mental representation of a particular thing that serves them as a mental model.

A mental model can be represented as a network as is shown in Figure 6-9. The structure of a given network contains our understanding of the world around us.

Managers use their mental models to find the answer to different questions they face in their daily activities. For example, What needs to be done to capture the market? How do organizations achieve a lower product cost? The answer may or may not be correct, but it is based on their mental model. Each question the managers face evokes a different network. The longer we think

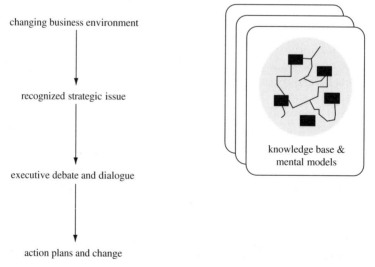

Figure 6-9 Strategic Change, Mental Models, and Debate

Source: Modeling for Learning Organizations by John D. W. Morecroft, John D. Sterman, editors, p. 6. Copyright © 1994 Productivity Press, P. O. Box 13390, Portland, OR 97213, 800-394-6868, www.productivity inc.com. Reprinted by permission.[29]

about a topic, the more facts and concepts we remember, and the more complex the network becomes. This is why a mental model should be viewed not as a static pattern of connections comprising a core network of familiar facts and concepts but as a vast matrix of potential connections that are stimulated by thinking and by the flow of conversation.

In the decision-making process, it is important to use the collective mind by having dialogues and taking the time to think together as a team. There is a strong relationship between our mental models and the knowledge we have. The mental models we possess are based on whatever knowledge and experience we have. Our knowledge, naive or sophisticated, whether it is based on real experience or imagination, remains the basis of our mental model. This is why our mental models may differ significantly from each other's. The amount of variation between the mental models of a team also depends on a number of other factors, such as the time spent working together as a team, the organizational culture, the diversity of the team, the education and background, etc. The differences in mental models can be a source of conflict or power. They may create problems in communications and the decision-making process, or if properly managed, these differences can also be a core capability of the organization. They can serve as a source of innovation and proper decision-making.

Knowledge managers should understand the nature of mental models and consider this in their management process. Mental models have a significant influence on the organization's strategy. On the basis of its mental model, an organization observes the changes in the external environment and makes its strategic decisions and future plans. Remember when world-renowned watchmakers did not recognize that a small quartz crystal could replace most of the complicated mechanism of a traditional timepiece, or when the large motorcycle makers did not recognize that a simple, easy to run and maintain Moped could take a large part of their market away from them, or when large computer makers did not recognize the potential power of small personal computers? These are simple examples of frozen collective mental models that did not allow management to observe the changes in the external environment and react properly and on time. This almost cost these organizations their existence as manufacturers.

Our experience at AMD has shown that for an organization to have a dynamic system of mental models that conform to the environment, the knowledge manager must become a mental model builder. The materials for building organizational mental models are the mental models of individuals who work in the organization. It is relatively easier to influence a mental model when it is only in the individual's mind and not a shared mental model that is already a part of the organization's culture.

As Forrester (1971) argued, "...Within one individual, a mental model changes with time and even during the flow of a single conversation. The human mind assembles a few relationships to fit the context of a discussion. As the subject shifts, so does the model.... Each participant in a conversation employs a different mental model to interpret the subject."[30]

From the time we started applying dialogue as a form of communication at AMD, we saw many instances in which, during an active dialogue among managers or engineers, their mental models changed in the process of expressing their thoughts. At the end of a productive dialogue, some participants may leave with a fuzzy, but different, mental model and may be unconscious of the change. However, when a mental model has become a part of the collective mind, or has been documented as a policy that has been used for many years, it is very difficult to challenge it. This does not mean that a mental model of an organization cannot be changed. It only means that it is more difficult and takes more effort and time. It requires work to bring the established assumptions to the surface and to challenge and test them by demonstrating their incompatibility with the current external environment.

As we mentioned above, mental models are based on the knowledge that individuals possess. To optimize the variation between individual mental models, we need to create an environment in which people can learn together and share their knowledge. When organizations are going through an intensive process of globalization, the art of the knowledge manager is to utilize the diversity of the subcultures to create an organizational culture in which diversity in thinking becomes a core competence. This can be accomplished with continuous dialogues, knowledge sharing, and the creation of an atmosphere of trust and free expression.

Here we have briefly outlined the nature of a mental model and its relation to individual and organizational knowledge. Summarizing what we mentioned above, we can say that:

1. Mental models are dynamic networks of facts and concepts that mimic reality. They are the sources from which knowledge managers derive their opinions and determine their course of action.
2. Mental models are based on the knowledge the individual possesses. To reduce the variation in mental models, collective learning and knowledge sharing is necessary.
3. The collective mental model is based on shared individual mental models through continuous dialogue.
4. The role of the knowledge manager is to continuously create an environment in which individuals share their knowledge.

References

1 W. Edwards Deming, *The New Economics*, Massachusetts Institute of Technology, Center for Advanced Educational Services, Cambridge, MA, 1994, p. 108
2 Peter F. Drucker, *Innovation and Entrepreneurship: Practice and Principles*, HarperCollins Publishers, Inc., New York, NY, 1985, p. 107
3 *International Science Yearbook*, Cahners Scientific Group, 2001, p. 8–10
4 Drucker, *Innovation and Entrepreneurship: Practice and Principles*, pp. 107–115
5 Drucker, *Innovation and Entrepreneurship: Practice and Principles*, p. 108

6. Ibid., p. 110
7. Ibid., p. 107, 111
8. Ibid., pp. 115–119
9. Ibid., p. 116
10. Corning Inc., "Retired Corning Specialist Recognized for Revolutionary Optical Fiber Research," 1999, http://www.corning.com/news/company/press_releases/1999/991102_draper.asp
11. Drucker, *Innovation and Entrepreneurship: Practice and Principles*, pp. 118–119
12. Albert Yu, *Creating the Digital Future: The Secrets of Consistent Innovation at Intel*, The Free Press, New York, NY, 1998, p. 3
13. Bill Gates, *Business @ the Speed of Thought*, Warner Books, Inc., New York, NY, 1999, p. 143
14. Clayton M. Christensen, "Exploring the Limits of the Technology S-Curve: Part I: Component Technologies," *Production and Operation Management* 1, Fall, 1990, p. 346
15. Deming, *The New Economics*, p. 108
16. Andrea Gabor, *The Man Who Discovered Quality*, Times Books, New York, NY, 1990, p. 10
17. W. Edwards Deming, *Quality, Productivity, and Competitive Position*, Massachusetts Institute of Technology, Center for Advanced Educational Services, Cambridge, MA, 1982, p. 7
18. Keki R. Bhote and Adi K. Bhote, *World Class Quality: Using Design Of Experiments To Make It Happen*, 1925 @ 2000 AMACOM Books, a division of American Management Association International, New York, NY, 2000, p. 7
19. Ibid.
20. Tom Peters, *The Circle of Innovation: You Can't Shrink Your Way to Greatness*, Hodder and Stoughton, London, UK, 1997, p. 30
21. Tom Peters, *Thriving on Chaos: A Passion for Excellence*, Alfred A. Knopf, New York, NY, 1987, p. 50
22. Ibid., p. 50
23. Ibid., p. 52
24. Ibid., p. 285
25. Richard J. Schonberger, *World Class Manufacturing: The Lessons Of Simplicity Applied*, The Free Press, New York, NY, 1986, pp. 229, 230, 232
26. Ibid., p. 229
27. Peters, *The Circle of Innovation: You Can't Shrink Your Way To Greatness*, p. 26
28. Ibid., p. 27
29. John D.W. Morecroft, "Executive Knowledge, Models and Learning" in John D.W. Morecroft and John D. Sterman (eds), *Modeling for Learning Organizations*, Productivity Press, Portland, OR, 1994, p. 6
30. J.W. Forrester, "Counterintuitive Behavior of Social Systems," Technology Review 73, No. 3, 1971, pp. 52–68.

Chapter 7

Knowledge Managers and Knowledge Workers

Remember the time when you could get plenty of responses to an advertisement you placed for an open position? When you, as a manager, had the problem of selecting one candidate from a pile of resumes? Those times are gone. Today you would be happy to have just one qualified applicant to fill your vacant position. You do not actually interview the applicants; they interview you. And why not, if the market is so hot that the applicants have many organizations and good opportunities to choose from? Companies are doing everything they can to attract talents. For example, the standard signing bonus for a graduating MBA is now $20,000 to $30,000. New motivational recruiting measures help, but not a lot.

The issue with attracting and retaining good people remains a problem, and we should probably expect it to get even worse. Why? Even though layoffs are now the lowest since 1991, there were still about a half a million people laid off in the U.S. in 1997 and there are still a relatively large number of unemployed people. What organizations need today is not just labor, but *knowledge workers*—a type of worker that makes up more than 50% percent of the workforce today.

Fifteen to twenty years ago, if you entered a semiconductor manufacturing plant, for example, you would have seen a process that consisted of a large number of operations performed by automated and semiautomated equipment and a large number of manual operations. At that time, direct production workers formed the main workforce in many manufacturing organizations. Over the years, we have observed a steady decline in the number of such workers, whose function was replaced by robots and fully automated machines. At the same time, we could see a fast growth of knowledge workers—engineers, accountants, programmers, researchers, statisticians, and others. How are these knowledge workers different from traditional workers? What is the difference in managing this kind of worker? What motivates people to become knowledge workers? What are their rights and responsibilities? How do we measure their productivity? We will try to answer these questions and others in this chapter.

7.1 TIME AND THE MEANING OF MANAGEMENT

The definition of management is changing with time. In his book *Post-Capitalist Society*, Peter Drucker relates, "When I first began to study management, during and immediately after World War II, a manager was defined as 'someone who is responsible for the work of subordinates.' A manager in other words was a 'boss,' and management was rank and power."[1] According to Drucker, this definition changed early in the 1950s when a manager was one who was "responsible for the performance of people."[2] This was a broader definition than the earlier one, but it did not correspond to the actual meaning of today's management activities. Drucker writes, "The right definition of a manager is one who 'is responsible for the application and performance of knowledge.'"[3] This definition explains the great attention that modern organizations are paying to knowledge management. It means that management today sees knowledge as the essential resource. The manager is responsible for the application and performance of knowledge. Particularly in a high-tech organization, the manager will need to have profound knowledge that includes a variety of soft and hard disciplines that are essential to managing knowledge work and knowledge workers.

7.2 THE KNOWLEDGE MANAGER

Knowledge workers are rapidly becoming the largest single group in the workforce of every developed country. For example, Peter Drucker stated in 1999 that in the U.S. 40% of the workforce were knowledge workers.[4] We have seen a tendency toward further growth of this category of workers. The failure to recognize this accelerating change in the workforce makes some managers think and act in a way that is not appropriate when dealing with knowledge work and knowledge workers. This may become a source of miscommunication and frustration, which can ultimately erect a barrier to innovation, knowledge creation, and application. This is why it is important for the contemporary manager to have a full understanding of what knowledge work and knowledge workers are and how to manage them in this new environment.

Managers themselves are also knowledge workers. In fact, they are the fastest growing group among knowledge workers. The main difference between a knowledge worker and a knowledge manager is in the knowledge and the authority a manager must have to lead knowledge workers.

The T shape of knowledge for a manager is different than that of other knowledge workers. (See more about T-shaped knowledge in Section 7.4.) The manager needs to know more disciplines related to process and human relations. Some examples of this are organizational psychology, systems dynamics, theory of variation, and organizational learning. Knowledge managers should know the general principles of managing process improvement and innovation and have a general understanding of globalization. So, if the T shape concept is applied, the T of knowledge managers would have a larger horizontal bar to reflect the knowledge of "soft" disciplines and the vertical bar, which is shorter than in the T of knowledge workers, would represent the specialized knowledge in the area they are managing in their organization. To be a successful manager in a knowledge-based organization, the knowledge manager should have a profound understanding of the difference between a knowledge worker and a traditional manual worker.

The next section is dedicated to some specifics that differentiate the knowledge worker from the traditional manual worker.

7.3 THE KNOWLEDGE WORKER

Peter Drucker coined the term "knowledge worker" in 1950. However, only in the last 5–10 years have organizations started to pay more attention to the meaning of this term. Time proved that Peter Drucker's prediction that knowledge workers were the fastest growing group of workers in any developed country was correct. But what is a "knowledge worker"?

The category of knowledge workers comprises various professionals such as accountants, engineers, technicians, computer experts, educators, and other highly skilled workers, such as line operators who use complicated equipment. As you can see, there are different levels of knowledge workers, but what holds them together is the fact that they all do more or less intellectual rather than manual work. For example, today's operators, who simultaneously control four or more processing machines in a semiconductor manufacturing environment, use statistical principles to analyze the output quality. They know how to use the computer installed in the equipment to control the complicated process. They perform diagnostics and some repairs on the equipment, which requires special knowledge—all of these are knowledge work.

Now let's move on to the professionals who are involved in a higher level of intellectual work. Is there any difference between a traditional engineer and an engineer of the twenty-first century? A contemporary engineer, as a knowledge worker, has a deep, specialized knowledge in a particular field. In today's environment, especially in large organizations, there is a need for knowledge workers who are specialized. New college graduates do not start in the organization as generalists but as professionals in their own field. With today's developments in science, it is difficult to be sufficiently knowledgeable in a number of disciplines. Knowledge engineers should have an understanding of concurrent engineering, statistical theory, theory of constraints, project management, and many other disciplines that together make them knowledge workers.

Take contemporary salespeople as another example. They are not just people who sell goods, but they work together with customers to introduce complicated solutions. A salesperson in the semiconductor industry is usually an engineer who has knowledge about semiconductors, the customers' products, economics, and cost. In a way, salespeople are also psychologists and good communicators. Again, besides a deep knowledge of their products, salespeople need some knowledge in a large number of disciplines. All these examples suggest that knowledge workers should have T-shaped knowledge, where the vertical bar is their specific knowledge, which needs to be strong and deep, and the horizontal bar is the extra knowledge they need to do the job.

Knowledge workers should not only know more in their specialized field than their managers, but should also take on responsibility. Peter Drucker wrote, "Since knowledge workers have considerable power, they must also assume responsibility. Their sole responsibility—and the one most sadly lacking today—is responsibility for the individual's own contribution."[5] It is the knowledge workers' responsibility to continuously upgrade their knowledge to fit the organization's needs and to direct themselves toward the organization's objectives and find where they can make the maximum contribution in achieving the organization's goals. Peter Drucker wrote, "It is not enough for the knowledge worker to apply his or her knowledge; that knowledge must be applied so that it redounds to joint performance. This requires that knowledge workers direct themselves toward the objectives of the institution."[6] A knowledge-based organization will depend on the willingness of its knowledge workers to take responsibility for their contributions to the whole and to understand and support the objectives and the values of the organization.

7.4 THE T SHAPE OF KNOWLEDGE: A REQUISITE FOR THE KNOWLEDGE WORKER

As a professional, what kind of knowledge and skills do you need to be successful in these turbulent times? Do you need to be a specialist with deeply specialized knowledge? Or do you need to be a generalist who knows a little bit about many disciplines? There is no question that you must know your profession well enough to be able to say that you know your job better than anyone else (at least in your organization). But what is "well enough," and what else do you need to know to consider yourself a knowledge worker or a knowledge manager?

In real life, it doesn't matter how much you know or how hard you study, there will always be a gap between your current knowledge and the knowledge you need to fulfill your and the organization's needs. As Figure 7-1 suggests, you must continuously learn to reduce this gap to a reasonable level.

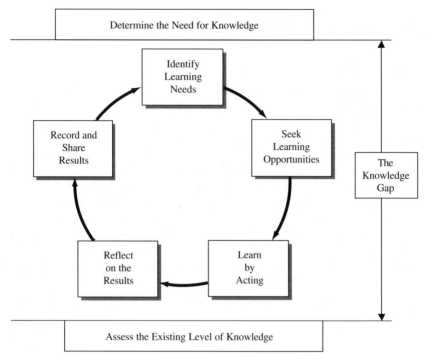

Figure 7-1 The Process of Learning to Reduce the Knowledge Gap

Every profession and every job has its own requirement for knowledge and skills. Today the knowledge and skills required for most knowledge workers, including managers, can be described in a T-shaped form, where the vertical bar represents the deeply specialized knowledge needed for a particular function and the horizontal bar represents the secondary skills and knowledge needed to communicate the specialized knowledge to others. This allows the worker to have a holistic view of things and to participate in cross-functional activities. We provide some examples below to illustrate the T-shaped knowledge concept.

An Example of a T-Shaped Set of Knowledge and Skills for an Engineer

Figure 7-2 is a T-shaped set of knowledge and skills for engineers who support semiconductor assembly lines. The vertical bar contains professional skills and knowledge that pertain to the engineering discipline. However, without the knowledge and skills contained in the horizontal bar, it would be difficult for an engineer to fully utilize the knowledge in the vertical bar.

As the T-shape suggests, the knowledge and skills in the vertical bar require deep fundamental experience, which makes an engineer a specialist in his or her field. The set of knowledge and skills in the horizontal bar are supporting tools, just as a microscope is for a biologist. Management should encourage and motivate engineers to develop a body of knowledge that contains specialized and supporting tools and techniques.

An Example of a T-Shaped Set of Knowledge and Skills for a Manager

What should be the fundamental knowledge and skills for a manager? Is accounting the fundamental discipline for an accounting manager? Is engineering the fundamental discipline for an engineering manager? Experience has shown that the main discipline for a manager is management. A manager should know how to practice management, just as a designer needs to know how to design. The main point we want to make here is that management is a discipline for a manager, just as economics is a discipline is for an economist. This certainly does not mean that a manager should not have an understanding of the technology of the process he is involved in, or not have an understanding of economics, statistics, or accounting; but these are complementary tools for a manager. In Figure 7-3, the vertical bar contains a body of knowledge

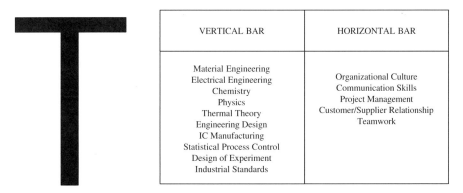

Figure 7-2   The T Shape of Knowledge for an Assembly Engineer

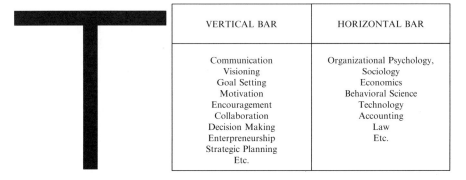

Figure 7-3   The T Shape of Knowledge for a Global Knowledge Manager

and skills that belong to the discipline of management. This is the stuff that makes him or her a manager. The horizontal bar also contains a body of essential knowledge and skills, but these are considered to be management tools.

The only difference between a manager's T-shaped chart and an engineer's T-shaped chart is that in the manager's case, the soft stuff, such as communications with organizations, visioning and goal setting, motivation and encouragement, comparisons and collaboration, decision making, etc., make up the major disciplines, which will fill out the vertical bar. The horizontal bar for a manager will contain practices from other disciplines such as psychology, sociology, economics, etc. In the engineer's case, all elements of knowledge and skills in the vertical bar are "hard stuff." Designing different sets of knowledge based on the T-shape concept is becoming more and more important for multinational, high technology organizations. This clarifies the curriculum of knowledge and skills for different individuals, and for an organization as a whole.

Today as never before, organizations will need different types of knowledge workers. Where will these workers come from? Colleges don't prepare them, and organizations don't provide enough motivation for the development of workers with a T-shaped set of knowledge and skills. When specialists come straight from college to work for an organization, they are motivated to concentrate on deep functional activities, to utilize the knowledge they obtained in college. As a result, the individuals are driven even deeper into specialization. In other words, the organizational environment often rewards narrow specialization. Organizations need to create an environment where the individuals will be motivated to develop themselves according to the requirements of the T-shape concept. Designing T-shaped "packages" of the skills and knowledge needed for every particular job function will allow management to improve the process of developing knowledge workers.

7.5 RECRUITING KNOWLEDGE WORKERS

It is no doubt getting more difficult to recruit someone to fit the bill. Every day there are fewer suitable applicants. During the interviewing process, we try to find out as much as we can about the applicant, and the applicant tries to find out about the organization and its environment. Both parties have something to offer, and the deal depends on both of them. If we analyzed the questions the applicants ask today and compared them with those asked 10 or 15 years ago, we would find a tremendous difference. Not only are applicants concerned with benefits and salary, but they also want to know whether they can apply their "portable" knowledge. They want to know whether the organization will provide them with the opportunity for information, risk taking, and knowledge upgrading. This does not mean that they are not interested in the benefits and salary; of course they are. Applicants are selling their knowledge. We call it "portable" knowledge now because in the knowledge area when people apply for jobs, they carry their knowledge with them like they carry their laptop computer. Their knowledge is their product, their property, their capital, and their asset. Previously, when we were looking for a perfect match to the job description, we lost so much time by rejecting someone who could have made a big contribution to the company but was not a perfect match. Now, in a cross-functional and high-performance team-based environment, it is not necessary to concern yourself with a perfect match, simply because you no longer hire a person only for a narrowly defined job function.

In knowledge-based organizations, what we are looking for is people with knowledge in a particular field, people who are capable of learning, who can become familiar with the work with limited supervision, who will do the job the way they feel is most efficient. It is not necessary to continue from year to year performing a function the same way just because the spec says so, or because "we always do it like that here."

One of the authors can remember an example of this, which took place many years ago when a very complicated technological task in aerodynamics was given to a group of scientists and engineers who were familiar with designing submarines. The task was completed successfully and in record time.

The art of management is not to replace a vacancy exactly the same way we would replace a device on a printed circuit board—by looking in the manual to find the right part number and replacing it. It is probably riskier, and will take more effort, to hire somebody who is well educated and knowledgeable in a related field than it would be to hire somebody who was a perfect match. However, with proper selection and good management, this risk can be minimized and the match can provide better results than we ever could have foreseen. Knowledge-based organizations make jobs dysfunctional. We need to look for knowledge workers who are capable of fast adaptation and not limit ourselves to people who just match the job description.

7.6 MANAGING KNOWLEDGE WORKERS

In real life, we can observe the tendency of managers to continue to treat knowledge workers as they did many years ago; forgetting that they are dealing with knowledge workers who have the capability and willingness to manage themselves. This certainly does not mean that the manager is totally detached from the responsibility for knowledge workers' activities. The manager acts more as a conductor in an orchestra. The conductor does not know how to play every instrument, but he knows how the whole orchestra should sound. He knows how to coordinate the efforts of every player to achieve a great performance. The same applies to a knowledge manager. He needs to have some understanding of the job content of every knowledge worker. He needs to know what he can expect from them. He assumes that he has the right players, and he tells them what he expects from them, but then he lets them play. So, metaphorically speaking, the knowledge manager is a conductor, and a conductor does not control the individual but coordinates, advises, and controls the whole orchestra so that the members work together as a whole. There is only one major difference between a knowledge manager and a conductor. The conductor uses a score written by a composer. He makes sure that the orchestra performs according to the score. In addition, he brings his own interpretation of the score to life by coordinating the collective performance of each individual. If he does this well, he is a great conductor who can make a score sound magical. A knowledge manager knows what senior management wants, and he has policies, procedures, organizational values, and beliefs that he should abide by, but he does not have all the "scores" in front of him. Hence, the knowledge manager is also a composer, who composes the music according to the company's strategy. At times, he needs to change the score on the spot, depending on environmental signals. To have a great performance, we need knowledge workers who know their part deeply and we need a knowledge manager who, as a composer and a conductor, facilitates the right performance.

Abraham M. Maslow writes, "People are growing and growing, either in their actual health of personality, or in their aspirations.... The more grown people are, the worse authoritarian management will work, the less well people will function in the authoritarian situation, and the more they will hate it."[7] This also relates to knowledge. The more people become knowledgeable, the less they will accept authoritarian management.

Maslow also wrote, "... If the Americans can turn out a better type of human being than the Russians, then this will ultimately do the trick. Americans will simply be more liked, more respected, more trusted, etc., etc."[8] Maslow's thoughts regarding the role of people in creating strong nations also applies to organizations. If your organization can turn out a better type of employee, who is more knowledgeable, more creative, and more dedicated to his responsibilities, than your competitors, you will win the competitive edge. Knowledge managers should use their knowledge, wisdom, and experience to create an environment that will allow knowledge workers to be creative and effective.

7.7 THE RELATIONSHIP BETWEEN THE ORGANIZATION AND THE KNOWLEDGE WORKER

The growth in the number of knowledge workers depends on the growth of organizations. Knowledge workers need an organization where they can apply their knowledge. In this respect, knowledge workers depend on the organization, and the organization's accomplishments depend on its knowledge workers. However, at the same time, knowledge workers own the "means of production"[9] (to use Drucker's term), which is their knowledge. While saying that knowledge workers have the "means of production," we need to clarify the point that knowledge workers still need the organization's capital investment in their tools to produce and apply new and existing knowledge.

It is important to note that what knowledge workers possess is owned by them and cannot be taken away. When an employee leaves the organization, he is leaving with his capital—the knowledge. This is one of the important differences between the manual worker and the knowledge worker.

An operator who runs a machine is doing only what the machine allows. A knowledge worker, who may still need some kind of equipment, will do whatever he would like to do with the machine. The computer or microprocessor does not dictate what job to do or how to do the job for a knowledge worker. For example, a technician who is responsible for running test equipment properly and performing precise measurements will not be influenced by the testers. The tester depends on her, not the other way around. The technician can use her knowledge to accomplish her task. The testers are just a means to do the work.

Knowledge workers need the organization as a place to work and continue to enhance their professionalism. At the same time, they have more mobility, because knowledge workers carry the means of production—their knowledge—with them. From a wider perspective, a good example is the former Soviet Union. Through the years, they developed a strong educational system that prepared a large number of knowledge workers who worked in the nuclear industry, at military institutions, and in other areas. When the Soviet Union collapsed, a lot of knowledge workers lost their jobs because many of the government institutions where they could apply their knowledge were closed. But because they carry their knowledge in their heads, you can find Russian knowledge workers all over the world today, working in organizations where they can apply their knowledge.

The main point we are trying to make here is that knowledge workers are very mobile. Because of this, knowledge managers who work with knowledge workers must learn how to attract and retain them. The expression, "No one is irreplaceable" needs to be used more carefully. In these times, when a large part of the organization is made up of knowledge workers, they are replaceable if they leave, but it may cost a lot of time and money to replace them, and the organization still loses.

How about loyalty? Where is loyalty in the knowledge worker era? As Peter Drucker wrote, "... 'Loyalty' from now on cannot be obtained by the paycheck; it will need to be earned by proving to knowledge employees that the organization which presently employs them can offer them exceptional opportunities to be effective."[10] For knowledge workers, even though money still remains a motivation to seek another place to work, when the paycheck reaches a certain level, its power of influence lessens. Knowledge workers need opportunities to be effective. They seek satisfaction in their work—satisfaction, in the sense that they can apply their talent, knowledge, innovative mind, and expertise to their work. Hence, to attract or retain knowledge workers, knowledge managers need to offer challenging job content, coupled with a job context that matches what the knowledge workers seek and an appropriate salary.

References

1 Peter F. Drucker, *Post-Capitalist Society*, Harper Business, New York, NY, 1993, p. 44
2 Ibid.
3 Ibid.
4 Ibid., *Management Challenges for the 21st Century*, Harper Business, New York, NY, 1999, p. 141
5 Ibid., *The New Realities: In Government and Politics, in Economics and Business, in Society and World View*, Harper & Row Publishers, Inc., New York, NY, 1989, p. 96
6 Ibid.
7 Abraham H. Maslow, *Maslow on Management*, John Wiley and Sons, Inc., New York, NY, 1998, p. 292
8 Ibid., p. viii
9 Peter Drucker, *Post-Capitalist Society*, p. 64
10 Ibid., p. 66

Chapter 8

Knowledge Transfer and Knowledge Management

"In the new economy, conversations are the most important form of work."
Alan Webber[1]

8.1 MANAGING KNOWLEDGE TRANSFER

Why has knowledge transfer gotten so much attention in the last 5–10 years? This is probably because organizations, more than ever, have finally recognized that knowledge is the main core capability that can make the difference between organizations. If we have the necessary knowledge earlier than anyone else, we will succeed. If we don't, we will lose. It's as simple as that. We need at least two things to accomplish this. The first is a commitment to share knowledge among employees and groups within the organization. The second is a network that will allow us to share knowledge and information with those outside the organization. This is especially important because the pace of globalization is accelerating quickly.

Today, with the development of computerization and networking capabilities, we have a significantly enhanced opportunity for sharing information and knowledge worldwide. And although we must give full credit to the Internet and other means of communications, more than this is needed to make people and organizations willing to exchange real, valuable information and, especially, knowledge.

We sometimes naively think that if we invest in a particular network, we will automatically succeed in transferring knowledge. However, what knowledge transfer really means is essentially sharing and understanding of particular levels of knowledge that have accumulated in our individual and organizational memory.

In addition to the opportunity for knowledge transfer provided by computers and networking, there is a human side that needs to be considered. How can we make individuals or groups within the organization willing to share knowledge? How do we make outsiders, in particular competitors, transfer their knowledge, when this is the only core capability they have that makes them different? We may think that knowledge transfer is just an information technology issue, but in actuality, it is also a psychological, cultural, and managerial issue—in short, a human issue.

Knowledge transfer is a difficult issue not only because individuals or organizations want to keep the knowledge to themselves and are not willing to share, but also because knowledge is embedded in context. Take knowledge out of context and it loses its meaning. Without context, knowledge is just information. Knowledge is like a living organism that continuously grows, and when the time comes, it dies (goes out of date). We may think that by accumulating information we accumulate knowledge. In a way, this is true. However, if we do not apply this information, it becomes obsolete. We should have a system that will select, update, share, and use information. Only information in action can create knowledge.

Knowledge transfer is a complicated multidisciplinary topic that requires a systemic approach. Here we will only discuss the preliminary issues of this topic, giving more emphasis to the human and organizational aspects. Figure 8-1 displays some sources and forms of knowledge sharing. If these are applied and managed properly, they can help us build a knowledge transfer culture, which is an important part of a knowledge-based organization. However, because of the space limitations of this book, we will only describe some of these elements.

8.1.1 Informal Knowledge Transfer

Most of the time, management considers the classroom to be the only place of learning. Although we can require adults to attend classes, we cannot force them to absorb the information. This is not within our control. They can sit there and still be free to think about anything they want. In general, management does not like to see people sitting in the cafeteria too long. It is obvious that they are "doing nothing." Alternatively, you can sit at your workstation and do nothing, but you are not as obvious and not as much under management's control. Actually, your lunchtime conversation is exactly where real and effective work may occur. Alan Webber shared the same thought in his article "What Is So New About the New Economy?" He said, "In the new economy, conversations are the most important form of work. Conversations are the way knowledge workers discover what they know, share it with their colleagues, and in the process create new knowledge for the organization."[2]

In an organization, knowledge is transferred from one person to another, or from one group to another, independent of how we manage this process. When a technician cannot fix a new machine and he requires consultation with a more

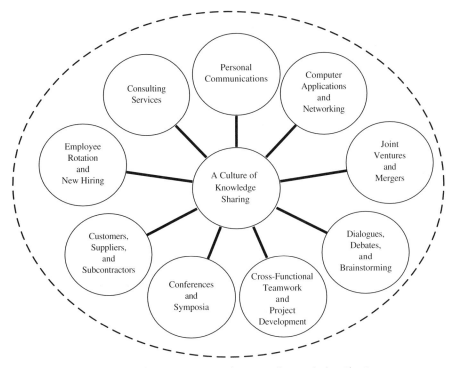

Figure 8-1 Some Sources and Forms of Knowledge Sharing

experienced technician or engineer, knowledge transfer occurs. When an engineer seeks a colleague's advice on how to deal with a particular problem, knowledge transfer occurs. When a manager in a Quality Engineering meeting gives a detailed report on how a task was accomplished, this is knowledge transfer. And when two people sit in the cafeteria and share some thoughts, they transfer knowledge to each other. These are just some examples of natural knowledge transfer that occurs in all organizations.

The manager's responsibility, in this case, is to nurture an atmosphere of trust that encourages the willingness to share knowledge. She needs to ensure that she has a sufficient group of people who know the business in the organization and to build an environment in which the employees can continuously talk with each other. This is probably the most effective and least expensive way to transfer knowledge from one worker to another. It sounds very simple, but it is not. It is easy to put a system in place to find and hire intelligent people, but to make them talk to each other and share their knowledge is much more time consuming. It takes a lot of management effort to build a sharing environment. The phrase "Let's stop talking and get back to work" is probably familiar to you, but when people stop talking with each other, it is the beginning of a catastrophe for an organization. In the right organizational environment, with a strong culture oriented toward innovation and getting things done, in a place where self-management and personal accountability are at work, the employees will know when to talk and for how long. In a knowledge-based organization it is probably more appropriate to say, "Let's start talking to each other." The knowledge workers will know when to start and when to stop and get back to work because they feel responsible for getting things done.

One of AMD's senior fellows, Raj Master, replying to a question on how he came up with an idea that brought many innovations to the company, said, "It was a tough time because of the pressure from the top, and I was sitting in my office seeking for the right solution. At the end of the day, exhausted from unsuccessful experimentation, I went to the cafeteria for a cup of coffee, and you would not believe, but this was it. I had a casual conversation with two of our engineers who worked on the same problem, we started to talk about the stock market, moved on to a conversation about our problem, and something clicked in my mind. The solution was found. Honestly speaking, I can't even tell you exactly when we came up with the right idea. We talked for about 20 or 30 minutes, and bingo! The solution was found not in my office, but in the cafeteria. This is why when we applied for the patent, we applied together—all three of us." And at the end of his presentation, he said, "You know, this is not the first time things like that have happened to me." This story is very interesting to us. Sometimes when walking past the engineers' cubicles, we can hear them talking about football or other non-work-related things, and we may think they should "get back to work." But if we were to listen to their whole conversation, we would probably notice an interesting phenomenon: People may start to talk about one thing, and sometimes nonsense, but they will usually end up talking about something that's bothering them. This could be an engineering problem, an interesting problem that needs an answer, or just frustration that needs to be expressed. All this is typical of people who feel responsible for their job.

Another example related to this issue occurred here at AMD. To keep the management informed about what is going on in our overseas plants and to share experiences, we regularly have management teams present their achievements for a particular period and express their needs for the future. In one of these presentations, the managing director of AMD Thailand, Mr. Yuthana Hemungkorn, in sharing his knowledge transfer experience, mentioned the expansion of their facilities and the construction of a learning center with computerized classrooms, a large conference room with electronic presentation equipment, and sports facilities, including a gymnasium, a soccer field, etc. At the end of his presentation, someone asked, "We understand the need for classrooms and conference rooms, but how do soccer and other sports relate to knowledge transfer?" Mr. Hemungkorn replied, "Any activity or any opportunity that allows people to get together and talk with each other can be related to the issue of knowledge transfer. And the more unplanned and unmanaged these activities are, the more sharing occurs. We now have a large cafeteria that is open 24 hours a day, and we eat and drink there, but the casual conversations between our employees at all levels are valuable. A learning process is actually going on there. It may be invisible, but it is there." This kind of everyday knowledge transfer is part of organizational life.

With this, it is also important to note that the unstructured knowledge transfer, with all its positive features, also has its weaknesses. When dealing with critical problems, or when there is a need for a fast and important decision, episodical conversations may not work. When there is a need for specific knowledge, the process of gaining this knowledge must be properly managed. To expect to always find the solution to a serious and complicated problem while eating your lunch in the cafeteria is naive. However, in general, the unstructured transfer of knowledge creates an opportunity for spontaneous meetings, which have the power of generating new ideas and finding new and unusual ways to solve existing problems.

8.1.2 Structured (Formal) Knowledge Transfer

All the knowledge transfer we mentioned above occurs between cubicles and offices within one plant. In other words, this type of knowledge transfer has a local character and is not sufficient for a global organization with employees located all over the world. Talking with colleagues in neighboring cubicles is great, but sometimes we need the expertise of someone who is not in our organization. How do we find this person? Obviously, the answer would be the Internet. The Internet brings us new opportunities for knowledge transfer. Now we can be in any part of the world and in seconds we can communicate with any person in the world. With a computer we can find an expert in any field we are interested in and get the information we need, but tacit knowledge is very hard to transfer at a distance. Experience shows that when we are dealing with the transfer of a mix of explicit and tacit knowledge, the best thing to do is to bring in the people who need the knowledge on a rotating basis to work with the originators of new knowledge, or vice versa if possible.

This happened when AMD licensed and adopted IBM's flip-chip technology. A group of U.S. engineers learned from IBM how to implement the new technology, which by itself is an example of knowledge transfer from outside the organization. Then engineers and technicians from our overseas plants came to AMD Sunnyvale to learn flip-chip technology. At the end of their assignment, they returned to their plants and transferred their new knowledge on a broader basis. This cascading form of knowledge transfer is probably the only way to deal with knowledge you cannot easily explain with words or just by writing procedures. Its methodology can be used when we need to transfer knowledge within the corporation, from one organization to another, or, as in the case we just described, from an outside organization to your own organization. This is a very motivational form of knowledge transfer, from which people learn more than just a particular, narrow part of the technology.

8.1.3 International Conferences and Symposia: A Form of Sharing

What we have discussed so far were examples of knowledge transfer between individuals and small groups. This is very important, because in large groups it is more difficult to achieve deep learning and proper knowledge transfer. But to accelerate the speed of the spread of knowledge throughout the whole organization, we need some other forms of knowledge transfer. This is why at AMD we usually hold international conferences and symposia once a year in different countries. We usually have 100–200 people, including representatives from our plants in different countries, participating in these activities. We usually also have one or two guest speakers who will speak on a particular topic. Presenters from different plants will also share their best practices. The presenters are people whose projects are selected as the best at the plants' internal conferences before the international event. On the second day of the event, we conduct workshops on a variety of topics to which people come in groups of 25. Participants share what they know on the workshop's particular topic. Three days of working together and having lunch and dinner together gives us a lot of time to understand each other better and to share knowledge. It is important to note here that we also invite representatives from different divisions and groups of the corporation to these conferences, as well as representatives from our major suppliers and customers. People gain a tremendous amount of information from personal contacts made during the breaks or the formal presentations. In addition, to help the participants get even more value from these events, we have a set of handout materials available and also videotape the 3-day event. In this way, the participants can share their impressions of the event with their peers back home, using the videotape as a facilitating tool.

This form of knowledge transfer is broad and covers both internal and external issues. Although it is difficult to estimate the effectiveness and savings we receive from these activities, every project presented at these events represents a hard figure of savings that, when combined, results in an impressive number.

8.1.4 The Culture of Knowledge Transfer

Knowledge transfer is a part of the whole organization's culture. We cannot successfully adopt forms of knowledge transfer that do not fit in our culture. For example, Japanese managers spend a lot of their after-office hours meeting with their peers, having fun, talking about business, and transferring knowledge. This would probably not work in other countries.

While visiting our suppliers, the authors noticed that the Japanese prefer face-to-face communication to conversing through E-mail. For example, if serious problems arise, they will prepare themselves on the issues by doing a lot of homework to understand the problem and visiting customers to discuss the issue face-to-face. Knowledge transfer occurs easily during face-to-face conversation with the Japanese. They would not understand the problem as clearly through E-mail messages.

In our AMD Singapore plant it has become a tradition for the managers to spend at least 2 days every year away from their routine work attending an off-site meeting, during which they will review past and current processes and make strategic plans for the coming years. They will make time during this meeting for some teambuilding activities as well. The purpose is to enhance their team skills and develop greater trust in each other. The results are tremendous. (Trust is the most essential component of knowledge transfer.) In addition to this, people in this meeting just talk to each other, express their feelings, and share knowledge. Such off-site meetings are very popular and fruitful in Singapore, but if you conduct this type of program in some American companies, it may not work.

There are many different types of celebrations in Thailand that have been incorporated into our AMD Thailand organizational culture. The authors were invited to participate in the Water Festival when it took place in AMD Thailand. The festival, by itself, has a lot of meaning, but we want to point out that during the full-day celebration, there were many unplanned, spontaneous conversations and dialogues with a rich sharing of knowledge. It would be difficult to overestimate the importance of such festivals in knowledge creation.

Mr. Teoh, the managing director of AMD Penang, told us, "What I will tell you may sound like a joke, and it probably is a joke, but I want to bring it up just to emphasize how important face-to-face conversations are. There was a time when smoking was not regarded as a bad habit, and was something men were supposed to do. You felt good about it. 'Let's have a smoke,' was an invitation to a friendly conversation, an invitation to share some thoughts. I remember at that time that the cafeteria was full of heavy smoke, but we didn't even notice it. We talked and shared our experience and knowledge. Now I have reduced my frequency of smoking and I have lost a lot of participants in my smoking conversations. I am happy and proud of my colleagues, that they understand the dangers of smoking; but at the same time, we lost the opportunity for many good fruitful conversations. You might say, why not continue the conversations without the smoke? The answer is that we are too busy. Before, the habit of smoking made us take the time to step out, and made time for our conversations. There is no way back. This was just a way to tell you that we need to make conversation a habit again, and a part of our work."

Knowledge managers should learn how to build a culture of knowledge transfer where informal and formal structured knowledge transfer will take place. Today, only a combination of face-to-face and electronic contacts on all levels of the organization will bring about successful knowledge transfer. As the organizational culture is the glue that holds the organization together, so does knowledge transfer hold together all the elements of the knowledge transfer system and make it work effectively.

8.2 THE SYSTEM OF KNOWLEDGE MANAGEMENT

As we all know from experience, there is no one way of managing an organization. However, there are some common sense elements that allow us to propose an outline of a management system. This may provide knowledge managers with an orientation to develop their own management system that will fit their organization's needs.

As Figure 8-2 suggests, the knowledge manager should organize the work starting from the assessment or reassessment of the current state of his organization. There is a lot of discussion about how to set a goal or develop a mission, with the assumption that we always know where we are. This assumption is not always true, and it is sometimes even more difficult to assess the current state of your organization than to develop a vision or a goal. Assessment of the current state directly impacts the success of the whole management process and the final results. Having determined the current state, the knowledge manager involves his colleagues in formulating the desired state. The desired state can be formulated in the form of a vision, mission, or goal depending on the level of the social group (team, department, division, plant, corporation, etc.). The desired state should be based on the organization's current capabilities, economic conditions, the external and internal environments, and the customers' needs.

Having properly developed the first two steps—the current reality and the desired state—we form a creative tension, which will create a force toward

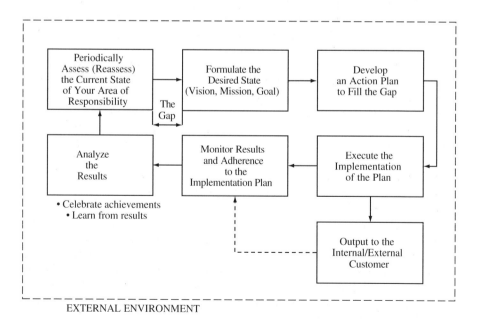

Figure 8-2 The System of Knowledge Management

achieving the desired state—tension resolution. It is very important here for the knowledge manager, as a good instrument player, to be able to design the right strength of structural tension. As in tuning a guitar, if you undertune or overtune the instrument, you will not achieve the desired tone. The goal, vision, or mission should include elements of challenge, innovation, and risk. All this will spice up the work, making it more interesting and meaningful.

The next step is to develop an action plan to fill the gap. It is important here to mobilize the strengths and capabilities available in the organization and to properly use the skills, knowledge, and talents of all the employees. A knowledge manager's talent is to make the right match between the individual or team capabilities and the work to be done. In real life, a manager will often avoid taking a risk by always giving the most challenging work to those who have already proven that they can do it, leaving other talent unused. This generates frustration and reduces the organization's capabilities. In developing the plan, it is very important to develop road maps, which will illustrate the long-term picture of what is to be accomplished and, at the same time, show the short-term details.

Planning from "the whole" to "the parts" is a very important systemic approach in the management process. In practice, when we take a three- or five-year goal, we are actually only capable of describing in detail the activities to be accomplished for the next one or two years and we leave the rest of the time blank. The concept "we will cross that bridge when we get to it" limits the vision, and sometimes we "get to it" and then find out that we took the wrong direction. It is not always easy to develop all the details, especially when we are dealing with high technology and building a long-term vision. But, seeing the whole as clearly as possible, and expecting some corrections on the way to the goal, is a very important component of knowledge management.

So far, we have described the three blocks of managing a process that have to do with visioning, goal setting, and planning. The next step is the execution of the plan. It is very important here to avoid micromanagement. If the plan is right and people are properly assigned to their jobs, and if they see the whole picture, they will do their best to accomplish the task. If they have a lack of knowledge, they will seek out new knowledge to help them complete the task.

The execution of the plan results in actual output that goes to the internal or external customer. This is the final place where your efforts are assessed, where profit occurs, and where the customer's and your satisfaction connect. But the system of knowledge management does not stop here. The next steps will be to monitor the results, assess the adherence, and determine how closely the actual results match the desired results. On the basis of the results of the analysis, we celebrate the achievements and learn from the results, and the process of improvement and progress continues. The management system we have just described is not only a system to achieve the desired manufacturing or operational results; in a way it is also a system of continuous organizational learning and knowledge creation. Knowledge managers should make sure that their management activities are not based on episodic actions but comprise a *system* of management.

References

1 Alan Webber, "What's So New About The New Economy?" *Harvard Business Review*, Boston, MA, Jan–Feb 1993, p. 28
2 Ibid.

PART III

Managing in a Global Environment

What is the meaning of globalization? Why it is becoming more and more important for organizations to go through the process of globalization? What should a manager know to be able to work in a global organization? How can the process of globalization be measured? All these questions and many other issues on this subject will be discussed in this part of the book. Special attention will be given to such issues as developing global strategies, forming different types of coalitions, working in a multicultural environment, and other organizational elements related to the subject of globalization. The main purpose of Part III is to familiarize the manager with the general concept of globalization and to develop an interest in this subject.

Chapter 9

On the Road to Globalization

"Ready or not, Americans, the global village has arrived."

Tom Peters, 1987[1]

In his 1987 bestseller, *Thriving on Chaos*, Tom Peters wrote, "I cannot overemphasize the need for the average firm to 'think (and do) international.' "[2] He warned the corporations by saying, "If you're $25 million or larger, and not doing 25 percent of your business overseas, and at least a little bit in Japan, you are avoiding today's realities and opportunities, and you risk being out of touch in general."[3] At that time, this warning may have sounded like a bit of an overstatement. However, it is no longer an exaggeration.

9.1 THE TREND TOWARD GLOBALIZATION

Today, a large number of organizations have increased their activities in an attempt to sell their product outside their home country. For example, in the semiconductor industry a large number of organizations exceed the 25% figure mentioned earlier (see Table 9-1).

The contents of Table 9-1 suggest that the semiconductor companies use international expansion as a vehicle for globalization and as a way toward economic growth. Further analyzing the globalization activities of semiconductor companies, we can observe a tendency toward direct investment in foreign countries (see Table 9-2).

By doing this, corporations receive a greater opportunity to use the "global mind" for developing new products and continuous innovation. It is also interesting to observe the increase of diversity in the semiconductor operations workforce (see Table 9-3).

Table 9-1 Sales Outside the U.S. for the Top 15 U.S. Semiconductor Companies*

Company	Revenues (K$)	Outside U.S. Sales
Hewlett Packard	47,100,000	54%
Intel	26,300,000	55%
Texas Instruments	9,700,000	67%
IBM Microelectronics Division	8,000,000	N/A
Motorola Semiconductor Sector	8,000,000	N/A
Philips	4,400,000	N/A
ST Microelectronics	4,200,000	71%
Micron	3,515,000	21%
Lucent Technologies	2,760,000	49%
Advanced Micro Devices	2,542,141	55%
National Semiconductor	2,500,000	N/A
NEC Electronics	2,000,000	0%
LSI Logic	1,490,000	N/A
Analog Devices	1,231,000	50%
Conexant	1,200,000	N/A

*Based on the SIA 1999 Annual Directory and arranged by total revenues.[4]

All of these steps strongly suggest that there is a globalization process going on in the semiconductor industry that allows the companies to grow and prosper. It is also interesting to observe the impact of globalization on technological processes. People who have been in the semiconductor business for a long time can remember the time when the design, fabrication, assembly, and test processes took place in one organization, under one roof. Now, to make, let's say, a competitive microprocessor, we need the participation of a number of countries (see Fig. 9-1).

The microprocessor, by itself, can be considered a global device because it is made and sold globally. We see the process of globalization coming like a tidal wave, and no one can stop it unless they want to go out of business. There is no shortage of data that substantiate the spread and development of globalization at both the macro and micro levels. The steps of globalization can also be seen in activities such as forming mergers, acquisitions, and joint ventures (see Table 9-2), licensing, supplier agreements, etc. So there is no doubt that organizations recognize the need for internationalization and globalization. But what is an international or global company? This may sound paradoxical, because we are trying to quickly become global but we do not clearly understand what a global organization looks like. Most of the time, we use the terms international, multinational, and global interchangeably. However, these terms have different meanings and require different mind-sets to understand the philosophy behind them.

Depending on what type of a company you want to create (domestic, international, multinational, global, or a combination), you need to design a different strategy and structure and align the organization's culture to the new

Table 9-2 Fabrication Facilities Outside the U.S. for the Top 15 U.S. Semiconductor Companies*

Company	Fabrication Facilities Outside the U.S.	
Intel Corporation	Leixlip, Ireland (2) Jerusalem, Israel	Kiryat Gat, Israel
Texas Instruments Inc.	Freising, Germany Hatogaya, Japan Hiji, Japan	Miho, Japan Oyama, Japan Taipei, Taiwan
IBM Microelectronics Division, IBM Corporation	Corbell-Essonnes, France (joint venture with Infineon) Mulhuddart, Ireland	Vermiccate, Italy Yasu-Gun, Japan
Motorola Semiconductor Sector, Motorola Inc.	Sendai, Japan East Kilbride, Scotland	South Queensferry, Scotland
Philips Semiconductor Inc.	Hsinhu, China Shanghai, China Hazelgrove, England Caen, France Limell, France	Böblingen, Germany Hamburg, Germany Nijmegen, The Netherlands Stadskanaal, The Netherlands
STMicroelectronics	Crolles, France Grenoble, France Rennes, France Rousset, France Tours, France	Agrate, Italy Castelleto, Italy Catania, Italy Singapore
Micron Technology Inc.	Avezzano, Italy Nishiwaki City, Japan (2) (joint venture with Kobe Steel Limited)	Singapore (2) (joint venture)
Lucent Technologies	Matamoros, Mexico Singapore (2)	Madrid, Spain Bangkok, Thailand
Advanced Micro Devices Inc.	Dresden, Germany	Aizu-Wakamatsu, Japan (3) (joint venture with Fujitsu Limited)
National Semiconductor Corporation	Greenock, Scotland	
NEC Electronics Inc.	None	
LSI Logic Corporation	Tsukuba, Japan	
Analog Devices Inc.	Limerick, Ireland	Taiwan (joint venture with TSMC)
Conexant Systems Inc.	None	

*Based on the SIA 2000 Annual Directory; companies arranged by total revenues.[5]

strategy. In their book *Managing Across Borders*, Christopher A. Bartlett and Sumantra Ghoshal wrote, "As the turbulent international competitive environment of the 1970s boiled over into the decade of the 1980s, it unleashed a rash of studies, reports, and recommendations telling managers how to run their business more effectively in the new 'global' environment...but like most fads,

Table 9-3 Percentage of Employees Working Outside the U.S. for the Top 15 U.S. Semiconductor Companies*

Company	Total Employees	Outside U.S. Employees
Hewlett Packard	124,600	46%
Intel	64,000	40%
Texas Instruments	44,140	50%
IBM Microelectronics Division	20,000	N/A
Motorola Semiconductor Sector	47,000	N/A
Philips	27,000	N/A
ST Microelectronics	31,000	90%
Micron	16,000	25%
Lucent Technologies	12,706	31%
Advanced Micro Devices	13,384	46%
National Semiconductor	10,500	N/A
NEC Electronics	2,800	0%
LSI Logic	5,500	N/A
Analog Devices	7,100	N/A
Conexant	6,300	N/A

*Based on the *SIA 1999 Annual Directory* and arranged by total revenues.[6]

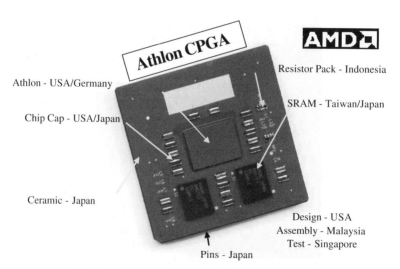

Figure 9-1 AMD Athlon™ CPGA Microprocessor

'globalization' soon became a term in search of a definition."[7] This comment was made in 1989, more than 10 years ago, but it still sounds right today. Even though there is no shortage of literature about internationalization and globalization, the question of what it really means, how to get there, and how to measure it still remains accurate. In his book *A Manager's Guide to Globalization*, Stephan H. Rhinesmith wrote, "Ask Andy Grove, the sage of Silicon Valley and founder of Intel, whether Intel is a global corporation. He will answer emphatically, 'Yes.' Ask Dov Frohman, director of Intel's Research Center in Israel, the same question and he will just as emphatically answer, 'No.' Ask Sharon Richards, cross-cultural coordinator for Intel and she will answer, 'Yes in strategy, but no in the skills and attitudes of the people and the corporate culture of the company.'"[8] On the question, "Who is right?" Rhinesmith elaborated, "All three are—and that's the first problem with globalization. Nobody, even in the same company, seems to have the total picture of how to make globalization work. In fact, Intel is working conscientiously to integrate and align its global strategy, structure, culture, and people."[9]

During the preparation for one of our seminars on globalization, we distributed a questionnaire to a cross section of employees who work in the U.S. and in offshore plants. The purpose of this questionnaire was to obtain information from the employees on how much they know about the subject and to assess their needs and interests. We received interesting feedback that helped us to design a comprehensive program (which we call Level F, Part 2). At the same time, the information we received also allowed us to draw the same conclusion as Stephen Rhinesmith—that different layers of people have a different understanding and see the company differently in respect to where it stands on the road to globalization.

Globalization is a process that, sooner or later, all organizations small and large will need to go through, and the sooner the organization takes the road to globalization, the greater will be the opportunities to enjoy the benefits from these efforts. In his article "The Nature of the Competitive Landscape," C.K. Praxald wrote, "Even businesses that until recently were considered purely local, such as cleaning services, will be subject to global competition.... Further, globalization will not be the concern only of very large, investment-intensive businesses. For example, smaller software firms are discovering that they can be global with revenues as low as two to three million dollars. They can sell around the world and often can outsource part of their work—developing code—worldwide."[10] This means that the subject of globalization relates not only to those large organizations in the list of global companies but to organizations of any size.

In the following sections of this chapter, we will elaborate on the differences in the terms related to globalization, describe different strategies and structures for globalization, and elaborate on the specifics of the organizational culture and how to align all these components of globalization together. In addition, a special section will be devoted to the subject of global management. In conclusion, we will summarize the material of this chapter by providing recommendations for action.

9.2 THE DRIVING FORCES OF GLOBALIZATION

Now that we have provided some data (see Tables 9-1, 9-2, and 9-3) that demonstrate a tendency toward globalization, we would like to try to answer some of the questions that are frequently raised by management.

The term "globalization" is often used and sometimes overused by managers. In his book *Total Global Strategy*, George S. Yip brought our attention to the issue of overusing the terms "global" and "globalize." He said, "Instead of being used to designate a particular type of international strategy, these terms are being used to replace the term 'international.' ... Everyone seems to want a global strategy rather than just an international one. As a result of the widespread use of the term 'global,' we are losing the ability to refer to different types of international strategy. More important, executives will find it easier to delude themselves that they have a global strategy if they are careless as to what they call their worldwide strategy."[11]

So, what is globalization? Well, without going into the history of its development, the term "globalization" started becoming popular in the late 1980s. Environmental changes forced businesses to consider global strategies and structures—to see the world as one marketplace. Loosely defined, the term "globalization" refers to the activities of designing standardized products, establishing worldwide operations, expanding market participation, and achieving competitive positions.

What external environmental forces influenced the shift to globalization? In their book *The Global Challenge*, Robert T. Moran and John R. Riesenberger define 12 major forces that influence an organization or industry to continuously work on the process of globalization (see Fig. 9-2).

Proactive environmental forces

1. Global sourcing
2. New and evolving markets
3. Economies of scale
4. Trend toward homogeneous demand for products/services
5. Lowered global transportation costs
6. Government interaction: tariffs, non-tariff barriers, customs and taxes
7. Increased telecommunications at reduced cost
8. Trend toward homogeneous technical standards

Reactive environmental forces

9. Increased competition from non-domestic competitors
10. Increased risks due to volatility in exchange rates
11. Trend of customers evolving from "domestic only" to global strategies
12. Increased pace of global technical change

Figure 9-2 Environmental Forces Influencing an Organization and its Industry's Degree of Globalization

Source: Reprinted with permission from Robert T. Moran and John R. Riesenberger, *The Global Challenge*, McGraw-Hill, 1996, p. 23[12]

For managers who are planning to get involved in the process of globalization, it is very important to have a full understanding of the meaning, dynamics, and stages of all these forces. This will allow the management to be better prepared for developing a global vision, strategy, and structure. In this section, we will provide a description of the major environmental forces and show their influence on the organization's success.

9.3 TECHNICAL GROWTH

It would be difficult in these days to find an industry that does not have something to do with globalization. There are also industries that, by their nature, must be global if they want to survive and prosper. One such example is the electronics industry. If you analyze the growth and development of this industry, you can see that there was a strong correlation between technological development and what was needed for globalization.

The design of new products and technologies facilitates the development of global industries. This can be colorfully demonstrated by an example of development in the electronics industry. By the mid-1950s, the replacement of vacuum tubes by transistors allowed manufacturers to expand the efficient scale for production of electronic components. In the 1960s, the introduction of integrated circuits further reduced the number and cost of components and increased the optimum manufacturing scale. The use of automated insertion machines, the replacement of the visual inspection of electronic components with robots, the introduction of "smart" testers, and the automation of the assembly and packaging processes allowed electronics and semiconductor manufacturers to reduce their costs and achieve low ppm (parts per million) quality levels. These and other innovations had a significant effect on the manufacturing scale. For example, the efficient scale for production of color TVs rose from 50,000 sets per year in the early 1960s to 500,000 sets in the early 1980s—a 10-fold increase.[13] At the same time, the scale of economies in R&D and Marketing were also increasing, and this factor also facilitated globalization. No single local, or even international, market could generate the revenue needed to fund the hotly sought after highly skilled scientists, electronic engineers, and knowledge workers. For example, Intel invested six billion dollars to produce the Pentium® microprocessor (Pentium® microprocessor is a registered trademark of Intel Corporation).[14] Considering the continuously shortening lifetime of electronic products, Intel would never have made the necessary revenue without being a global organization.

Another factor that facilitated globalization in the electronic industry was the growth of product reliability and quality, which brought a progressive decline to the importance of local repair and maintenance services. The development of replaceable service boards practically eliminated the barrier to globalization. This factor—improved quality and reliability—was and is as important for globalization in the electronics industry as some other factors that are more obvious. Further development of the electronics industry will generate a greater need for globalization than there is today.

By introducing new products, such as microprocessors, corporations like Intel and AMD generated a new industry within the electronics industry. In his 1998 book, *Creating the Digital Future: The Secrets of Consistent Innovation at Intel*, Albert Yu wrote, "The invention of the microprocessor created a new microprocessor industry.... In the process of creating this new industry, we also nurtured and stimulated a new infrastructure that supported new markets for other companies, such as independent hardware and software vendors, distribution channels, etc.... Together we created a whole new industry."[15]

The continuously increasing demand for new products, which increased the need for globalization, generated a requirement for high investments in research and development (R&D) and other capital expenditures. Thinking of this as a cycle, the customer demand for new products leads to an increasing demand for R&D, which leads to an increase in capital investment, which will require new markets, which will increase the need for globalization. And the cycle repeats (see Fig. 9-3).

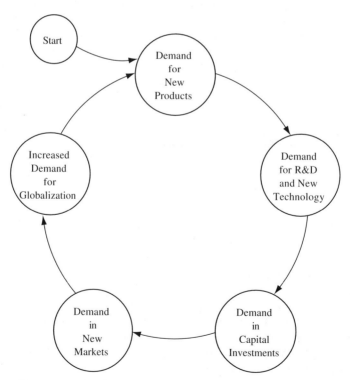

Figure 9-3 The Cycle For Increased Demand for Globalization

9.4 THE EVOLUTION OF GLOBALIZATION

As we mentioned above, globalization is an evolutionary process. However, this does not mean that an organization must go through all the stages—from international to multinational to global.

Here we want to introduce a concept that was developed many years ago and still remains applicable. It describes how an organization may choose to develop its international structure. One of the most comprehensive studies was done by John Stopford, who included 187 of the largest U.S.-based multinational corporations in his research. Even though this research was done in late 1960, his work still deserves attention at least to observe the thought process given to globalization. His work resulted in a "stage model" of intentional organizational structure that became the benchmark for most of the work that followed.

In their 1972 book, *Managing the Multinational Enterprise*, Stopford and Wells describe a model that suggests that at different stages of international expansion, corporations typically adopt different organizational structures. Figure 9-4 is based on two variables that capture the organization's complexity.

The variable that represents the number of products sold internationally is shown on the vertical axis as "Foreign Product Diversity." The second variable represents the importance of international sales to the company and is shown on the horizontal axis as "Foreign Sales as a Percentage of Total Sales." The model shows that at an early stage, when both foreign sales and the diversity of

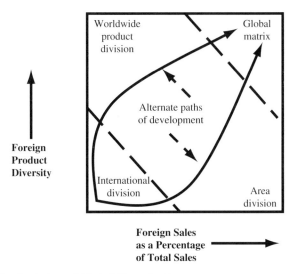

Figure 9-4 The Evolution of Global Organizations

Source: From *Managing the Multinational Enterprise* by John M. Stopford and Louis T. Wells, Jr., Copyright © 1972 by Basic Books, Inc. Reprinted by permission of Basic Books, a member of Perseus Books, L.L.C.[16]

Table 9-4 The Structural Stages Toward Globalization

Structural Stages	Foreign Sales	Foreign Product Diversity	Notes
International division structure	Limited	Limited	An international approach
Area division structure	Expanded	Limited	A multinational approach
Worldwide product division structure	Limited	Expanded	A global approach
Global matrix structure	Expanded	Expanded	A global organization

products sold are limited, companies usually manage their international operations through an international division. With time, as the corporations expand their foreign sales but do not significantly increase product diversity, they may adopt an area approach—the multinational model. However, when the companies' development results in a substantial increase in foreign product diversity, they tend to adopt the worldwide product division structure. When companies achieve high volumes of foreign sales and a significant increase in product diversity, they move to a *global matrix structure* (see Table 9-4). This model of structural stages may help organizations to better understand the evolution toward globalization.

Although Stopford's ideas were presented as a descriptive model, it was perceived and applied by consultants and practitioners as a prescriptive model. Organizations confronted with increasing complexity and diversity were looking for optimal ways of restructuring, and the global matrix seemed to be the right solution. However, very soon this global matrix resulted in disappointment for many organizations. Dual reporting led to conflict and confusion. Overlapping responsibilities resulted in less accountability. Organizations that were separated by barriers of distance, language, and culture found it difficult to clarify the sources of confusion and impossible to resolve the conflicts. As a result, the communication and management process was complicated and costly. The need for constant travel and frequent meetings introduced difficulties and duplication of efforts. The initial attractiveness of the global matrix faded into the recognition that something different was needed.

9.4.1 Going Beyond Classic Globalization

An organization continuously searches for a better structure to fit its needs. The basic problem underlying an organization's search for the best structure is that it usually focuses on only one organizational variable—formal structure. This simple variable cannot capture the complexity of the strategic task that global organizations are facing today. Management recognizes that the formal structure is an important but insufficient tool to manage strategic change. What is important now is to have a strategy that will allow us to manage organizational flexibility. To develop multinational and flexible strategic capabilities, an organization must go beyond structure and expand its fundamental organizational capabilities. The key to this task is to learn how to bring our assumptions about a global organization to the surface and reorient our way of thinking about organization systems.

In this section, we will introduce four models for an organization, which will help you understand the process of gradually developing a model that best fits today's internal and external environments. Having an understanding of the evolution of a contemporary model of a global organization will allow you to develop an organizational model that will fit your own organizational needs.

In their book *Managing Across Borders*, Christopher A. Bartlett and Sumantra Ghoshal described a series of organizational models that provided a historical perspective of how organizations went through a gradual process of changing their mental model of forming effective organizations. What attracted us the most to these models is the fact that the authors viewed the process of globalization mainly as a mental process. During different periods of time, management has held different mind-sets in regard to organizational structure. The authors named the first organizational model the "multinational organizational model." This structure was well suited to the European companies that expanded abroad in the pre-World War II period. The multinational organization is defined mainly by three characteristics:

- A *decentralized federation* of assets and responsibilities,
- A management process defined by simple financial control systems, and
- A dominant strategic mentality that views the company's worldwide operations as a portfolio of national business (see Fig. 9-5).

The approach in this model was literally multinational—each national unit was managed as an independent entity whose objective was to optimize its situation and achieve maximum influence in the local environment.

The second structure was named the "international organization model." This model became popular in the early post-World War II period. The main task for organizations at that time was to *transfer knowledge*, skills, and expertise overseas where technology and markets were less developed. This model fit the U.S.-based organizational cultures very well. While retaining overall control through a complicated management system, a delegation of some responsibility could be observed. The structure of this model provided the necessary channels for a regular flow of information to the center. In this design, the top management had the opportunity to control and manage the independent subsidiaries. The main difference between the multinational and international models is that the international subsidiaries were more dependent on the center for the transfer of knowledge and information (see Fig. 9-6).

DECENTRALIZED FEDERATION

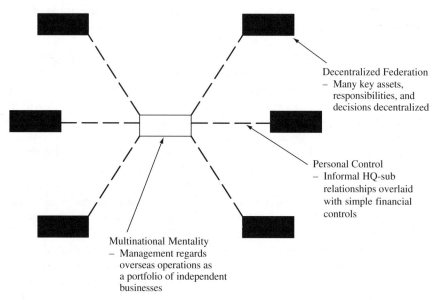

Figure 9-5 The Multinational Organizational Mode

Source: Reprinted by permission of Harvard Business School Press. From *Managing Across Borders – The Transnational Solution* by Christopher A. Bartlett and Sumantra Ghoshal. Boston, MA, 1991, p. 50. Copyright © 1989, 1991 by Harvard Business School Publishing Corporation, all right reserved.[17]

And, finally, the third type is the "classic global organization model." This model can be considered one of the earliest organizational forms of globalization. It was adopted by Henry Ford and John Rockefeller, who built global-scale facilities to produce standard products and ship them worldwide under a strong centralized control strategy. The concept of the global organization model is based on the centralization of assets, resources, and responsibilities and the use of overseas operations to reach foreign markets. The major purpose here was to build up to a global scale.

In this model, the role of offshore subsidiaries was mainly limited to sales and services. The development of local assembly plants was dictated not so much by economic resources but by political pressure. The subsidiaries in a global structural configuration usually had much less operational freedom compared with the subsidiaries in multinational and international organizations. The dominant management mentality was that in global organizations the world could and should be treated as a simple integrated market in which similarities were more important than differences. So the classic global organization could be viewed as an entity of tightly controlled subsidiaries with a management mentality that saw the world as a single economic entity (see Fig. 9-7).

COORDINATED FEDERATION

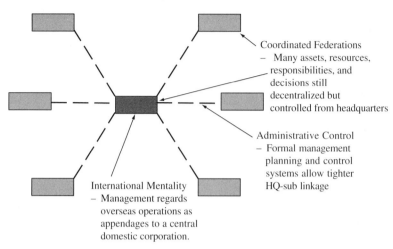

Figure 9-6 The International Organization Model

Source: Reprinted by permission of Harvard Business School Press. From *Managing Across Borders: The Transnational Solution* by Christopher A. Bartlett and Sumantra Ghoshal. Boston, MA. 1991, p. 51. Copyright © 1989, 1991 by Harvard Business School Publishing Corporation, all rights reserved.[18]

CENTRALIZED HUB

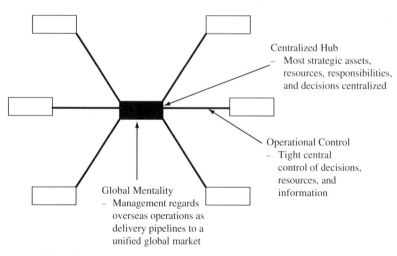

Figure 9-7 The Global Organization Model

Source: Reprinted by permission of Harvard Business School Press. From *Managing Across Borders: The Transnational Solution* by Christopher A. Bartlett and Sumantra Ghoshal. Boston, MA. 1991, p. 52. Copyright © 1989, 1991 by Harvard Business School Publishing Corporation, all rights reserved.[19]

This global strategic approach reflects Professor Theodore Levitt's argument that the future belongs to those companies that make and sell "the same thing, the same way, everywhere."[20] This organizational configuration requires significantly more central coordination and control than the other configurations. In such global organizations, R&D and manufacturing activities are usually managed from the headquarters and strategic decisions are mainly made from the center.

As we can see, these three models have different strategic capabilities: The multinational model allows organizations to be *responsive and sensitive to local markets*. The international model creates *an environment for knowledge and skills transfer* and uses them for local development. The global model allows organizations to coordinate strategies and to capture *global-scale* efficiencies.

The models described above characterize the linkage between environment, strategy, and structure. The models suggest that great performance comes from the argument between corporate strategy and environmental demands and between the organizations structure and its strategy. This means that organizations need to make sure that there is always a strong fit between strategy, structure, and environment—and this is not easy to do. In this regard, Bartlett and Ghoshal express their findings, "Some researchers imply that firms can change their strategy or their formal organizational structure to regain fit. Our findings, however, suggest that such changes are extremely difficult to achieve, since both strategy and structure are products of a company's unique and ingrained administrative heritage."[21] To achieve a successful implementation of change in strategy and structure, organizations must make sure that those changes match the changes in the organization's values and management's behavior. Bartlett and Ghoshal note that "...change in competitive strategy or in formal organizational structure is difficult to implement and rarely effective unless it is accompanied by matching changes in the company's values and management processes."[22] The three models—multinational, international, and global—demonstrate how the external environment influenced organizational change. With time, management gradually adjusted their strategies and structures to adapt their organizations to the external environment.

Each of these three models has some implicit assumptions about how best to achieve global competitive positions. The global configuration is based on the assumption that scale and the resulting cost leadership are the key sources of competitive advantage. The international configuration is based on the assumption that innovation, created at headquarters, is the main core competency that will allow them to reduce cost and increase revenue. The multinational configuration is based on the assumption that differentiation is the primary way to enhance organizational performance.

Even though these three models serve the purpose for a worldwide organization, there was still a need at this time for a new model to simultaneously achieve global efficiency and national responsiveness and the ability to develop and deploy knowledge on a worldwide basis. At this time, Bartlett and Ghoshal proposed a new organizational configuration that they named the "transnational model," which goes beyond the traditional global organization (see Fig. 9-8). However, because the term "global" has penetrated the business vocabulary for so many years, we prefer to preserve this term by using "contemporary global" meaning "transnational."

This new model is based on a diversified mind-set. Instead of emphasizing a particular core capability such as responsiveness or learning, the contemporary global model is based on the assumption that all three core capabilities should be achieved simultaneously for a global organization to remain competitive. Table 9-5 summarizes the major characteristics that differentiate the four models. By revealing the organizational characteristics for every type of organizational model separately, we can clearly see the problems each of the first three configurations—multinational, international, and global—faces in responding to the new external environment. For example, an organization that is built on the traditional global concept achieves efficiency primarily by exploiting potential scale of economics in all its activities. At the same time, when the capabilities are concentrated at the center, the lack of resources and

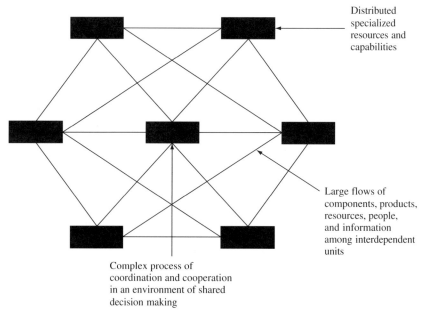

Figure 9-8 The Contemporary Global (Transnational) Organization Model

Source: Reprinted by permission of Harvard Business School Press. From *Managing Across Borders: The Transnational Solution* by Christopher A. Bartlett and Sumantra Ghoshal. Boston, MA. 1991, p. 89. Copyright © 1989, 1991 by Harvard Business School Publishing Corporation, all rights reserved.[23]

Table 9-5 The Organizational Characteristics of the Contemporary Global Model

Organizational Characteristics	Multinational	International	Traditional Global	Contemporary Global
Configuration of assets and capabilities	Decentralized and nationally self-sufficient	Sources of core competencies centralized, others decentralized	Centralized and globally scaled	Dispersed, interdependent, and specialized
Role of overseas operations	Sensing and exploiting local opportunities	Adapting and leveraging parent company competencies	Implementing parent company strategies	Differentiated contributions by national units to integrated worldwide operations
Development and diffusion of knowledge	Knowledge developed and retained within each unit	Knowledge developed at the center and transferred to overseas units	Knowledge developed and retained at the center	Knowledge developed jointly and shared worldwide

Source: Reprinted by permission of Harvard Business School Press. From *Managing Across Borders: The Transnational Solution* by C.A. Bartlett and S. Ghoshal. Boston, MA. p. 65. Copyright © 1989, 1991 by Harvard Business School Publishing Corporation, all right reserved.[24]

responsibilities at the subsidiary level may make them lose their motivation to respond to the local market needs. The lack of resources and the limited participation of the overseas units in the implementation of new technology prevent the whole organization from utilizing the global brain in knowledge creation and learning. These are examples of problems that a classic global organization cannot overcome without jeopardizing its trump card of global efficiency. This is why the contemporary global model became so interesting and important. The attributes of the contemporary global configuration are internally consistent and mutually reinforced. The integrated network, which is essential for work configuration, differentiation of the subsidiary roles and responsibilities, and the simultaneous management of multiple innovation processes, constitutes an integrated system that is capable of sustaining and adapting to new environmental changes.

9.4.2 The Levels of International Development

Jay R. Galbraith, describing the weakness of the traditional measure that companies use to determine how global they are, proposed another method of measuring the level of globalization. He proposed using the proportion of assets and employees outside of the home country as the main criteria to determine the type of international organization (see Table 9-6).

Table 9-6 The Levels of International Development

Level	Proportion of Assets and Employees Outside of the Home Country	Role of the Subsidiary	Type of International Organization
I	Zero	None	National company
II	Low	Startup	International geographical division
III	Moderate	Implementer	Multidimensional network
IV	High	Leader	Transnational organization

Source: Jay R. Galbraith, "Structuring Global Organizations," Chapter 4, from *Tomorrow's Organization* 1998, p. 105. Reprinted by permission of Jossey-Bass, Inc,[25] a subsidiary of John Wiley & Son, Inc.[25]

Galbraith divided the way to globalization into four levels, with the first level being a national organization and the remaining three levels reflecting different types of international organizations. We can also see that on different levels of internalization, the role of the subsidiary is changing. For example, if at Level II—international geographic division—the subsidiary has the role of a startup organization. On Level IV—transnational organization—the role of the subsidiary is as a leader. Describing the four types of international organizations, Galbraith notes that this does not mean that an organization must go sequentially through all these levels. He writes, "A company may start at Level I and move sequentially through the other levels, but no assumption is made here that this is the only or necessary sequence to be followed. For example, many new companies starting up today begin at Level IV...."[26] In recent years, we have seen a strong tendency of young companies to adapt the transnational model of organization without going through the other levels. The four types of international development are another way of measuring the progress toward globalization.

9.5 THE MEANING OF GLOBALIZATION IN THE TWENTY-FIRST CENTURY

We can see that for many years there was an attempt to establish a clear understanding of what globalization is all about, to determine the differences between multinational, international, and global organizations, and to find a generally accepted methodology to measure the progress toward globalization. Academicians and practitioners proposed different methodologies to assess the level of globalization based on the percentage of sales or amount of assets outside the home country or the number of subsidiaries and employees worldwide. All attempts to determine the globalization level have their strengths and weaknesses, but how do these fit into today's environment? Is globalization still a fuzzy subject? Is there any established methodology as to how to get global, or how to measure the level of globalization?

At the 1999 meeting of the World Economic Forum in Davos, Switzerland, Jacques Manardo, Deloitte Touche Tohmatsu's European chairman, said, "Globalization is a hollow word. There is no generally accepted understanding of what it means."[27] Interestingly, this statement was made at a time when more and more organizations were going global, and the term globalization was becoming more and more frequently used.

That reminds us of the story about the elephant and the blind men, who touched different parts of the elephant and expressed what they thought it was, even though none of them had seen what an elephant really looked like. Managers need a clear understanding of what a global organization is all about and how to assess achievements toward globalization.

This is probably why the World Economic Forum and Deloitte Touche Tohmatsu (DTT) at Davos, Switzerland came up with a comprehensive study on globalization, which was reported to the Forum's annual meeting that took place in Davos on January 29, 1999. The aim of the report was to give organizations an understanding of what globalization actually is, how it should be measured, and how to identify innovative practices that will best position organizations for global expansion.[28]

The World Economic Forum did not come up with a formal definition of globalization or with a set of terms that can be used as a base for building a vocabulary that can be used in global work. However, they suggested a framework for assessing corporate globalization. This framework significantly differs from earlier attempts in this direction. In this new framework, globalization is not measured by the percentage of the organization's sales outside its host country or the number of subsidiaries located around the world. According to the proposed framework, an organization can be tested against six broad organizational capabilities: governance and responsibility, strategy and planning, marketing and services, operations and technology, research and development, and organization and human resources management (see Fig. 9-9 and Insert 9-1).

Comparing organizational achievements to these six criteria will help an organization better understand the requirements of a global organization, and at the same time, assess how close the organization is to globalization. As Figure 9-9 suggests, the central measure of globalization success is enterprise value. However, at the World Economic Forum mentioned above, this criterion was the subject of debate. In the 1999 World Economic Forum Annual Report we read, "Shareholder value remains a subject of intense debate. Although some academic and business leaders believe that all activities should ultimately be driven by value to shareholders, other believe that shareholders are but one set of stakeholders, all of whom must be given equal weight in the decision making process."[30] The World Economic Forum also suggested a set of questions that are based on the proposed framework (see Fig. 9-9). Besides being

THE MEANING OF GLOBALIZATION IN THE TWENTY-FIRST CENTURY

Figure 9-9 The Global Framework
Source: World Economic Forum / Deloitte Touche Tohmatsu. http://www.deloitte.com/davos/findings.html[29]

Insert 9-1

The Global Diagnostic

- **Independence and planning.** Is my company's board of directors sufficiently well-diversified with regard to region of origin and experience in global issues to allow for thought that is independent of our region's prevailing business models?
- **Disclosure and accountability.** Does my company disclose financial and operating data in a manner that enables global investors to understand its business and financial performance?
- **Social responsibility.** Are my company's ethical standards communicated in a manner that is understood to have the same meaning by all managers and staff worldwide?

2 **Strategy & Planning**

- **Value-based management.** Do managers within my company accept or reject projects according to their relative potential to create value for shareholders, customers, or employees?

(continued)

Insert 9-1 (*continued*)

- **Partnership development.** Can my company develop mutually beneficial relationships with established customers, government officials, suppliers, and distributors?
- **Global vision.** Do product and service managers within my company see the world as a single economic and operating unit?

3 **Marketing & Service**

- **Organization of products and services.** Does my company assign responsibility for profitability and sales on a global basis?
- **Customer service.** Are my company's procedures and policies designed to serve local needs, measured against a standard of global excellence?
- **Local market development.** Can my company develop an understanding of a targeted market's local culture and practices necessary for integrating my company's messages and building a global brand-and getting close to the customer?

4 **Operations & Technology**

- **Cost efficiency.** Can my company source materials strategically from the most cost-efficient supplier, regardless of its location?
- **Operational effectiveness.** Can my company standardize core processes around the world while allowing sufficient flexibility at the assembly level for tailoring products to local markets?
- **Technology integration.** Can my operations optimize global production and distribution capabilities by using sophisticated decision-aid tools to incorporate the most recent information about all costs in the supply chain?

5 **Research & Development**

- **Innovatory process.** Does my company foster innovation by enabling knowledge-sharing across the R&D function? Does it provide researchers with information about global markets and consumer preferences?
- **R&D partnerships.** Can my company form alliances with strategic research partners to promote new markets and opportunities for its products and services?
- **Innovative capacity.** Can my company's R&D structure leverage talent and knowledge anywhere in the world? Does my company operate only out of centralized R&D centers?

6 Organization & Human Resources Management

- **Leadership development.** Does my company provide the opportunities and tools necessary to develop future leaders wherever my company operates? Can my company identify talent globally?
- **Human resources processes.** Do the human resources processes in my company move people to geographical areas where they can both disseminate knowledge and absorb it?
- **Culture.** Does my company's culture encourage and support managers in their endeavors to gain global experience by managing products, services, operations, and people in a foreign market?

Source: World Economic Forum/Deloitte Touche Tohmatsu. http://www.deloitte.com/davos/global.html[31]

very instrumental in assessing the global level of an organization, these questions can help form a framework of areas on which organizations can focus to achieve success with globalization. They may also serve as guidelines to develop concrete plans toward globalization and as a framework to design a more instrumental methodology for assessing the organization's level of globalization.

References

1. Tom Peters, *Thriving on Chaos*, Alfred A. Knopf, New York, NY, 1987, p. 151
2. Ibid., p. 164
3. Ibid., p. 165
4. *SIA 1999 Annual Directory*
5. *SIA 2000 Annual Directory*
6. *SIA 1999 Annual Directory*
7. Christopher A. Bartlett and Sumantra Ghoshal, *Managing Across Borders: The Transnational Solution*, Harvard Business School Press, Boston, MA, 1989, 1991, p. 19
8. Stephen H Rhinesmith, *A Manager's Guide to Globalization: Six Skills for Success in a Changing World*, McGraw-Hill Companies, New York, NY, 1996, p. 13
9. Ibid.
10. C. K. Praxald, "The Nature of the Competitive Landscape", *The Organization of the Future*, Frances Hesselbein, Marshall Goldsmith and Richard Beckhard (eds.), Jossey-Bass Publishers, San Francisco, CA, 1997, pp. 160–161
11. George S. Yip, *Total Global Strategy: Managing for Worldwide Competitive Advantage*, Pearson Education, Inc., Upper Saddle River, NJ, 1992, p. 10
12. Robert T. Moran and John R. Riesenberger, *The Global Challenge: Building the New Worldwide Enterprise*, McGraw-Hill Companies, New York, NY, 1994, p. 23
13. Bartlett and Ghoshal, *Managing Across Borders: The Transnational Solution*, p. 23
14. Susan Albers Mohrman, Jay R. Galbraith, Edward E. Lawler III, and Associates, *Tomorrow's Organization: Crafting Winning Capabilities in a Dynamic World*, Jossey-Bass Publishers, San Francisco, CA, 1998, p. 108

15 Albert Yu, *Creating the Digital Future: The Secrets of Consistent Innovation at Intel*, The Free Press, New York, NY, 1998, p. 30
16 John M. Stopford and Louis T. Wells, Jr., *Managing the Multinational Enterprise*, Basic Books, New York, NY, a member of Perseus Books, L.L.C., 1972
17 Bartlett and Ghoshal, *Managing Across Borders: The Transnational Solution*, p. 50
18 Ibid., p. 51
19 Bartlett and Ghoshal, p. 52
20 Theodore Levitt, *The Globalization of Markets*, Harvard Business Review, Boston, MA, May-June, 1983, pp. 92–102
21 Bartlett and Ghoshal, *Managing Across Borders: The Transnational Solution*, p. 54
22 Ibid.
23 Ibid., p. 89
24 Ibid., p. 65
25 Jay R. Galbraith, "Structuring Global Organizations," Chapter 4, *Tomorrow's Organization* by Susan Albers Mohrman, Jay R. Galbraith, Edward E. Lawler III, and Associates, Jossey-Bass Publishers, San Francisco, CA, 1998, p. 105
26 Ibid., p. 106
27 Thomas A. Stewart, Going Global, "Part II—A Way To Measure Worldwide Success", *Fortune*, March 15, 1999, p. 196
28 Deloitte Touche Tohmatsu (1999), http://www.deloitte.com/davos/background.html
29 Ibid. (1999), http://www.deloitte.com/davos/findings.html
30 Ibid. (1999), http://www.deloitte.com/davos/global.html
31 Ibid.

Chapter 10

Managing Mergers, Acquisitions, and Other Strategic Alliances

There are many different activities that managers perform periodically. They develop and implement strategies, set goals and create long-term visions, design and manufacture products, etc. But managing mergers, acquisitions, or other strategic alliances remains an infrequent activity for many organizations. This is probably why most organizations don't have the necessary experience when it comes to managing strategic alliances. They usually treat this activity not as a process that they must know, but rather as a one-time event, and so it may not be in their management portfolio of knowledge and skills. At the same time, many studies suggest that in the near future managers will frequently be involved in activities related to organizational integration and collaboration. This means that knowledge in this area has become more significant.

For some people, the term "acquisition" is usually associated with layoffs, changes in the structure and culture, losing positions, diminishing power, etc. So when an acquisition is announced, it may create stress and frustration from insecurity and the unknown. This is probably why we often use the term "merger," even when we are actually talking about an acquisition. These two terms actually mean different things (see Table 10-1).

Having the mindset that merger or acquisition means "bad news" makes some people frustrated and less supportive of this activity. We need to challenge the existing mental models to make sure that people think positive about mergers, acquisitions, and other strategic alliances. Those who have worked through a successful acquisition-integration, or were involved in other alliances that had positive results, realize that collaboration through strategic alliances provide a great opportunity for individual and organizational growth and should be perceived as "good news."

Many studies have shown that with the continuous development of globalization and further technological progress, the importance of mergers,

Table 10-1 Some Forms and Definitions of Strategic Collaborations

Acquisitions	A corporate *acquisition* is a process by which the stock or assets of a corporation come to be owned by a buyer. The transaction may take the form of a purchase of stock or a purchase of assets.
Merger	A *merger* occurs when one corporation is combined with and disappears into another corporation. Note: The word *merger* has a strictly legal meaning and has nothing to do with how the combined companies are to operate in the future.
Strategic Alliance	A *strategic alliance* is a substitute for a merger or an acquisition, and in U.S. law is treated as such for antitrust and other legal purposes. Note: The most common strategic alliance is the *joint venture*. Other forms of strategic alliance are: *licensing agreements, technology exchanges, marketing agreements, cooperative agreements*, and others (see Fig. 10-2).
Joint Venture	A *joint venture* is a collaboration in which two different corporations set up a third, jointly-owned enterprise in corporate form. Note: If more than two corporations are involved in a venture it is called a *consortium* rather than a joint venture.

Source: Based on *The Art of M&A* by Stanley Foster Reed and Alexandra Reed Lajoux, McGraw-Hill Companies, 1998, pp. 4,5,825[1]

acquisitions, and other strategic alliances will grow quickly. This is especially important for the high technology companies, as they cannot perform effectively without integration and collaboration between them, with their customers, suppliers, competitors, and other organizations.

Strategic alliances, which include mergers, acquisitions, joint ventures, licensing agreements, and other forms of collaboration, are part of a comprehensive discipline that requires special attention. In this chapter, however, we will only touch on the surface of this matter to generate some interest in this subject. In the references you will find suggested literature that may help you with further study of this important, increasingly popular discipline.

10.1 MERGERS AND ACQUISITIONS

Managers often use the terms "merger" and "acquisition" interchangeably, but as we mentioned earlier there is a significant difference between these two terms. Mergers involve a much higher degree of cooperation and interaction between the partners than do acquisitions, in which one organization takes over another. Most of the time, mergers occur between relatively equal-sized organizations, while in acquisitions one organization tends to be larger and more powerful than the other. The key factor here is not so much the size of the organization as it is the capabilities and competencies of the merging organizations, and how they can complement each other. The capability of the merged organization must be greater than the sum of the combined parts. For example, in February 2002, AMD completed the acquisition of Alchemy Semiconductor, Inc. Alchemy was a privately held company that designed, developed, and marketed low power, high performance microprocessors for personal connectively devices such as personal digital assistants (PDAs), web tablets, and portable and wired internet access devices and gateways. This was a small, innovative company that had a lot to offer. Hector Ruiz, at that time the COO of AMD, said, "The founders of Alchemy have created some of the most innovative and commercially successful processors in the industry, including the Alpha and StrongARM processors. By joining forces, AMD and Alchemy can supply the building blocks of connectivity—computing solutions from Alchemy coupled with wired technologies and flash memory devices from AMD. Alchemy's MIPS-based solutions provide the ideal combination of high performance and low power for the Internet access device market."[2]

The most important thing here is not what we name the transaction, but how we think about it. It is very important to see an acquisition as a comprehensive process with different types of subprocesses. It is also very important to recognize that a merger or acquisition does not end with signing the papers; the combination is accomplished only after a lot of effort directed towards integrating two or more separate parts into one.

10.1.1 The Acquisition-Integration Process

If we want a merger, acquisition, or any other strategic alliance to succeed, we need to treat it as a process and not as an event. Determining exactly why and what we want to acquire, finding the right candidate to be acquired, assessing the capabilities and competencies of the candidate are just one segment of the process that takes place before the transaction actually takes place. The whole process also includes the post-merger integration activities that take care of all the problems related to merging two or more companies, such as changes in strategies, structure, culture, etc. We recognize that it is impossible to develop a process that will fit the needs of every acquisition-integration. Every combination is unique and requires the development of a process that will fit the specific requirements of that particular combination. However, it would be very helpful to find a model that would give management a framework, a direction that would make it easier to design their own process.

In the Jan/Feb 1998 issue of Harvard Business Review, we came across an interesting article, "Making the Deal Real: How GE Capital Integrates Acquisitions," written by Ronald Ashkewnas, that describes a model of acquisition-integration. We believe this is an excellent framework that could be used by any organization to design a more detailed process of acquisition-integration (see Fig. 10-1).

174 MANAGING MERGERS, ACQUISITIONS, AND OTHER STRATEGIC ALLIANCES

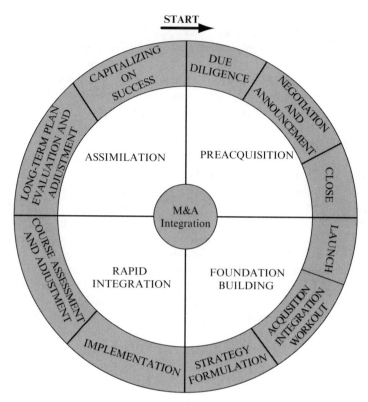

Figure 10-1 The Model of Acquisition-Integration

Source: Used by permission of Harvard Business Review. Based on "Making the Deal Real: How GE Capital Integrates Acquisitions" by Ronald N. Ashkewnas, Lawrence J. DeMonaco and Susan C. Francis, *Harvard Business Review*, Jan-Feb 1998, p. 167. Copyright © 1998 by the Harvard Business School Publishing Corporation, all rights reserved.[3]

In this model, the whole process of acquisition-integration is divided into four major segments, starting with Preacquisition and going through Foundation Building, Rapid Integration, and ending with the Assimilation segment. The output of this process is the actual operational, economic, and financial results from the combination.

Every segment of this model includes two or three process elements. For example, the Preacquisition segment contains three major elements: (1) Due Diligence, (2) Negotiation and Announcement, and (3) Close of the transaction. The model also describes some practical steps for every segment that managers can take to facilitate the process (see Table 10-2).

This model can be used as an introduction to our educational programs on managing acquisition-integration. It allows the participants to see the whole picture and helps them understand that the work on acquisitions doesn't end

Table 10-2 Some Practical Steps to Facilitate the Process of Acquisition-Integration

Segment	Steps For Action
Preacquisition	• Begin cultural assessment • Identify business/cultural barriers to integration success • Select integration manager • Assess strengths/weaknesses of business and function leaders • Develop communication strategy
Foundation Building	• Formally introduce integration manager • Orient new executives to business rhythms and nonnegotiables • Jointly formulate integration and communication plans • Visibly involve senior management • Provide sufficient resources and assign accountability
Rapid Integration	• Use process mapping, and other tools to accelerate integration • Use audit staff for process audits • Use feedback and learning to continually adapt integration plan • Initiate short-term management exchange
Assimilation	• Continue developing common tools, practices, processes, and language • Continue longer-term management exchanges • Utilize corporate education center • Use audit staff for integration audit

Source: Used by permission of Harvard Business Review. Based on "Making the Deal Real: How GE Capital Integrates Acquisitions," by Ronald N. Ashkewnas, Lawrence J. DeMonaco and Susan C. Francis, *Harvard Business Review*, Jan-Feb 1998, p. 167. Copyright © 1998 by the Harvard Business School Publishing Corporation, all rights reserved.[4]

after all the parties achieve agreement and the formal papers are signed. Rather, the hard work begins at the post-merger integration where the combination of elements results in wholeness.

We cannot describe every element of the model in detail in this book, so we will limit ourselves to describing the Due Diligence subprocess as an important part of the whole process.

10.1.2 Merger and Acquisition Due Diligence

When a human organ transplant is undertaken, the success of the operation depends greatly on the selection of the right donor. In addition to the surgeon's qualifications, there must be a perfect match between the donor and the patient. After the operation, preventative measures must be taken to ensure that the patient's body does not reject the organ. It is also understandable that an organ transplant is always done only as a last resort. If the human body could recover by itself and function properly, no one would undertake the complicated and risky procedure of an organ transplant.

The same thing applies to a merger or acquisition. Before acquiring a company, management needs to make sure that there is clearly a need to do this. If the organization has an internal capability to grow and achieve the desired state, then there is no need to risk an expensive collaboration such as a merger or acquisition. However, when the need is determined, the next important step in the process is to find the right 'donor.' There should be strong evidence that the candidate selected has the capabilities and culture to satisfy the acquirer's needs. The process of due diligence comes into play here.

The Process of Due Diligence The main purpose of due diligence is to conduct a review of acquisition candidates. This is to make sure that purchase poses no unnecessary risks to the acquirer's shareholders.

When an organization, after assessing the gap between its current reality and the desired long-term vision and goals, concludes that an acquisition or merger is a viable strategy to fill the gap, a search for a candidate begins. It is very difficult to find the right candidate. Even the most promising candidates have a high risk of failure. In an article in the Nov/Dec 2000 issue of *Harvard Business Review* titled, "Integration Managers: Special Leaders For Special Times," we read, "Less than half of all mergers and acquisitions ever reach their promised strategic and financial goals, yet companies spent more on M&A last year than ever before. According to investment bankers J.P. Morgan, companies worldwide spent $3.3 trillion on mergers and acquisitions in 1999—fully 32% more than was spent in 1998. Basically, that means those companies failed to get the value they expected from a whopping $1.6 trillion in investments. That's a very expensive irony indeed."[5] Due diligence will not eliminate the high risk, but if properly applied, the acquirer will have a greater chance to succeed.

Companies that succeed in mergers and acquisitions are those that go beyond the usual strategic, financial, and legal checks that, by themselves, are very important. They focus on long-term capabilities to ensure that the candidate's existing products reflect their knowledge and expertise, and are not a result of reverse engineering. Successful acquirers learn how to look beyond the products to make a conclusion about the candidate's core competencies. To do this, in addition to business development experts, they involve experienced engineers and technologists who have the necessary knowledge and skills to assess potential design and engineering capabilities that are not apparent on the surface. These days, when the lifetime of a product is measured in months, assessing the intellectual capital of the candidate-organization is more important than the value of the existing processes and products.

10.1.3 The Cultural Aspect Of Due Diligence

If we tried to analyze the reasons for the high failure rate of mergers and acquisitions we would find that one of the major reasons is cultural compatibility. Although companies that don't attend to the cultural aspect of due diligence may not fail right away, they gradually lose the people who hold the organization's intellectual capital. Cultural matches are especially important when two large companies merge together.

Sometimes two CEOs who are in the process of merging their companies talk about the compatibility of the two cultures, about the perfect match of their values, beliefs and assumptions; but after the merger they still have a crash. The senior people of the acquired organization leave first, and gradually most of the other key players also leave. It can be expected that there will be cultural difficulties after a merger. Because of this, the management must create an environment where the two cultures will absorb the best parts of each other and gradually transform the culture to a higher level that is in sync with the vision, strategy, and goals that are shared by both organizations.

We participated in a combination when AMD acquired NexGen in 1996. We believe that this is a good example of how an organization should act in order to succeed in the process of acquisition. It is also an example of how the external and internal environments can influence the process of making a decision about an acquisition. After many years as a second-source semiconductor manufacturer, AMD had invested heavily in the development of the AMD-K5 microprocessor. The intent was to design and produce a microprocessor that would compete against Intel's Pentium® microprocessor. While the AMD-K5 did not produce all the results we expected, it did provide good experience towards becoming an independent microprocessor producer. The AMD-K5 results strongly suggested that AMD needed to take a new approach toward designing and building the next generation of microprocessor. Because of time constraints, AMD was seeking a candidate to help accelerate the process of building a subsequent generation of microprocessors. When NexGen was considered as a candidate for acquisition, AMD did not rush the decision. Due diligence was taken to make sure that NexGen had capabilities that would match AMD's needs. Competent people from both sides started to evaluate NexGen's technological capabilities, their intellectual capital, and their existing culture. In this early stage, excitement and the willingness to merge were expressed from both sides. While it's usually somewhat difficult to assess the intellectual capabilities of engineers and other professionals, AMD representatives were satisfied that NexGen could contribute to the acceleration of a process to produce the next generation of AMD's microprocessor. AMD's representatives also realized that the NexGen culture, which obviously had its own pattern, was nevertheless similar, and could gradually be embedded in AMD's culture.

The positive attitude of NexGen employees was based on their belief that if AMD acquired NexGen they would be better able to materialize their ideas. At that time, NexGen had limited financial capabilities and was unable to fully utilize the intellectual capital available to them.

AMD and NexGen joined forces on January 17, 1996. At that time AMD chairman and CEO, Jerry Sanders, said, "...The engineering resources of NexGen and its sixth-generation microprocessor design, combined with

AMD's 8-inch wafer, sub-0.35-micron process megafab in Austin, Texas, and our recently concluded intellectual property agreement with Intel, further strengthens our capability to provide personal computer manufacturers with high-performance, high-volume alternatives to Intel-based technology... The union of AMD and NexGen enables us to take advantage of the forthcoming NexGen Nx686 microprocessor, a device that will dramatically outperform a Pentium and be competitive with Intel's sixth-generation Pentium Pro."[6]

Over time it was proven that the decision to acquire NexGen, based on due diligence, was right. The efforts of the combined engineering teams and many other departments led to the successful launch of the AMD-K6® microprocessor, which consequently led to an even greater success with the new architecture of AMD Athlon™ microprocessor.

AMD's acquisition of NexGen is an example that demonstrates how due diligence can help make an acquisition a success. It also demonstrates how a successful acquisition can help improve the competitive position of the acquiring organization.

10.1.4 The Integration Manager

Who in the organization is actually responsible for the whole process of acquisition integration? When the need for a merger or acquisition occurs, organizations usually form a due-diligence team that is responsible for the process, but that team acts only up to the moment the deal is closed. After that, most of the time the merged organizations are left alone to naturally integrate. This may create the loss of knowledgeable people who contain the acquired organization's intelligence. This is why some progressive organizations developed a relatively new position—the integration manager. A candidate for this position must have experience in solving complex problems related to collaboration, change, communication, etc. The position also requires knowledge and skills in cross-functional and multi-cultural project management. In the article "Integration Managers: Special Leaders for Special Times," published in the Nov/Dec 2000 issue of *Harvard Business Review*, Ronald N. Ashkenas and Suzanne C. Francis wrote about the experience of some pioneering organizations that first introduced positions such as integration manager. They wrote, "We found that integration managers help the process in four principal ways: they speed it up, create a structure for it, forge social connections between the two organizations, and help engineer short-term successes that produce business results."[7]

As the authors of the article suggest, there are two critical periods in the life of most acquisitions. One is the time between the deal's announcement and its close, and the other is the first 100 days after the close. The major roll of the integration manager is to move integration through these two critical periods as quickly as possible.

The position of the integration manager is a full-time job. However, most of the time this position is needed only from time to time, when the organization is acquiring another organization. Experience has shown that people who take this position and succeed usually end up with a high position in the organization. As soon as a new need arises, the organization already has experienced people who have proved that they can perform this complicated job. So what does the integration manager actually do? As we mentioned earlier, every acquisition requires its own management approach. However, in their article, Ronald N. Ashkenas and Suzanne C. Francis propose an outline of what an integration manager is all about (see Fig. 10-2).

It is very important to recognize that acquisition integration is not just a technical or economical activity; it is also an activity that is strongly related to people. The integration manager should be a very well educated and experienced person with a human touch. Unless the human side of the combination is understood and recognized, the integration will not succeed.

Inject Speed

- Ramp up planning efforts
- Accelerate implementation
- Push for decision and actions
- Monitor progress against goals, and pace the integration efforts to meet deadlines

Engineer Success

- Help identify critical business synergies
- Launch 100-day projects to achieve short-term bottom-line results
- Orchestrate transfers of best practices between companies

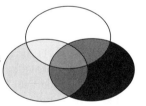

Create Structure

- Provide flexible integration frameworks
- Mobilize joint teams
- Create key events and timelines
- Facilitate team and executive reviews

Make Social Connections

- Act as traveling ambassador between locations and businesses
- Serve as a lightening rod for hot issues; allow employees to vent
- Interpret the customs, language, and cultures of both companies

Figure 10-2 What Integration Managers Do. Every acquisition is different, demanding a different balance of efforts from the integration manager. But in a single integration project, the manager may use any or all of the following four strategies.

Source: Used by permission of Harvard Business Review. From "Integration Managers: Special Leaders for Special Times" by Ronald N. Ashkenas and Suzanne C. Francis, *Harvard Business Review*, Nov/Dec 2000, p. 115. Copyright © 2000 by the Harvard Business School Publishing Corporation, all rights reserved.[8]

10.2 OTHER STRATEGIC ALLIANCES

The most visible indication of the growing role of collaborative strategies lies in the phenomenon often described as *strategic alliances*: the increasing success of multinational organizations to form cooperative relationships with their global competitors.

The term strategic alliance is often used to describe a variety of different interfirm cooperative agreements, ranging from sharing research to forming joint ventures (see Fig. 10-3). In this section, we will describe some of the most frequently used strategic alliances.

10.2.1 Joint Ventures

The popularity of joint ventures (JV) as a form of strategic alliance is rapidly growing from year to year. In Reed and Lajoux's book, *The Art of M&A*, we read, "The number of new joint venture announcements has risen dramatically to rival the number of merger completions—a correlation of nearly 100 percent. Based on this ratio, we can estimate that in 1998, there have been 8,000 joint ventures—about the same number as completed mergers anticipated for the year. Just two decades ago, the ratio of joint ventures and mergers was only about 1 to 4, and overall JV and merger activity was lower. For example, in 1978 (shortly before *Mergers & Acquisitions* stopped tallying JVs), only about 500 JVs were announced, compared to 2,000 mergers."[9]

Many years ago, a traditional joint venture was usually formed between a senior multinational corporation based in an industrialized country and a junior local partner based in a less-developed country. The main goal of this alliance was to gain access to the new markets for existing products. This was an arrangement in which the senior partner provided existing products and the junior partner provided the local market expertise. In this kind of contractual agreement, both parties benefited: The senior organization increased sales revenues, and the junior organization gained access to the new products and learned new technologies from its partner.

Figure 10-3 Frequently Used Forms of Strategic Alliances

The scope of strategic alliances has become significantly broader. Christopher A. Bartlett and Sumantra Ghoshal[10] recognize three major trends that can be seen in the modern formation of strategic alliances:

1. The formation of strategic alliances is between two firms located in industrialized countries.
2. The focus is on creating new products and technologies.
3. Strategic alliances are often formed during industrial transitions in which competitive positions are shifted and the very basis for building and sustaining competitive advantage is being redefined.

These three major trends significantly changed the meaning of joint ventures, and the alliances became strategically more important than the classic joint venture.

Bartlett and Ghoshal also recognize four key motivators[11] that are driving the formation of strategic alliances. These motivators are:

- Technology exchange
- Global component
- Industry convergence
- Economies of scale and reduction of associated risks

Technology exchange, R&D, and technology transfer have become major objectives for most of the strategic alliances formed in recent years. This is because it is very difficult today for one organization to absorb the necessary capital investment and to have the necessary skills and knowledge to cope with global requirements and the speed of change.

For example, FASL, a joint venture formed by AMD and Fujitsu Limited in 1993, operates integrated circuit manufacturing facilities in Aizu-Wakamatsu, Japan, to produce flash memory devices. This joint venture is continuously upgrading its capabilities by building fabrication facilities with state-of-the-art technology and equipment. The cost of such equipment grows repeatedly from year to year. In AMD's 2001 Annual Report, it states, "We expect FASL JV2 and FASL JV3, including equipment, to cost approximately $2.4 billion when fully equipped. As of December 30, 2001, approximately $1.5 billion of these costs had been funded by cash generated from FASL operations..."[12]

One company, standing alone, would have difficulties maintaining the state-of-the-art capabilities without engaging in a joint venture. But, as we mentioned earlier, there are also many other reasons for forming joint ventures such as technology exchange, knowledge transfer, global competition, and economies of scale. In today's environment, with the lifetime of products and technologies continuously getting shorter, it is very important to have partners to share capital investments so that you can all stay technologically competitive. This is why joint ventures have become a popular form of strategic alliance.

10.2.2 Supplier Agreements

Establishing long-term agreements with suppliers is another effective form of coalition. It allows suppliers to adjust their activities to meet the customer's unique requirements. The customer then treats the other party as a preferred supplier, which reduces the supplier's risk in taking extra steps to better satisfy the customer's needs. This two-way relationship also involves a great deal of communication about future plans and needs, which requires an "open door" on both sides. This, in turn, requires full trust from each other. This kind of sourcing relationship opens a great opportunity for both sides to learn from each other by working on joint projects and sharing information. Supply agreements, as a form of collaboration, are especially popular in the semiconductor industry. For example, Hewlett-Packard (HP), Nortel Networks, and many other companies have entered into long-term agreements with AMD, who will supply them with flash memory products. A supply agreement is an example of a win-win activity. As Ken Bradley, the Chief Procurement Officer of Nortel Networks said, "This agreement with AMD, our 1999 Supplier of the Year, assures supply of the critical flash memory products that are required to meet our continued growth in the Optical, Access, Wireless and Enterprise businesses."[13] "Nortel's pioneering technology in areas such as optical and wireless Internet solutions and AMD's leadership in flash memory technology make for a winning combination," said Walid Maghribi, at that time the Group Vice President of AMD's Memory Group. He continued, "AMD's relationship with Nortel highlights its commitment to supply the networking market with extremely reliable, high density flash memory devices. This agreement further solidifies AMD's position as the leading supplier of flash memory devices to the major networking companies of the world."[14]

Suppliers involved in a long-term supply agreement will feel more secure to concentrate on the customer's unique needs. Being a partner gives the supplier a greater feeling of association than just being a provider. Walid Waghribi said, "We are thrilled to be working with HP, a worldwide leader in desktop computing and PC peripherals. Our high density and Page Mode flash memory devices offer an ideal complement to HP's innovative products."[15]

Usually every supplier is, at the same time, a customer that has its own suppliers with whom it develops supply agreements. So both sides can assess the importance of developing supply agreements. As you can see, there are many forms of strategic alliance that allow a company to move toward organizational transformation and globalization.

10.2.3 Licensing Technology

Another popular form of coalition is licensing technology from other companies. Licensing is a great approach to getting access to a technology that has been developed by and is owned by another organization. This may allow the borrower to develop and introduce a new product to the market faster than his competitor. For example, when AMD was in the process of developing and designing the technology for its AMD-K6® microprocessor, AMD needed a new packaging technology to fit the new requirements. Licensing flip-chip packaging technology from IBM was the best solution.

Using a license to promote your business faster is always a good thing to do if it is economically justified. There is almost no risk involved for the buyer. However, selling licenses should be treated as a risky transaction and should be done only under special conditions. Licensing fees are not usually large enough to offset a loss of competitive advantage, but for an organization that has a competitive technology, it may be strategically convenient to award licenses. Michael Porter[16] recognized many circumstances to consider when awarding licenses. However, we will limit ourselves to four circumstances and describe them in relation to the electronics industry.

Circumstance 1: The Inability to Exploit the Technology Sometimes, for economic or other reasons, an organization is unable to fully exploit the technology by itself. This happens very often, especially in the electronics industry where innovative start-up organizations lack the capability to fully exploit their creative ideas. When this is the case, failure to license will create the opportunity for competitors to invent around an organization's technology. However, by licensing, competitors gain a cheaper, faster, and less risky alternative to investing in their own technology. Thus the technology holder, instead of being imitated, may be able to set the standard and collect licensing royalties in addition to the profit it makes from the application of its own technology.

Circumstance 2: Tapping Unavailable Markets By licensing the technology, an organization may gain new opportunities for extra revenue from markets otherwise unavailable to it or from geographic markets where it cannot or does not want to participate.

Circumstance 3: Rapidly Standardizing the Technology Licensing may accelerate the process by which the industry standardizes on a firm's technology. If several firms are pushing the technology, licensing will not only legitimize it, but may also accelerate its development.

Circumstance 4: Quid Pro Quo A company may award a license in return for a license of another firm's technology. Organizations should award licenses only to noncompetitors or a "good" competitor. By a "good" competitor, we mean one who can later play a variety of important roles, such as stimulating demand, blocking entry, and sharing the cost of pioneering. A firm ideally should license noncompetitors that may become good competitors if they later decide to enter the industry. Considering the fact that a licensee may eventually become a competitor, the licenser should make sure that licenses contain renewal clauses to avoid an ongoing commitment to turn over technology.

10.3 SOME FORMS OF GLOBAL COMPETITIVENESS

To be able to compete in a global environment, an organization will need to implement a set of factors that will form the right environment to attract global customers. There is a large range of factors that interact with each other and work as a system to keep the global organization competitive. Below we describe a simplified system of factors that can serve as the core of forming a competitive environment. Not in order of priority, these factors are:

1. *High quality*: the entry ticket into the global market
2. A large *variety* of products and services
3. *Customer convenience* when dealing with the global supplier
4. On time innovative products
5. *Competitive cost* of products and services
6. *Global mind-set*

These six factors are strongly interrelated and interdependent. They can be viewed as a system, as seen in Fig.10-4. For example, to achieve a competitive cost, we need to have a high quality level. A variety of marketable processes can

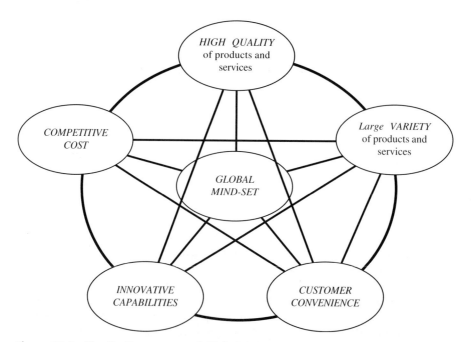

Figure 10-4 The Six Components of Global Competitiveness

Based on Stephan H. Rhinesmith, "A Manager's Guide to Globalization," McGraw-Hill, 1996, p. 48[17]

only be achieved when we have innovative capabilities. The six peripheral elements in the figure depend on the organization's global mind-set.

This system can be effectively applied in any global organization. Here is a short description of every element of the system in relation to the electronics industry.

10.3.1 High Quality: The Entry Ticket into the Global Market

In the last 10 years or so, quality has ceased to be a means of going global. It is now a minimum entry point into global competition. However, keeping the quality of products, processes, and service at a competitive level remains an important requirement without which an organization cannot be considered a global enterprise. For example, in the semiconductor industry, if the finished product does not have a low ppm defect level it cannot be considered a global product. A semiconductor global organization must have processes that are capable of running at a very low ppm defect level.

10.3.2 A Large Variety of Products and Services

Variety is becoming a very important component of customer demand. This tendency can be seen especially in high-technology industries. We will use motor vehicles as an example. In Stephen H. Rhinesmith's book, *A Manager's Guide to Globalization*, he said, "At the 1995 auto show in Detroit, there were 671 different types of motor vehicles, including cars, trucks, and vans."[18] All these vehicles require different types of microprocessors, radios, meters, and other components that are produced in high-technology industries. The increasing variety not only is a result of customer demand, but also comes from the producers who are trying to compete for a higher market share. With the growth of technological capabilities, organizations bring new products to market that the customers could not even dream of before. Take, for example, the electronic navigator that is now installed in many types of motor vehicles. This requires a sophisticated network and equipment, but customers are willing to pay extra money for this kind of service.

The process of globalization, the increase of people's wealth and education, the growth of technological capabilities, and other factors will continue to have influence on the demand for product variety. Organizations in the high-technology industry need to be prepared for this and to be able to foresee customers' needs.

10.3.3 Customer Convenience in Dealing with the Global Supplier

Miniaturization Making devices smaller and smaller makes the end products more convenient to consumers. To satisfy the users of semiconductor components, we need to continuously reduce the size and increase the portability of our products. For some companies, miniaturization is a core competence to gain market share. This tendency is putting pressure on the semiconductor industry to shrink the size and weight of their components.

Speed Computer users want a situation with which they can just push a button on their machine and have the necessary information come up on the screen immediately. This requires a high-speed microprocessor. To make a higher-speed microprocessor, we need to increase the number of transistors per die and reduce the die size for economical manufacturing and equipment miniaturization purposes. For comparison, in 1980 semiconductor technology allowed us to build 40,000 transistors on a chip, and today we use more than 30 million transistors per die (see Fig. 6-7 in Section 6.4). For an organization to be global and stay in business, all these elements of convenience become core competencies that lead to success.

10.3.4 On-Time Innovative Products

Organizations that want to be in the global market must be able to continuously develop and introduce new products. A new product should be ready before the current product reaches its performance limitations. Figure 10-5 illustrates this important requirement. For example, the AMD Athlon™ microprocessor was introduced while the AMD-K6® and K6-2® microprocessors were still selling very well in the market. It is predicted that the AMD Athlon™ will have a longer life than our previous products, but AMD Hammer™, a new microprocessor, is already on the way and will be on the market at the end of 2002. This is the only way for an organization to satisfy the needs of the market and hang on to customers.

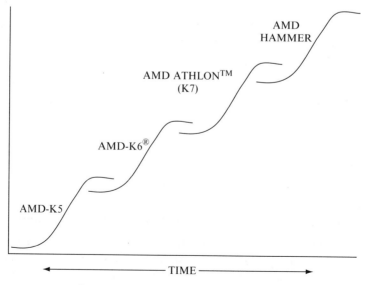

Figure 10-5 The Timeliness of Innovation

For an organization to have the capability of continuous innovation, it must invest heavily in research and development. Table 10-3 shows 20 semiconductor companies investment in R&D in relation to their revenue. As you can see, the average spending on R&D as a segment of total revenue by these 20 companies is 13%.

10.3.5 Competitive Cost

The time when organizations can ask for premium prices for higher quality is almost gone. The best way to gain a bigger portion of the global market is to produce products with the highest quality and performance and the lowest price. This puts great pressure on global organizations to seek areas where the cost of the product can be reduced to a minimum without jeopardizing quality levels.

Organizations are continuously reducing the number of their suppliers, which allows them to develop long-term relationships with key suppliers who

Table 10-3 Some Semiconductor Companies R&D Costs in 2000 (Arranged by Descending Order of Percentage to Total Revenues)

Company Name	Total Revenues ($ Millions) in 2000	R&D Costs ($ Millions) in 2000	Percentage of R&D Costs to Total Revenues
Broadcom	1,090.6	249.7	22.9%
Conexant Systems	2,004.0	394.8	19.7%
National Semiconductor	2,380.0	428.4	18.0%
Maxim Integrated Products	1,073.3	177.1	16.5%
Analog Devices	2,968.2	460.1	15.5%
Texas Instruments	11,875.0	1,745.6	14.7%
Cypress	1,287.8	184.2	14.3%
Advanced Micro Devices	4,644.2	640.9	13.8%
LSI Logic	2,744.3	378.7	13.8%
Altera	7,287.6	911.0	12.5%
Atmel	2,012.7	251.6	12.5%
Xilinx	1,559.0	188.6	12.1%
Lucent Technologies	33,490.0	3,985.3	11.9%
Motorola	37,580.0	4,434.4	11.8%
Agilent Technologies	11,368.0	1,330.1	11.7%
Intel	33,726.0	3,912.2	11.6%
ATI Technologies	1,309.0	128.3	9.8%
Fujitsu	49,807.9	3,237.5	6.5%
Micron Technology	7,584.2	439.9	5.8%
Fairchild Semiconductor International	1,783.2	83.8	4.7%
Average Percentage			13.0%

Source: Based on Cahners Research, EB300: In Lean Times, Only the Strong Survive (Top 300 Electronics Companies Ranked on Revenue), *Electronic Business*, v27, n8, p.8, August 2001[19]

will collaborate with them to achieve material cost reductions. Organizations redesign their structures to reduce product and material transportation costs and seek knowledge workers in any place that they can be found at a relatively lower cost. For example, some companies are now going to India and Russia for computer software programmers who will work for much lower wages than the same specialists in the companies' home countries. The globalization process allows companies to reduce the burden of R&D on product costs by expanding the markets for their products. Joint ventures and other alliances also allow companies to reduce the ultimate cost of the product. To be competitive, the process of cost reduction should be ongoing.

10.3.6 Global Mind-Set

Given the importance of all the above factors, the major factor for success in globalization is the manager's mind-set. To become a global organization, we need to learn how to develop a global organizational mind-set that will allow us to develop and utilize all the rest of the factors. Global thinking should become a major part of the organization's culture (see Section 9.5, The Meaning of Globalization in the Twenty-First Century).

References

1. Stanley Foster Reed and Alexandra Reed Lajoux, *The Art of M & A: A Merger/Acquisition/Buyout Guide*, McGraw-Hill Companies, New York, NY, 1998, pp. 4, 5, 825
2. AMD Press Release, "AMD Acquisition of Alchemy Semiconductor Expands Market Force," February 06, 2002, http://www.amd.com/us-en/corporate/virtual pressroom
3. Ronald N. Ashkewnas, Lawrence J. De Monaco and Suzanne C. Francis, "Making the Deal Real: How GE Capital Integrates Acquisitions," *Harvard Business Review*, Boston, MA, Jan-Feb 1998, p. 167
4. Ibid.
5. Ronald N. Ashkewnas, and Suzanne C. Francis, "Integration Managers: Special Leaders for Special Times," *Harvard Business Review*, Boston, MA, Nov-Dec 2000, p. 108.
6. AMD Online Dialogue, February 1996, http://amdonline.amd.com/dialog/dialog-arch/merge.html
7. Ronald N. Ashkewnas and Suzanne C. Francis, "Integration Managers: Special Leaders for Special Times," *Harvard Business Review*, Boston, MA, Nov-Dec 2000, p.110
8. Ibid. p. 115
9. Stanley Foster Reed and Alexandra Reed Lajoux, "The Art of M&A: A Merger/Acquisition/Buyout Guide," McGraw-Hill Companies, New York, NY, 1998, p. 825
10. Christopher A. Bartlett and Sumantra Ghoshal, "Transnational Management," 2nd edition, The McGraw-Hill Companies, Inc., New York, NY, 1995, p. 370

11 Ibid, pp. 370–372
12 AMD 2001 Annual Report
13 AMD News Release, "AMD signs three year flash memory supply agreement with Nortel Net-works," October 31, 2000
14 Ibid.
15 Ibid., "AMD announces flash memory agreement with HP," September 25, 2000
16 Michael E. Porter, "Competitive Advantage," The Free Press, a division of Simon and Schuster, New York, NY, 1985, pp 191–193
17 Stephen H Rhinesmith, *A Manager's Guide to Globalization: Six Skills for Success in a Changing World*, McGraw Hill Companies, New York, NY, 1996, p. 48
18 Ibid., p. 50
19 Cahners Research/EB300: In Lean Times, Only the Strong Survive (Top 300 Electronic Companies Ranked on Revenue), *Electronic Business*, v.27, n8, p. 8, Aug 2001

Chapter 11

Globalization and Culture

What happens to a culture when the organization becomes a global enterprise? In other words, what impact does globalization have on the organization's culture? How do subsidiaries located in different countries perceive the culture of the parent organization? How do organizations involved in mergers, acquisitions, or joint ventures share their culture? These and other questions arise when organizations are on their way to globalization, and this chapter will attempt to address these concerns.

11.1 CULTURAL CONDITIONING

You are probably familiar with the well-known experiment used in organizational behavior courses that involves showing the class an ambiguous picture—one that can be interpreted in two different ways. Figure 11-1 represents either an old man with glasses or a rat, depending on the way you look at it. Dr. Geert Hofstede used an old woman/young woman picture in an experiment to introduce a discussion on cultural conditioning. The experiment shows that in five seconds you can condition half of the class to see something different from what the other half sees.

We reproduced this experiment several times, using other pictures because the "old/young woman" picture was already familiar to many participants. It was interesting to see how different people in the same situation could perceive things quite differently. Half of the seminar participants were asked to close their eyes while the instructor showed the other half of the participants a slightly altered version of the picture—one in which only the old man can be seen—for only five seconds. Then the instructor asked those who just saw the

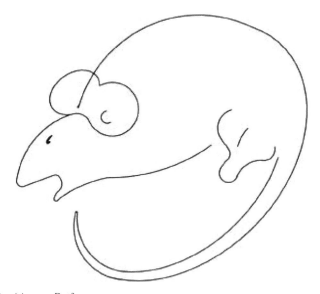

Figure 11-1 Man or Rat?
Source: Copyright (1992) from "Can You Believe Your Eyes?" by J.R. Block and Harold E. Yuker, p. 20. Reproduced with permission from Taylor & Francis, Inc., http://www.routledge-ny.com.

picture of the old man to close their eyes while the instructor showed the other half of the participants a picture of the rat for five seconds. After this, the instructor showed the ambiguous picture to all the participants at the same time. The results are always amazing–most of those "conditioned" by seeing the picture of the old man first only see the old man in the ambiguous picture, and those "conditioned" by seeing the picture of the rat first only see the rat in the ambiguous picture. Even after the secret is revealed, each group usually finds it very difficult to see the "other" picture.

This experiment is a very effective way to demonstrate the power of conditioning. If it is possible to achieve such an effective result in five seconds, think how much stronger the differences in perception of the same reality would be between people who have been conditioned for many years by their family, school, religion, and country's culture before they became a member of a global organization. This conditioning effect should be considered by managers who work in multicultural organizations.

In Part V of this book, we provide some definitions of organizational culture. Here, however, we will introduce another definition of culture that we feel better applies to our discussion. Dr. Hofstede defines culture "as the collective mental programming of the people in an environment."[2] According to his definition, culture is not a characteristic of individuals; it encompasses a number of people who were conditioned by the same education and life experience. From this perspective, organizational culture refers to the collective mental programming that people have in common, the programming that is different from other organizations. Seeing organizational culture as collective mental programming, we recognize that as organizations grow they also gradually develop a mental model, which may be very difficult to change. The difficulty comes from the fact that the mental model not only is programmed into the people's minds, but is also collectively shared and becomes institutionalized in the organization's structure, policies, and regulations, which reflect common beliefs derived from the shared culture.

Having described organizational culture as collective mental programming of people in an environment, and having mentioned that this programming is almost impossible or very difficult to change, the main question is, How do we form a global culture?

11.2 CULTURAL STRATEGIES

When the structure of the organization becomes more complex, it is very important to challenge the existing assumptions in the relationships between all elements of the global enterprise, its subsidiaries, strategic alliances, global customers, suppliers, subcontractors, and the other elements of the organizational macrosystem that make the organization work. Because of the implicit character of the cultural relationship among all elements of the multinational organization, we sometimes think that "business is business," and that we don't need to pay much attention to the fact that when we make, buy, and sell products we are dealing with a large variety of cultures. Not paying sufficient attention to the cultural differences between organizations creates conflicts and may negatively impact the corporate results. Susan C. Schneider and Jean-Louis Barsoux differentiate three cultural strategies—ignore, minimize, and utilize—that in their opinions have implications for relationships between headquarters and subsidiaries. Table 11-1 reflects the difference between these three strategies.

These strategies are based on three different assumptions of culture. One assumption is that culture is irrelevant because business is just business, and it does not matter if you are dealing with the Japanese, the Chinese, or any other nationality. The second strategy is based on the assumption that the difference in culture can create problems or can be a threat to the business. The third strategy is based on the assumption that the cultural differences can create an opportunity or source of competitive advantage. In reality, there is probably no clear-cut way that organizations think about cultural differences and their influence on the organization's business. But understanding the difference between these three strategies may help managers assess where they stand in relation to this issue. Challenging the organization's assumptions about multinational culture can also help the organization form the right mental model of global thinking. We will use our experience of working with different countries to elaborate on these three strategies.

Table 11-1 The Differences Among Three Cultural Strategies

Ignore	Minimize	Utilize
Assumptions: culture as Irrelevant	A problem/threat	An opportunity A source of competitive advantage
Headquarter/subsidiary relationships:		
Ethnocentric	Polycentric/regiocentric	Geocentric
Expected benefit:		
Standardization Global integration	Localization Responsiveness	Innovation and learning
Performance criteria:		
Efficiency	Adaptability	Synergy
Communication:		
Top down	Top down Bottom up reporting	All channels
Major challenge:		
Gaining acceptance	Achieving coherence	Leveraging differences
Major concern:		
Inflexibility Missed opportunities	Fragmentation Duplication of effort and loss of potential synergy	Confusion Friction

Source: Susan C. Schneider and Jean-Louis Barsoux, *Managing Across Cultures*, p. 211.[3] Reprinted by permission of Pearson Education Ltd., U.K.

11.2.1 Ignoring Cultural Differences

When AMD's Manufacturing Services Group was just beginning to form, some managers had a tendency to ignore the cultural differences between the subsidiaries. They assumed that the relationship between a person who works in the U.S. and a person who works in one of our Asian subsidiaries was just a business relationship that was based on policies and specifications. These managers thought that culture had nothing (or almost nothing) to do with doing the job right. If there were a problem, we would try to fix it by enhancing the policies to make sure that the person from the host country would understand what it was all about. We were sure that to maintain or improve the process quality levels, it was good enough to follow the technological standards. This is not to say that organizational culture was not considered one of the most important elements of the organization. Culture was not ignored; what was ignored was the cultural differences. Maybe the word "ignore" is not even appropriate to use here. Usually we ignore something that we know exists in nature but don't want to pay attention to. In our case, at the beginning of our organizational development as a multinational enterprise, we didn't even see or feel this difference.

Cultural differences are implicit. When AMD formulated the corporate core values, we organized "Walk The Talk" training programs and activities that involved open discussions of those values to help every AMD employee, regardless of what country they were in, understand and share these core values. The objective was to have their behavior aligned with the corporate values and beliefs. Using the Corporate interpretation of the values as a foundation, each international site deployed and translated these values and behaviors according to the local culture of their country and organization. We understood that some underlying assumptions are unique to specific sites. Once we had accepted the possibility that subsidiaries can have their own set of cultural values that are not contradictory but complementary to the corporate core values, the employees were more satisfied with the corporate culture and started thinking about their deep-seated assumptions and challenging them.

We cannot advocate or publish a comprehensive set of behaviors for the employees and expect them to behave likewise. Behavior is dynamic and spontaneous. It occurs naturally and is based on the values that are embedded in our minds, starting from our first days of life. Some countries have cultures in which people like to be told exactly how to behave, but most nationalities like to behave according to their own beliefs and judgment. For a multinational organization, it should not matter how people behave, as long as their behavior allows them to get things done and is in sync with the corporation's core values. In general, people may agree with the behaviors their company recommends, but they usually resent being told in detail what they should do and how to do their job. This is especially true with knowledge workers, who know their job better than anyone else in their organization.

Although we are critical of ignoring cultural differences when working in a multinational organization, we also recognize that there are many examples of successful global organizations that do not pay much attention to cultural differences. This is especially true in high-technology industries such as electronics, where the technological process can be "just like in America" in any subsidiary, independent of the country it is located in. What we are talking about here is not the technology, but the culture.

11.2.2 Minimizing Cultural Differences

Minimizing the cultural impact on the organization's behavior is another strategy that some companies use. With this strategy organizations recognize the importance of managing the cultural differences, but they look at it as a problem that needs to be resolved. To minimize the influence of cultural differences on performance, managers in these organizations segregate employees into homogenous teams, train them to create sameness everywhere they can, and take other measures to reduce potential conflict. Managerial efforts are directed at creating a "global" corporate culture in which the strategy is to allow some autonomy, but at the same time, rely on rigorous systems of reporting and financial control. Companies that adapt a strategy of minimizing cultural differences rely mainly on two assumptions, that it is possible to create a strong corporate culture by mixing up all the cultural differences and achieving a homogenous culture, and that it is appropriate to allow subsidiaries to do things "their way" as long as they deliver the required results. This strategy works in some organizations, but experience has shown that it is very difficult to introduce standardized systems, policies, and procedures or to create a real global corporate culture that does not reflect the home organization's national practices and culture. In this case, there is often resistance to the global corporate practices, systems, and core values.

We conducted a workshop on organizational culture in five countries with a total participation of 600 employees. We designed a special method to assess how people of different nationalities perceive the concept of multinational culture in a global organization. Instead of using a traditional questionnaire, we decided to design a process that would metaphorically demonstrate the influence of the corporate culture on the culture of the subsidiaries.

For the demonstration, we divided the participants into five groups, each of which represented a different nationality. Representatives from each group brought a box of M&M candy (representing the core values of the culture) up to the podium. They placed their "core values" in a large jar of hot water (representing the length of time the employees had worked under one organization). As we expected, the M&Ms melted and, after a little mixing, they formed a homogenous "culture" of a multinational organization. When the participants were asked how comfortable they felt with this metaphorical experiment demonstrating the formation of a corporate culture, the response was 3.5 on a scale of 1 to 10, where 10 is very comfortable.

We then repeated the experiment, but this time we mixed up different-colored jelly beans with little plastic beans. The result of this experiment was also a melted mixture of different beans, but the plastic beans remained unmixed, even when we added more boiling water (time). We asked the same question we had asked at the end of the last experiment, and this time we got a 100% approval, which means that all the participants felt that the second experiment best represented the process of forming a multinational culture.

We repeated this demonstration in five countries—the U.S., Singapore, Malaysia, Thailand, and China—and the results were the same. What this second experiment demonstrated was that it is impossible to obtain a melting pot of values composed of cultures from different countries. These values have been programmed for many years and will stay with the people from these countries. Some people may change after being influenced by time and place, but their core values will stay with them. These individual core values are imbedded in the corporate culture and will form a multinational culture but not a homogenous mixture as we demonstrated in the first experiment.

We paid a lot of attention to this particular subject because the mind-set exists in many organizations that if you are an employee of a U.S.-based company, you must carry a pure culture that comes only from the mother country (U.S.). In a global environment, we must change our mental model and accept reality, which shows us that a multinational culture is more like a beautiful art magazine that is made up of separate blocks of color. Every block is represented in the whole picture, and the picture is reflected in its parts.

11.2.3 Making Cultural Differences a Core Competence

It is evident that more organizations are moving toward globalization. Following this trend, organizations are building new subsidiaries, creating joint ventures, and arranging long-term global suppliers. Hence, greater integration between national companies is inevitable. This creates a great opportunity for pooling talents from all over the world, with people and organizations of different cultures. Although, as we mentioned above, some firms adopt the strategies of ignoring or minimizing cultural differences, others choose to capitalize on cultural differences and use them as a core competence.

There are some preconditions necessary for cultural differences to become value-added. A special organizational environment is required to create a balance between the corporate leadership and local initiatives, deliver appropriate education, expose employees to other nationalities and culture, etc. Organizations that are moving toward globalization introduce organizational structures that support the development of this kind of environment, which allows them to use diversity as a core competence.

At AMD, for example, we practice various approaches that allow us to capitalize on cultural differences. We mentioned the power of dialogue above. This form of thought-sharing is exactly what is needed to boost the expectations of people of different culture backgrounds, to share their thoughts and learn from each other.

There was a time when we thought that only in a debate or brainstorming session could we really learn something. There was a time when at the end of a meeting, we could hear the boss say, "No more time for debate. Let me now tell you what you should do." When a statement such as this is made, everything else becomes hard to express. In such a situation, the home country will tend to impose its way of doing things without giving other sites the opportunity to share their opinions.

Today, people participate in meetings in which they can freely express their point of view and create a collective opinion. This is not to say that managers never take personal responsibility and come up with their own conclusion. Rather, most of the time, their conclusion is based on the shared opinions of people from different backgrounds. Hence, even in situations where a true dialogue did not actually take place, the atmosphere of sharing thoughts and opinions is implicit in the many varied daily exchanges.

An example of a continuous dialogue can be seen in AMD's World-Class Supplier Committee meetings, where representatives from different countries come together to have business discussions on material and equipment supply issues. It is interesting to observe the process of learning from each other in these meetings. They express the same problems and proposed solutions from different cultural perspectives and mind-sets. This can happen only in a multicultural working environment. At the end of the meeting, the solutions and decisions are wiser because of the richness of shared opinions.

Another example is AMD's International Engineering Conferences conducted by the Manufacturing Services Group. These conferences are conducted at least twice a year. Engineers and managers come together to develop innovative plans and initiatives. The people who need to fulfill the projects participate from the very beginning of the planning stage in creating a road map that includes the time frame of the project, the strategy of execution, and other components necessary to make the project a success. Even though the participants return to their routine functions in different sites, after the conference there is still something that binds the whole engineering community together— the valuable time they have spent planning the project right from the start, the common goal they have, the feeling of working as a team to develop and introduce technology, which is the most competitive edge in the semiconductor environment. Having International Engineering Conferences is certainly good exposure and a breeding ground for much deeper global contacts and partnerships. It is also a developmental and learning opportunity for the representatives. Through the project, the participants learn to support and integrate AMD's corporate values into the mainstream while appreciating and respecting the unique differences of each location.

The aim here is not to eliminate the cultural differences, but rather to build on those differences to make the multinational organization even stronger. The differences are capitalized to view problems from different perspectives, thereby using a global brain to resolve, solve, or dissolve the problems.

For many years, every AMD site functioned independently and kept the "secrets" of its success to itself. The excuse for not sharing was "We are different. What is good for us would probably not be feasible for another country." Now, under the assumption that diversity is a strength and not a problem to be resolved, the "cultural walls" between the offshore plants have collapsed and they developed different ways to communicate and share knowledge. They conduct international conferences, have regular meetings at different levels, participate in group planning, and work together as one great multinational team.

It would be difficult to estimate the value returned from "breaking the walls" between the cultures, but the overall results, spirit, and organizational mood are obviously higher. Continuous dialogue, sharing and learning from each other, helping each other, and respect for each culture are a basis for working together in a global culture.

We have briefly discussed the idea that to capitalize on the cultural differences as a core competence, managers need first of all to recognize that diversity is not a problem but a great value. Secondly, we need to make sure that managers at all levels, both domestic and international, share the global perspective and have a global mind-set. All managers who are involved in the process need to be engaged in formulating global plans and goals and enlarging their span of responsibilities. This exposure could be an expansion of their career advancement as well as a lubricant in the process of utilizing cultural differences. In the next section, we will describe in greater detail what a global manager is.

11.3 THE GLOBAL MANAGER

Is there such a thing as a global manager? This question was asked by William Taylor in an interview with Percy Barnevik, the head of ABB (Asea Brown Boveri), and his answer was, "Yes, but we don't have many. One of ABB's biggest priorities is to create more of them; it is a crucial bottleneck for us."[4] This is the opinion of a leader of a global organization that employs 36,000 people and generates an annual revenue of more than four billion U.S. dollars. It is an organization that has 50 business areas in different countries.

So, there is no question that preparing managers to work in a global environment is a critical issue. It will certainly take a lot of time, effort, and resources to fulfill this task. But another question that arises here is, Do we need every manager in a global organization to have the knowledge of a global manager? If not, how many of them are needed? Continuing his response to the original question he was asked, Barnevik said, "... On the other hand, a global company does not need thousands of global managers. We need maybe 500 or so out of 15,000 managers to make ABB work well—not more. I have no interest in making managers more 'global' than they have to be."[5] Roughly, this means that out of every thirty managers, one global manager is needed. Certainly every global organization has its own specifics, which depend on the industry, size, culture, structure, and other factors. The ratio of 1 to 30 is not proposed as a standard but as an orientation as to how many global managers will need to be educated in a "global way."

A more detailed way of looking at the issue of preparing different managers in a global organization has been suggested by Enrico Auteri and Vittorio Tesio of Fiat.[6] They have proposed four categories of positions operating in different levels of a global organization.

- *Transnational positions* operate over the whole geographic area pertaining to the business without segmentation or limitations.
- *Multinational positions* operate in the context of several countries defined by specific limitations.

- *Open local positions* operate within the context of a single nation, with significant links, reference points, and dependence on elements outside the country.
- *Local positions* operate within the context of a single nation on the basis of locally determined variables, without significant interaction with other countries.

As Figure 11-2 suggests, the glue that holds all four categories of managers together is their global mind-set. Independent of the managers' functions or positions, all of them who work in a global organization should think and act in line with the organization's global culture.

With the use of these four categories at Fiat, it was found that more than 40% of the managerial positions worked with international interaction. The most exposed areas were commercial, administrative, planning, and personnel. Analyzing the four categories, we can see that the first three need to have knowledge in global management. Perhaps the program of education can be slightly different in these categories, but it is apparent that all three categories need to have special knowledge to be able to work in a global environment.

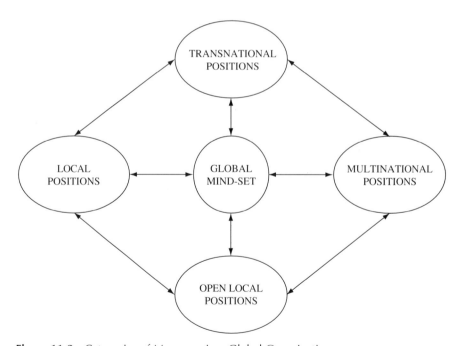

Figure 11-2 Categories of Managers in a Global Organization

Source: Based on Enrico Auteri and Vittorio Tesio, "The Internationalization of Management at Fiat," Journal of Management Development 9, 6, MCB University Press Ltd., p. 10, © MCB University Press Ltd.[6]

Although we recognize the importance of all four categories of management, from a global perspective, the third category—open local position—deserves special attention because it involves national managers who constantly communicate with someone outside their country. This category of managers probably forms the largest part of all managers in the majority of global corporations and therefore will need more attention, recognition, and help to develop global knowledge and skills.

The first two categories—transnational and multinational positions—are a smaller portion of the whole body of management, and they need to be carefully selected and developed under a special program that includes not only formal education, but also a program that rotates these managers to work in different countries and in different positions. Preparing people for transnational and multinational positions will take years. Providing them with a course in globalization alone cannot achieve this.

By emphasizing the importance of the first three categories, we do not mean to undermine the importance of the fourth category—local position. Here, the managers are the employees who get things done locally. They are equal partners in the management team. They are the source from which global managers are selected and groomed. Because of this, the local managers also need to be exposed to information and knowledge that will facilitate them to build a global mind-set. We liken this to a ship on the ocean, where all the positions on board the ship make it stay on course.

11.4 SHAPING THE CULTURE IN A GLOBAL ENVIRONMENT

When things go wrong, we always fear that it is necessary to change the organization's culture, under the assumption that values and attitudes influence behavior. However, values and attitudes are embedded in our minds and cannot be easily changed. It is much easier to influence our behavior by ensuring desired measurable results and leading the members of the organization toward achieving the results. On the basis of the values of the organization, reward for the right behavior and good results, punish (if necessary) for bad behavior and poor results. The drive toward achievement of the desired results will generate the proper behavior, which will eventually shape the organization's values (see Fig. 11-3).

We'll illustrate this with the example, "Driving for Perfection" in our list of corporate core values. We can "Walk The Talk" for hours, but if it is not an element of our attitude, it will not become one of our core values. If you set a goal for the organization to achieve six-sigma quality, develop the strategy, build up a sense of urgency, explain to the employees what will happen if this goal is not achieved, measure and monitor the results, and reward (or punish) according to the results, people will change their behavior. Gradually, your drive for perfection will become a mind-set, an attitude, a core value, and a part of the organization's culture. The only thing that management should do is

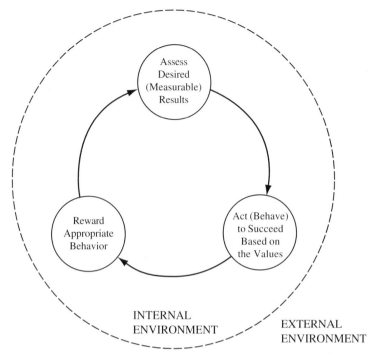

Figure 11-3 Shaping the Organization's Culture

read the signals from the external environment to create an internal environment and then facilitate people toward achieving the desired effects. As soon as people feel that their changed behavior is benefiting them and the organization, they will accept this behavior as a part of their attitude. From then on, less management and supervision will be needed; the culture will work; and the values will influence behavior. When newcomers join the organization, the new cultural environment will force them to behave accordingly. The alternative would be to leave the organization.

This way of shaping the organization's culture is very important when we think about a multinational culture, where people work in different countries with a variety of values and attitudes. What we have described above has come from the real practice of working with subsidiaries located in different countries. In particular, AMD has implemented the six-sigma quality philosophy in all four overseas plants, and it has gradually become a part of the culture.

11.5 CULTURAL INFLUENCES ON MERGERS AND ACQUISITIONS

When a merger or acquisition is being prepared, a lot of research in organizational, financial, and other activities goes on to make sure that both parties feel

202　GLOBALIZATION AND CULTURE

comfortable in reaching an agreement. But less consideration is paid to the cultural aspect of the transaction. Organizations usually assume that there must be a total assimilation of the different cultures for the merger or acquisition to be successful. Another assumption is that this assimilation will occur by itself. This misconception is probably one of the major reasons why a large number of mergers and acquisitions fail or do not produce the planned results. Our experience has shown that, because of the cultural clash, a lot of the people involved in the combination leave the company in the first two to three years, and they take with them the knowledge that the acquirer or both sides of the merger were counting on. To reduce the cultural tension during the period of combination and prevent the loss of talented employees, it is very important to understand what is going on in the combined organization from a cultural point of view.

Because we described different types of mergers and acquisitions earlier in the chapter, we will only describe different types of cultural mergers and acquisitions here. This will help you better understand the cultural processes that are going on during the period of organizational transformation.

As shown in Figure 11-4, there are at least four main types of cultural merger and acquisition outcomes that reflect typical organizational and operational merger implementation strategies.

11.5.1　Cultural Pluralism

The basic assumption underlying cultural pluralism is that "strength comes from diversity." Under this assumption, it is possible to allow the maximum flexibility for the acquired organizations to operate autonomously. It is also possible to allow cultural diversity and cultural subgroups within the context of a shared strategy, which represents the whole organization.

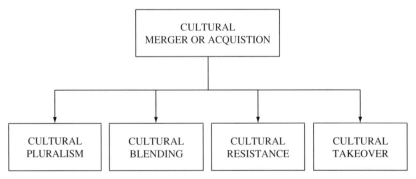

Figure 11-4　Four Major Types of Cultural Merger or Acquisition
Based on Anthony F. Buono and James L. Bowditch, *The Human Side of Mergers and Acquisitions*, 1989, p. 143–149. Used by permission of Jossey-Bass Inc., a subsidiary of John Wiley & Sons, Inc.[7]

11.5.2 Cultural Blending

The cultural blending concept assumes "mergers of equality." Organizations that participate in this transaction focus on assessing the strengths on each side and combining these strengths to form a competitive organization. On the surface, this way of forming the culture looks rational and simple, but in real life it is sometimes very complicated. The process of "blending together" usually creates cultural tension, which then needs to be resolved. However, if properly handled, the cultural blending works very well.

11.5.3 Cultural Takeover

As the name suggests, a cultural takeover merger requires that the culture of the acquired organization be replaced with the dominant culture of the acquiring firm. To succeed in the implementation of this form, we need to have strong leadership and skillful management. A major factor here is the reputation of the acquiring organization before the merger or acquisition.

11.5.4 Cultural Resistance

This type of cultural merger and acquisition refers to situations in which the transaction results in severe cultural conflicts, which leads to a high level of management turnover, market share shrinkage, and difficulties in achieving the desired effects. This kind of situation usually arises when there is a lack of understanding between the cultures of the merger partners. This may also occur when the participants in the transaction do not have a clear understanding of what the leaders of the firms are planning to accomplish. If the merger survives, the tension of resistance slowly reduces, which eventually leads to a cultural takeover by the more dominant of the participating firms.

11.6 CHANGING THE CULTURE DURING A MERGER

Depending on the form of strategy and structure we choose, the employees who participate in an organizational merger will experience different cultural tensions and different degrees of cultural conflict. How do we reduce this tension? And to what extent can we change the culture?

When two organizations with different cultures merge, there will always be a cultural clash. The significance of this clash depends on the strategy and structure selected, which in turn depends on the actual situation of the merge. Sometimes a significant cultural change needs to occur to make the merge or acquisition work. Anthony F. Buono and James L. Bowditch[8] recognize two fundamental ways to affect cultural change in an organization. One way is to get organizational incumbents to "buy into" a new configuration of beliefs and values, and the other way is to recruit and socialize new people into the organization (with an emphasis on new beliefs and values). This process is carried out simultaneously with that of removing existing members where necessary.

Figure 11-5 shows that there are five key intervention points and processes that can be utilized to create such change.

1. Changing the behavior of organizational member
2. Justifying the behavioral changes
3. Communicating cultural messages about the change
4. Hiring and socializing new members who fit in with the desired culture
5. Removing existing employees who deviate from the desired culture

Some examples of these processes can be seen in General Electric and IBM. To succeed in cultural change with fewer losses and less pain, a knowledge manager involved in this process should follow these main points. Even though the steps described in Figure 11.5 are self-explanatory, we will provide a short description of every step.

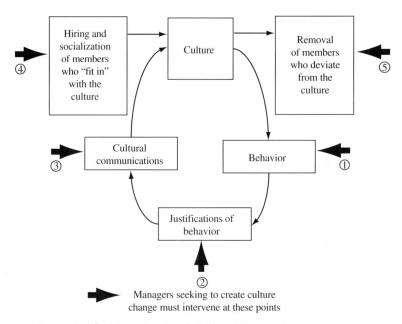

Figure 11-5 A Model of Organizational Culture Change

Source: *Culture and Related Corporate Realities* by V. Sathe, p. 385. Copyright by Richard D. Irwin, 1985. Reprinted by permission of The McGraw-Hill Companies.[9]

11.6.1 Behavioral Change

Values influence behavior. Most of the time when we want to introduce a major change in the organization, we begin by assessing and then trying to change the employees' attitudes. This is because we believe that the values our employees possess influence their behavior. This is true, but it is also correct to think that behavior influences values (see Fig. 11-6). Research shows that one of the most effective ways to change values and beliefs is to begin with changes in related behaviors.

It is very difficult to directly change deeply embedded values in people. Comparatively, it is easier to see and manage the organizational behavior. It is important to note that values do not exist as a set of separate things. You cannot, for example, pull out one particular value from the approved set of corporate values and replace it with another. It makes no sense, as some others recommend, to arrange a list of values by their level of importance. What is important is to see the organization's values as a system where they interconnect and interrelate.

By setting explicit expectations and rewarding appropriate behavioral changes, management can influence the behavior. This, however, does not mean that changes in organizational behavior will immediately translate into cultural change. Employees influenced by the new motivational system and other organizational factors may change their behavior and perform in the desired manner without having changed their values, beliefs, or attitudes. If the employees can see the inherent value of the change, they are much more likely to accept and identify with what the organization is attempting to accomplish.

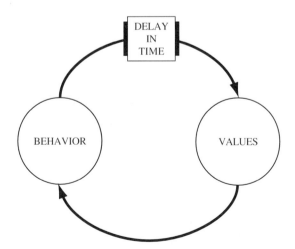

Figure 11-6 Values and Behavior Influence Each Other

In short, it is much easier to influence an individual's behavior than to try to change all the employees' values. If this is done persistently, the employees' values will also gradually change. This will finally lead to a change in the whole organizational culture.

11.6.2 Justifying Behavioral Change

We pointed out above that not only do values influence behavior, but behavior also influences values. It is also important to recognize that for real cultural change to take place, the desired new behaviors for the organization need to be explained, so the employees will know why they should act in a new way and what the consequences will be for retaining the former behavior. Only when the behavioral change has been intrinsically motivated can management expect to see real change in the values and beliefs—when people start to act in a new way because they want to, not because it is expected from top management. For this reason, it is very important that the cultural change be explained and justified to all organizational members. It is also very important to create a need for change by challenging the operating assumptions. When the employees can see that maintaining the status quo is a threat to the organization's survival, they will support the change (which may be a merger or acquisition). As soon as the need for change is created, members of the organization will not only change their behavior but also intrinsically accept the new set of core values and beliefs—the new or reshaped organizational culture.

11.6.3 Communicating the Cultural Change

As soon as we achieve results in changing the organizational behavior and stabilize those behaviors by intrinsic motivators, it is time to communicate these changes through explicit and implicit means. The explicit means include various types of announcements, memos, speeches, and other direct forms of communication. It is important that these forms of communication cascade from the top through all levels of employees. Now, with the availability of the Internet and E-mail, explicit communications can be disseminated faster and over a much broader range. The implicit means of communication include activities such as rituals, ceremonies, stories, metaphors, logos, and other symbolic actions.

Managers usually pay more attention to the explicit forms of communicating the organization's culture, but combining and aligning both explicit and implicit forms of communication will deliver better results. Knowledge managers need to learn how to incorporate the implicit forms in the overall system of communication. The major purpose of applying these two forms is to induce people to adopt the new cultural beliefs and values.

Given the importance of communicating via implicit and explicit means, it can be of little value, or even detrimental, if deeds and actions do not support the things that are communicated to the employees. This is especially true after a merger or acquisition has occurred. During this testing period, the employees are more sensitive to the message and the sender. In forming or shaping an organization's culture, speeches made by upper management are important; but even more important are the actions and symbols. These are better understood and can have a more potent impact on people involved in the merger/acquisition.

11.6.4 Hiring and Socialization

The merger and acquisition process usually creates a higher turnover rate. The manager's role is to implement a system of hiring and socializing new employees who have the best fit to the existing culture or to the culture we want to create. It is very important here to immediately introduce the newly hired employees to the policies and requirements. If the right selection is made, the new employees will adapt themselves faster to the new culture than the employees who were merged or acquired.

11.6.5 Removal of Deviants

The organizations involved in the combination must do everything they can to preserve the best people in the organization, independent of how well they fit the new culture. However, if an employee continuously resists the cultural change and is not willing to accept the new changes, it is better to help the employee find another job. When a large number of people are willing to leave the organization, management needs to be prepared for different types of readjustment. You are probably familiar with the situation in which the withdrawal of one key person leads to increased dissatisfaction and lower morale. This may also lead to a chain reaction in which employees who did not initially plan to leave the organization submit their resignations. All this can lead to a lot of financial and knowledge losses.

The model described above may help you design a proper plan for culture change.

References

1. J.R. Block and Harold E. Yuker, *Can You Believe Your Eyes*, Brunner/Mazel, Inc., New York, NY, 1992, p. 20. Reproduced by permission of Taylor & Francis, Inc., http://www.routledge-ny.com
2. Geert Hofstede, "Motivation, Leadership, and Organization: Do American Theories Apply Abroad?" *Organizational Dynamics*, Elsevier Science Oxford, UK, Summer 1980, p. 43
3. Susan C. Schneider and Jean-Louis Barsoux, *Managing Across Cultures*, Pearson Education Limited, Harlow, UK, 1997, p. 211
4. William Taylor, "The Logic of Global Business: An Interview with ABB's Percy Barnevik," *Harvard Business Review*, Boston, MA, March-April, 1991, p. 94
5. Ibid.

6 Enrico Auteri and Vittorio Tesio, "The Internationalisation of Management at Fiat," *Journal of Management Development 9,6*, MCB University Press Ltd., Bradford, West Yorkshire, UK, p. 10
7 Anthony F. Buono and James L. Bowditch, *The Human Side of Mergers and Acquisitions*, Jossey-Bass Inc., Publishers, San Francisco, CA, 1989, pp. 143–149
8 Ibid., p. 115
9 Vijay Sathe, *Culture and Related Corporate Realities*, McGraw-Hill Companies, New York, NY, 1985, p. 385

PART IV

Some Aspects of Managing Quality

Managing quality requires knowledge in many different disciplines. In this part of the book, you will have the opportunity to refresh your knowledge in managing variation, become familiar with the concept of Baldrige Award criteria as a method for measuring quality, and learn about managing changes. In addition to this, you will also have the opportunity to take a second look at some quality initiatives such as Kaizen, Total Quality Management, Reengineering, and others. We make a comparison of different initiatives to allow you to decide how to develop your own quality management system.

Chapter 12

Some Fundamental Concepts of Managing Quality

Every organization has its own purpose, but if you needed to formulate one universal purpose that would fit any business enterprise, what would it be? In *The Essential Drucker* (2001), we read, "There is only one valid definition of business purpose: to create a customer."[1] Creating a customer is more than customer satisfaction; it implies that in addition to the satisfaction of a customer's particular needs, organizations should also propose new offerings that may be beyond the customer's expectations. It also means that an organization is obligated to develop new technologies and design new products that will enhance the customer's life and make it more productive.

Creating a customer could be a good topic for an entire book, but in this chapter we will just touch the surface of the subject. We will show the relationship between quality and customer satisfaction, what level of quality is required to make it a core competence, and the importance of innovation in creating a customer. We will demonstrate that innovation comes not only from inside the organization, from the creativity of those who create the product, but also from getting as close as possible to the customers, to be able to hear them and understand their needs.

In this chapter, we will also demonstrate that to make a significant improvement in quality, you also need to reshape the existing culture. The chapter concludes with an offering of some forms of quality measurement, emphasizing the importance of applying Baldrige criteria to assess the organization's level of achieving performance excellence.

12.1 QUALITY AND CUSTOMER SATISFACTION

In *A History Of Managing For Quality*, Dr. Juran wrote, "There are now strong indications that the United States will become a world leader in quality during the twenty-first century."[2] As never before, American corporations have brought the quality of their products up to levels we never had before. Expressions such as "quality is now an entry ticket to the market" or "quality is a requirement for any organization to stay in business" are now heard often. This is understandable. In today's highly competitive environment in which customers have many choices of what to buy and where to buy it, quality has become a must. So does this suggest that quality is no longer a competitive advantage? The answer to this question depends on what we mean by the term "quality." Today, quality is not just "zero defects" (Philip Crosby), "Fitness For Use" (Juran), or even Motorola's Six-Sigma concept. Today, customers want much more. So what is the quality that can be considered as a core competence? What is the quality that helps to create and hold the customer? Here we will introduce you to a concept that will lead you to better understanding of how to make quality a core competence. The concept we will discuss is based on the work of Professor Noriaki Kano of Tokyo Rika University.

The Kano Model

Professor Kano developed a model that demonstrates the relationship between customer satisfaction and quality (see Fig. 12-1).

The horizontal line is a scale of the level of product quality, and the vertical line is a scale of the level of customer satisfaction. Usually we think that there is a direct relationship between customer satisfaction and the level of product quality. In other words, we assume that the lower the quality of the product, the less satisfied the customer will be, and the higher the quality of the product, the more satisfied the customer will be. Line (1) represents this kind of relationship. The customer is more satisfied with higher product quality and less satisfied with lower product quality. Such customer requirements are known as "one-dimensional" customer requirements. For example, the customer satisfaction with a personal computer is proportional to its speed of performance (or with the amount of memory). These relationships are sometimes called "satisfiers"—that is, the more fulfilled the requirement is, the more satisfied the customer is.

As Figure 12-1 suggests, one-dimensional customer requirements do not describe the complexity of the relationship between quality and customer satisfaction. Curve (2) indicates the "must-be" quality. This is when the customer is less satisfied or totally dissatisfied when the product doesn't conform to the expected quality level but is not more satisfied when the product quality is higher. For example, when buying a laptop computer, the customer choosing a particular price range expects a particular size screen, or a particular software package, etc. The customer will be dissatisfied if any of these expected quality

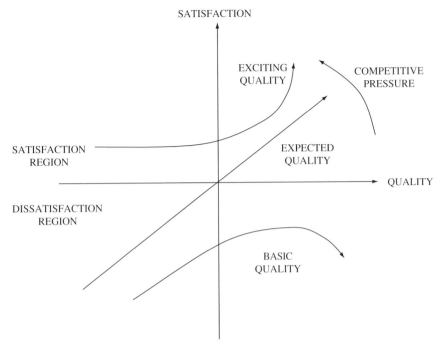

Figure 12-1 The Kano Model of Quality
Source: Quoted from Dr. Noriaki Kano[3]

attributes are missing, but he or she will not be more satisfied if all these must-be components are there. Sometimes the must-be components are called "dissatisfiers." They can dissatisfy, but they cannot increase the satisfaction.

Curve (3) indicates the "attractive" quality. This reflects a situation in which the customer is more satisfied when the product is more functional but not less satisfied when the product is less functional. For example, the customer is not dissatisfied when the computer doesn't come with a digital camera, but the customer is more satisfied when the computer does come with this feature. These attractive elements are sometimes also called "delighters"—they do not dissatisfy if absent, but they can delight the customers when present. This is why when constructing questionnaires to study customer satisfaction levels, it is not enough to simply ask questions like, "Are you satisfied with our product?"

The model gives you the opportunity to see a broader picture of what we mean by customer satisfaction. It shows that the must-be elements can dissatisfy the customer but cannot increase the level of satisfaction. The one-dimensional requirements (or "satisfiers") can satisfy customers only to some extent. From Figure 12-1, we can see that only the "satisfiers" (to some extent) and "delighters" (to a greater extent) can be considered as core competencies that differentiate one organizational from another. For example, if a semiconductor manufacturer, in collaboration with his major suppliers, comes up with a new device that beats all the records in speed and power, this semiconductor manufacturer has a "satisfier" or "delighter" quality level, which can be considered a core competence for the manufacturer. But as we know, a "satisfier" or "delighter" quality level doesn't last long. Competitors usually come up with the same or even better quality levels, so that after the passage of time, "satisfier" and "delighter" quality levels become "must-be" quality. This means that for an organization to have quality as a core competence or, in other words, to always have attractive quality, it must continuously improve and innovate. This means that improvement and innovation become an indisputable part of quality management. We will discuss continuous improvement and innovation as an important part of quality culture further in Section 12.6.

12.2 A DUALISTIC LOOK AT QUALITY AND COST

When an organization, after a lot of innovative efforts and capital investment, finally creates an environment in which all their processes and services run at levels close to zero ppm, can you say that this organization has achieved high quality? Does higher quality cost more (or less)? The answer to both questions is yes and no; it depends on how we perceive quality.

As we discussed in Section 12.1, there are two major aspects of quality: "attractive" and "must-be." In his book *Planning For Quality* (1990), J.M. Juran summarized the features of these two types of quality (see Table 12-1). The left part of Table 12-1 shows that quality means those features of products that must satisfy the customer. From this perspective, quality is oriented toward income. In other words, the organization is doing its best to satisfy customers' needs, and by doing so the organization hopes to increase its income. For example, providing a computer with greater speed, memory, and capacity would make the customer happier, but at the same time this requires an investment and, because of this, an increase in cost. From this point of view, the old assumption that "higher quality costs more" is correct. This is one side of the dualistic meaning of quality and its relationship to cost.

The right side of Table 12-1 demonstrates that quality is freedom from deficiency. This means freedom from nonconformance, rework, retest, scrap, inspection, field failures, etc. From this perspective, quality is oriented toward costs, and "higher quality produces less loss." So, the new definition "high quality costs less" is also correct. When an organization invests in creating an environment in which a nonconforming unit is a rare event, then scrap, rework, and inspection are reduced to a minimum and the product usually has high reliability. The overall product quality goes up, and the cost to produce it goes down. This is the other side of the dualistic look of quality and cost.

Understanding the dualistic meaning of quality allows managers to better see the relationship between quality and cost. This, in turn, will help them understand that the cost of achieving a high quality level is not an expense but a great investment with the objective of improving the organization's profitability

Table 12-1 The Meanings of Quality

The *Attractive* Quality Product features that meet customer needs	The *Must-Be* Quality Freedom from deficiencies

Higher quality enables companies to:
  Increase customer satisfaction
  Make products saleable
  Meet competition
  Increase market share
  Provide sales income
  Secure premium prices
	Higher quality enables companies to:
	  Reduce error rates
	  Reduce rework, waste
	  Reduce field failures, warranty charges
	  Reduce customer dissatisfaction
	  Reduce inspections, tests
	  Shorten time to put new products on the market
	  Increase yields, capacity

The major effect is on sales.	Improve delivery performance
Usually, higher quality costs more.	The major effect is on costs.
	Usually, higher quality costs less.

Source: Adapted with permission from *Planning for Quality*, 2nd ed. (1990) Juran Institute, Inc., Wilton, CT pp. 1–10[4]

and increasing customer satisfaction. A high quality level is a foundation to achieving fewer returns and complaints, shorter cycle times, and improvement in customer-supplier relations. It is also a great source for creating more net income that can be used in the research, development, and design of more attractive products.

12.3 MANAGING THE INTANGIBLE PART OF THE PRODUCT

Let's assume that you have successfully implemented the six-sigma concept and your customer receives a product with zero defects. Can you be sure that the customer is happy? Let's also assume that, in addition, the product is always delivered on time and its price is the lowest in the market. How confident are you that the customer will not switch to your competitor? There are many examples of where a customer, having all the components mentioned above, moves to another supplier. Why? There are many small things that are, by their nature, intangible and make the customer decide whether to stay with your organization or not. Management pays great attention to the "hard" stuff of the product, invests a lot of capital to make it right, and does not leave enough time and money to pay attention to the small things that sometimes make the difference. In this regard, Tom Peters, in his book *Liberation Management* (1992), illustrated the importance of intangibles by using a hotel in Florida as an example. He wrote, "Take the great location away from that resort hotel in Florida, and nobody would have come in the first place.... But having spent a ton on location, the owners proceeded to risk the whole kitty by ignoring the access to the morning newspaper.... We normally put 90 percent of our effort into location, 10 percent into things like newspapers at breakfast. In this crowded, fickle, madcap world, it ought to be darn near the other way around. In the end, it's simple, OK?

$$(1) \text{ "product"} + \text{intangibles} > \text{"product"}$$

Clean ashtrays and readily available newspapers make that Florida resort into a *new resort*. That is,

$$(2) \text{ "product"} + \text{intangibles} = \textit{new } \text{product"}^5$$

These equations make sense for any industry. Take computer support as an example. Having the right computer support, especially for those novice computer users, is a very important component in customer satisfaction. If a computer maker has a reputation for good service and support (intangibles), this may be the component that causes the user to purchase from that computer maker.

Management needs to learn how to manage the intangible part of product quality that will create customer delight. So we can add one more equation to Tom Peters' formula:

$$\text{"product quality"} + \text{intangibles} = \text{customer delight}$$

The Wall Street Journal (October 1991) printed a story that illustrates the importance of intangibles (see Insert 12-1).

As you can see from this story, the same doctor who performed the same service obtained different results in patient satisfaction only because of a small intangible—a letter to the patient. This letter strongly influenced the patient's perception about the doctor's work. Management of the high-technology industries can and should find a large variety of small things that will make their products more value-added. Value is not seen only in the physical things customers can buy for their money; it is also seen in the emotional and psychological things we can use to satisfy customers.

12.4 LISTENING TO THE CUSTOMER'S VOICE

How many times have you heard or said phrases such as "listening to the customer's voice, "focusing on the customer," or "customer satisfaction"? These terms express the importance of customers to the producer's success. To understand the customers' needs, different effective tools are in place, such as Quality Function Deployment (QFD), Failure Model and Effect Analysis (FMEA), and Customer Corrective Action Request (CCAR) Analysis. But how close are we to the customers to be able to listen to their voices? How much

> **Insert 12-1**
>
> **A Story About Intangibles**
>
> If doctors wrote letters to their patients repeating what they told the patients in the office, the patients would be a lot happier with their medical care.
>
> That's the implication of a test by two cancer specialists in Sydney, Australia, who were looking for ways to increase patients' satisfaction with their care.
>
> The researchers reported that they tested the idea on 48 cancer patients who came in to see one of the specialists, M.H.N. Tattersall, for a follow-up consultation after being treated for cancer. Half of the patients were randomly chosen to receive a letter dictated by Dr. Tattersall immediately after the patient's visit. The letter summarized what had been discussed during the visit....
>
> In the ensuing three weeks, all 48 patients were interviewed and asked how satisfied they were with the visit, on a scale of one to five. The patients who received the letters expressed a higher degree of satisfaction than those who didn't receive a letter. Of the 24 letter recipients, 13 reported, "complete satisfaction" with their visit, while only four of those who didn't get a letter were completely satisfied.
>
> The letter-receivers were more satisfied with the doctor's explanation of their condition and the amount of information given. These patients also said they felt they remembered everything they had been told and that they had been able to ask all the questions they wanted to.
>
> *Source*: Reprinted from "The Wall Street Journal", October 11, 1991, p. B4[6]

do the customers know about what they want? For example, is the navigation system that is now in place in many automobiles the result of listening to customers' voices? Are all the features on a modern camera (auto focus, etc.) the direct result of customers' requests?

Science and technology today offer solutions and features that customers could not even dream of in the past. Today, customers require high-speed computers to be able to produce as many calculations, graphics, and other things as possible in the shortest amount of time. But this appetite comes from the producers' offerings, which are based on their technological capabilities. Customer requirements and technological capabilities work together, and sometimes it is difficult to say which comes first. This is important, because it is essential to ensure that the producer is creating all of the possible innovative solutions to help "people live more productive lives" (from AMD's Vision Statement).

It is important to note here that not only do innovative ideas come from the producer's laboratories, or from the people who work there, but the customers are also a great source of innovative ideas if we learn how to use this source properly. To illustrate the point, we adapted a story from the book *Mastering the Digital Marketplace* (1999) by Douglas F. Aldrich. Lloyd Word, the CEO of Maytag explained,

> "...We put researchers into consumers' homes and watched them interact with our appliances. And through these observations, we gain insights that then provide a direction for us to pursue in terms of our innovation. I'll give you a simple example. We're watching a lady load her dishwasher. Before she closes the door and runs the cycle, she walks to the family room that's just adjacent and picks up two durable toys. She comes back, she puts them in the dishwasher, and runs the cycle. Later, the ethnographer says, 'Ma'am, I noticed you put the toys in the dishwasher. Why?' She says, 'Oh, you must not have little kids at home, because if you did, you'd know exactly why. Everything my little Suzie picks up, she puts in her mouth. And I put those in the dishwasher to sanitize them clean.' And he says, 'Well ma'am, do you realize that dishwashers don't sanitize your dishes?' She says, 'Of course they do. Every time I open up the door the steam hits me in the face.' He says, 'No, they clean your dishes but they don't sanitize your dishes.' She said, 'They don't?' And so we get two insights from that. One insight is that the consumers expected their dishwashers to sanitize their dishes. We confirmed that in focus groups and further research. Secondly, that consumers were using dishwashers to clean things other than their dishes. Two important insights. We worked with the National Sanitation Foundation. We developed a sanitation cycle that was time/temperature dependent. We then built it into our high-end Jenn-Air and Maytag dishwashers. It's the kind of innovation that we can charge for at retail because it adds real value for customers, and the product is flying off the shelves. Consumers get an LED visual read out that tells you your dishes have been sanitized clean when you select that cycle."[7]

We used the whole story here because it shows a number of important things related to customer satisfaction, from which we can draw a number of conclusions.

1. To really understand customers' needs, we need to get as close to them as possible. What we have described here probably would not have been picked up by performing a QFD project or by other forms of technological or marketing research.
2. Using the science of ethnography to study how customers use our products can help us find out not just what the customers say but also what they really need, things that they don't even think are possible.
3. Customers can be a great source of new ideas, which if materialized can serve both sides: the customers will receive more value for their money, and the producers will receive more customers. This will make both sides happy.
4. To gain a greater market, producers need to enlarge their responsibility and activities to educate the potential customer about what is going on in their laboratories and R&D departments and what the customers may

expect in the future. This will activate the customers' minds and make them dream of the impossible. Bill Gates is, in our opinion, an excellent model of sharing his thoughts with people—even ideas that are not yet in R&D. This is an excellent way of preparing the potential consumers' minds for a product or service that doesn't currently exist but is technologically possible.

As Figure 12-2 suggests, producers need to get closer to customers to understand their needs for today and use this information as input to the system of innovation for tomorrow. Then they should feed back to the customer (or potential customer) information about what will be going on in their laboratories in the future. This will create new demands from customers, which will make the customer input more valuable to the producer. This continuous cycle is the cycle of progress in customer satisfaction.

12.5 CREATING A QUALITY CULTURE

In Part V of this book, we discuss culture that is related to almost all aspects of the organization. However, considering that quality is a discipline that belongs to all parts of the organization's activities, we decided to write a separate section on quality culture. Below is a short description of some peculiarities that need to be considered in developing and sustaining a great quality culture.

In a meeting related to supplier qualification, a manager shared her observations. She said, "The two potential suppliers have the same process flow, equipment, and technology. They both obtain raw material from the same source, and they have almost the same technological process, but the quality of the product is significantly different. Why?" The meeting participants brought up different suppositions, but if we were to summarize them, we could conclude that the difference in outcome from the two similar processes is the difference in their quality culture. We sometimes don't consider culture to

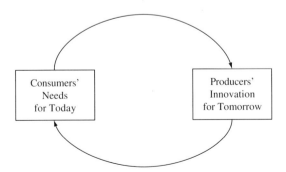

Figure 12-2 The System of Innovation for Tomorrow

be an important factor of quality, but in reality, culture and technology are inseparable components required to produce superior quality.

This is why it is important for management to stimulate a culture throughout the whole organization that continually views quality as a primary objective. In their book *Quality Planning and Analysis*, J.M. Juran and Frank M. Gryna define quality culture as "the pattern of human habits, beliefs, and behavior concerning quality."[8] In this context, culture is the human side of quality (the soft stuff), and together with the right technology (the hard stuff), human beings make quality (see Fig. 12-3).

For an organization to become superior in quality, it needs to work in at least two directions:

1. Develop technologies capable of producing competitive products and processes that meet and exceed customers' needs and expectations.
2. Create a culture that continuously views quality as a primary objective.

One of the reasons why management doesn't totally recognize the impact of culture on quality is probably because it is difficult to quantify. Quality culture is related to our assumptions and perceptions and what we think and feel should be done to achieve superior quality. For example, many years ago, the organization's mind-set was that incoming products must go through receiving inspection. It was also believed that internal processes should be inspected at every major operation and that, at the end of the process, an AQL sampling plan should be utilized before shipping the product to the customer. All this was embedded into the organization's mental model, and we felt good about these activities. Inspection was a part of the process for a long time. Even when the process was capable of producing superior quality, organizations still addicted to inspecting could not stop this activity. It took years to change this mental model in high-technology organizations.

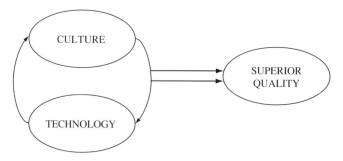

Culture and technology work together and complement each other to create superior quality.

Figure 12-3 Culture and Technology Work Together

Today, in a process-centered culture, many organizations take care of the process quality and the product quality occurs automatically. Customers started to trust their suppliers, and receiving inspection became obsolete in many cases. Certainly, the progress in the development of new technology and process automation was a major factor that allowed organizations to reach higher quality levels and achieve a greater trust in their processes. But at the same time, the focus on quality and the belief that organizations can produce products close to perfection are a result of cultural change, which was influenced by a change in technology. This is why we see the concept of six-sigma quality as not purely a technological issue but also a psychological issue—a cultural issue.

One more thought: When we inspect or prepare policies and procedures on how to build high-quality products, we always keep the customers in mind. We want to make them happy. But how much of our efforts is reflecting the customers' needs? Going back to the inspection activities, we remember that there were a lot of process requirements formulated by the inspection departments that the customer didn't even care about. Organizational culture sometimes requires perfection, even though this has no value to the customer. This is also culture. To illustrate this point, we will use a story from Dorothy Leonard-Barton's book *Wellspring of Knowledge* (1995). She wrote,

> "In the 1940s, Ellery Boss gave his factory floor operators the unprecedented right to personally reject any pen that was visually flawed, even if the imperfection was tiny and did not affect performance at all. Visitors touring the factory were shown with pride the bins of reject products that had in some very minor way failed to meet the stringent quality requirements of the people who produced them. Over the years, as ever more sophisticated equipment allowed the identification of even more microscopic flaws, the workers pursued their obsession with perfection with unwavering dedication. However, in the early 1990s, Cross managers recognized the need to define quality through the eyes of the users rather than the producers, and their research into consumers' first interactions with the products produced some surprises. Consumers did not visually inspect the pen as they took it out of the box. Rather, they hefted it, felt the surface, and then checked its function. They were not concerned about minute variations in surface color or finish."[9]

This example shows that the core values of the organization have not changed and the company remains dedicated to quality. What *have* changed are the operational values that materialize the core value—satisfying the customer. This allows the organization to reduce the production cost and align the quality culture to the customers' needs.

Challenging our assumptions about quality and customers' needs, and changing the operational values and cultural mind-sets, is equally as important as changing the technology. Culture and technology work together to achieve superior quality.

In an organization with a strong quality culture, cultural issues apply not only to the Quality Department but also to all levels of the organization, from the CEO to the operator working on the manufacturing floor.

Maslow's Hierarchy of Needs and Quality Culture

We described above how culture and technology work together to create superior quality. However, although technology can be purchased from outside the organization, culture cannot. It cannot be bought. Culture is an internal thing that belongs only to the organization that created it. Culture is the human side of quality, and because of this motivation is an important component that influences the quality culture of the organization. People need to be motivated to create superior quality. From this perspective, it is interesting to see how Maslow's classic hierarchy of human needs associates with the usual forms of quality motivation (see Table 12-2).

This connection makes it easier for management to understand the relationship of Maslow's theory to quality. In Table 12-2 you can see that improving quality is not just something that the customer needs; it is a way of satisfying *your* personal needs. Quality provides the opportunity to satisfy your psychological needs by increasing your earnings, receiving more benefits, etc. Higher quality allows you to stay together with the team you belong in (no layoffs). Higher quality raises your esteem and promotes you to the top of the hierarchy of human needs—self-actualization, a place where you have the opportunity for creativity and innovation.

To create a quality culture environment, management needs to use Maslow's theory in their work with their subordinates to demonstrate that quality is not

Table 12.2 Maslow's Hierarchy of Needs

Maslow's hierarchy of human needs	Forms of quality motivation
Physiological needs: i.e., things that money can buy—food, clothing, books, etc. In an organizational environment this translates into minimum earnings	Quality work provides more opportunities for an increase in salary, bonuses, and other monetary awards
Safety/security needs: i.e., a desire to protect oneself from losing what has already been achieved, a need to remain employed at the currently achieved level	The higher the quality, the larger the market share. This means more sales, more work, and a greater opportunity to be employed
Social/affiliation needs: i.e., the human desire to belong to a group that can satisfy his or her social needs	The desire to belong to the team motivates you to deliver a quality performance and not let the team down.
Esteem/recognition needs: i.e., the need for self-respect and for the respect of others	Workmanship and personal mastery brings you rewards and recognition
Self-actualization needs: i.e., the internal need of a person for creativity and self-expression. For Maslow, self-actualization stems from a personal realization that "What I can be, I must be and I will be."	Quality needs creativity and innovation; so quality is an opportunity for a worker to propose creative and innovative ideas

Note: See *Maslow on Management* (1998) by Abraham H. Maslow; John Wiley & Sons, Inc., 1998, p. xx. Copyright © 1998 by Ann R. Kaplan, all rights reserved.[10]

only something that the customers or organization's need but also something that everyone who works in the organization needs. While satisfying customers' needs by creating high quality, at the same time employees satisfy their own needs. This is the main message we want to convey here.

Summary

To become superior in quality, an organization needs to develop and continuously upgrade their technologies, to be able to create processes and produce products that meet customer requirements. At the same time, organizations need to stimulate a culture that continually views quality as one of the primary goals of the organization. Culture and technology work together to serve the customers' needs. These two elements must be integrated with the methodologies, strategies, and structures for quality. Management needs to remember that the quality culture of an organization cannot be changed overnight. Our habits of doing things, our assumptions that have been formed for many years, and the mental models that influence our decisions require time, patience, and effort to be changed. Because of this, if an organization wants to introduce changes in the culture of quality, it needs to consider this change as a part of the overall strategic planning process.

A final note: Quality is created by people. Because of this, motivation is one of the important components of the quality culture system. However, it is important to remember that creating superior quality by itself is the best source of employee motivation. Through quality we create and hold new customers, and this is the source of fulfillment of our material and morale needs, including the needs of self-actualization, which means the urge for creativity and self-expression.

12.6 CONTINUOUS IMPROVEMENT AND INNOVATION AS CORE COMPONENTS OF A QUALITY CULTURE

Continuous improvement should be considered as a core component of the quality culture system. It is very important for management to distinguish between the two types of improvement—improvement that results in increasing customer satisfaction and improvement that results in reducing customer dissatisfaction. Note that these two types of improvement are not opposites and do not contradict each other. In *A History of Managing for Quality* (1995),[11] Dr. Juran describes these types of quality improvement as:

1. Improvements directed at increasing customer *satisfaction* through product and process innovation. If properly implemented, such improvements result in greater customer satisfaction and greater profit.
2. Improvements directed at reducing customer *dissatisfaction* through reducing chronic waste in field failures, process scrap and product rework,

inspection, test, and so on.... If properly implemented, improvements of this type result in cost reduction and increased productivity.

These two types of improvement can be illustrated as in Figure 12-4.

As Figure 12-4 suggests, the definition of quality improvement has been broadened. However, many managers still consider quality improvement as an activity of creating processes and producing products that will meet the producer's or customer's specifications. The use of conformance to specifications or conformance to standards as a definition of quality is very dangerous when applied at managerial levels. At those levels, what is important is that products satisfy customer needs and that products and services will delight the customer. Conformance to specifications should be considered as only one of many means to achieve customer satisfaction.

In *Quality By Design* (1992), J.M. Juran broadens the definition and expands the scope of quality by introducing two new terms: the Big Q and the Little Q (see Fig.12-5).

As you can see from Figure 12-5, quality is no longer coordinated by the quality manager but is an issue for all upper management. Quality is no longer viewed as only a technological problem but also a business problem. Quality is not related to incremental improvement but is achieved through breakthrough-type activities, innovation, and creativity. This broadened concept of quality is in sync with the major quality initiatives, including TQM.

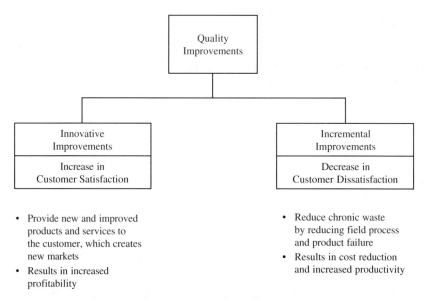

Figure 12-4 Two Types of Quality Improvement

Topic	Content of Little Q	Content of Big Q
Products	Manufactured goods	All products, goods, and services, whether for sales or not
Processes	Processes directly related to manufacture of goods	All processes; manufacturing support; business, etc.
Industries	Manufacturing	All industries; manufacturing; service; government, etc., whether for profit or not
Quality is viewed as:	A technological problem	A business problem
Customer	Clients who buy the products	All who are impacted, external and internal
How to think about quality	Based on culture of functional departments	Based on the universal trilogy
Quality goals are included:	Among factory goals	In company business plan
Cost of poor quality	Costs associated with deficient manufactured goods	All costs which would disappear if everything were perfect
Improvement is directed at:	Departmental performance	Company performance
Evaluation of quality is based mainly on:	Conformance to factory specifications, procedures, standards	Responsiveness to customer needs
Training in managing for quality is:	Concentrated in the Quality Department	Companywide
Coordination is by:	The quality manager	A quality council of upper managers

Figure 12-5 Contrast, Little Q and Big Q

Source: Reprinted with permission of The Free Press, a division of Simon & Schuster, Inc., from *Juran on Quality by Design: The New Steps for Planning Quality into Goods and Services*, by J.M. Juran. Copyright © 1992 by Juran Institute, Inc.[12]

12.7 QUALITY CULTURE MEANS FOCUS ON QUALITY

To stimulate a quality culture, organizations obviously need to continuously keep quality as their primary focus. Organizations have always paid attention simultaneously to a number of characteristics such as productivity, cost, and so on, including quality. But what is always changing is the order of priority. At different times, quality has different priorities. When pressure from the external environment increases on cost, organizations give a greater priority to raising productivity by focusing on direct labor reduction. When customers are oriented toward inventory reduction, organizations pay greater attention to scheduling and cycle time reduction. Recognizing the importance of keeping a finger on the pulse of the external environment, an organization with a strong quality culture should always keep quality in focus, not only because it is important itself, but also because it has a positive influence on productivity, cycle time, and other economic and organizational characteristics. Another reason to focus on quality as a high priority is global competition. High quality is one of the major components that allow organizations to hold on to their customers and retain market share.

When we say quality here, we do not mean products built to the specifications, but rather products perceived as having high quality by the market, quality that is key to the customers' valuation. Figure 12-6 demonstrates the

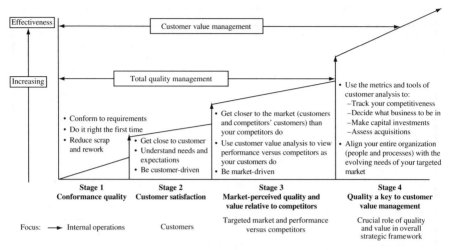

Figure 12-6 Making Quality a Strategic Weapon-the Four Stages
Source: Reprinted with permission of The Free Press, a division of Simon & Schuster, Inc., from *Managing Customer Value: Creating Quality and Service that Customers Can See,* by Bradley T. Gale. Copyright © 1994 by Bradley T. Gale.[13]

different stages of quality. An organization with a high quality culture should be in stage 2 and on their way to stage 3 and should have a plan to produce quality that is key to customer value management.

12.8 MEASURING QUALITY IN A CONTEMPORARY ORGANIZATION

Statements related to quality are most frequently used in management presentations. But what do you think is the reaction of the knowledgeable and dedicated workers in hearing a statement like "We must improve quality"? The workers may conclude that the manager feels they are not doing well enough. The point we want to make is that instead of frequently using the term "quality" without being sure that the people you are talking to have the same understanding of what it actually means, it is better to talk about creating an environment for performance excellence. Most employees have a common understanding that there is always room for improvement. People who have been in the business for a long time may remember different movements for quality improvement. "Zero defects" was perhaps the earliest formal quality improvement program whose aim was the involvement of people from all layers of the organization.

Zero defects started in the Martin Company in 1961, applied to missiles for delivery to the U.S. government, and it was formalized by Philip B. Crosby. In his book *Quality Is Free* (1979), Crosby wrote, "...Zero defects is not a motivational program. Its purpose is to communicate to all employees the literal meaning of the words 'zero defects' and the thought that everyone should do things right the first time."[14] Crosby's philosophy survived for many years, and even today it deserves the attention of management as a quality movement. But in recent years, the definition of quality has become significantly broader and "zero defects" does not fit the needs of measuring quality anymore.

In 1987, Motorola came up with an original philosophy that has become popular all around the world. In a way, it resembles the zero defects concept because both are aimed at excellent performance. However, the six-sigma approach has an absolutely different foundation. It is based on reducing the variation around the target and creating processes capable of producing products that will consume only 50% of the specification range. In other words, it is based on creating processes that can produce products close to perfection (3.4 nonconformities per million opportunities). Motorola's concept became popular because of its way of measuring quality. It is a method that allows people to see how the organization gradually progresses toward achieving excellence.

However, it is important to note here that the term "performance excellence" has a much broader meaning than just making the process close to perfect. Motorola, General Electric, and many other organizations use the term "six sigma" more as a philosophy, a symbol of high performance that includes many other requirements of which the six-sigma process is just a part. Organizations need a comprehensive measurement system that will reflect the multifaceted definition of contemporary quality. In our opinion, such a system is reflected in the Baldrige criteria. In this section we will provide a brief overview of the Baldrige criteria, and discuss their importance as a measure of quality.

12.8.1 Applying the Malcolm Baldrige Criteria for Measuring Quality

The Malcolm Baldrige National Quality Award (sometimes called the Baldrige Award) is an annual award to recognize U.S. organizations for performance excellence and quality achievements. The application of Baldrige Award criteria helps leaders identify organizational strengths and key opportunities for further improvement and innovation. The Baldrige Award promotes the awareness of quality as an increasingly important component of organizational competitiveness. Baldrige principles not only can be used when the management of the organization decides to apply for the Baldrige Award, but also can be used successfully as an internal instrument of measuring quality and assessing the organization's progress toward performance excellence.

Regardless of the criticism from some authors and practitioners who suggest that the Baldrige criteria do not contain contemporary measurements, we believe that the Baldrige measurement system is one of the best systems that can be offered to those in organizations who are looking for a way to measure their progress in performance excellence on a broader scale. This certainly does not mean that the Baldrige criteria cannot be further improved. Using Baldrige principles as a foundation, management can develop their own systems tailored to their organizational needs.

The intent of this section is not so much to encourage management to apply for the Baldrige Award (although this is a good thing to do) but mainly to suggest the Baldrige principles for use in the organization as an excellent instrument for periodic self-assessment. However, once you get results by using the Baldrige criteria, you as a leader will probably decide to apply for a Baldrige Award to formalize your achievements. Below is a brief introduction to Baldrige Award criteria.

12.8.2 The Baldrige Award Criteria

12.8.2.1 Criteria Purpose The main purpose of the Baldrige criteria is:

1. To give the organization a basis for self-assessment,
2. To give feedback to applicants, and
3. To award organizations who meet the Baldrige requirements.

This three-fold purpose suggests that the Baldrige criteria can be used as internal criteria for benchmarking the existing performance level, to receive external feedback from highly qualified specialists, and, finally, to apply for the Baldrige Award.

12.8.2.2 Core Values and Concepts The Baldrige criteria are based on the following value system and a set of concepts that can be observed in high-performance organizations.

Visionary Leadership: The role of a visionary leader is to set directions, maintain customer focus, articulate clear and visible values, and create great expectations. All these should balance with the needs of the organization's stakeholders. The visionary leader should also ensure the formulation of short- and long-term strategic plans, performance systems, and contemporary methods directed toward the achievement of high performance. A visionary leader's work is also to inspire and motivate people by personal example. He or she should serve as a model through ethical behavior and personal involvement.

What is important to emphasize here is that visionary leadership is concerned not only with the satisfaction of the shareholders but also with balancing the needs of all stakeholders (customers, employees, society, and others).

Customer-Driven Excellence: This means that the level of quality and performance in an organization is not only measured and controlled by internal specifications and policies but is also determined by the customer. It also means that satisfying today's customer needs is important but not sufficient for the organization's success. "Customer-driven" implies anticipating customers' desires and marketplace offerings. Organizations need to create products and services that will go beyond the customers' expectations (see Section 12.1). While recognizing the importance of having processes capable of making reliable, defect-free products, customer-driven excellence also requires more value for less cost. Finally, customer-driven excellence should be viewed as a strategic concept. High quality levels create customer loyalty and allow organizations to commit to global markets and gain market share, which requires great sensitivity to the changes in the external environment and the ability to react to change. Therefore, customer-driven excellence requires the introduction of new technology, continuous improvement, and periodic innovation. Customer-driven excellence is an important component of the Baldrige value system.

Organizational and Personal Learning: It is impossible to achieve customer-driven excellence or to develop visionary leaders without creating an environment of continuous learning. Organizational learning can be achieved through continuous incremental improvement of the existing processes and through the introduction of breakthrough-type improvement, which requires more creativity and innovation. Organizational and personal learning cannot be achieved only through classroom activities or other forms of education; something more is needed. According to Baldrige principles, learning needs to be embedded in the way your organization operates. Leaders should create an atmosphere of knowledge and experience sharing, an opportunity for employees to participate in cross-functional activities, an environment where risk taking is encouraged, etc. Organizations should use all kinds of sources from which new information can be taken and transferred into organizational knowledge. Learning from customers, suppliers, competitors, and other countries should be encouraged. Organizations can learn as fast as individuals. This is why leadership should be concerned with creating a learning environment for personal learning: Teamwork, job rotation, empowerment, participation in innovative projects, encouraging learning in colleges, and participation in conferences and symposia are just examples of promoting personal learning. Personal learning should be seen as a way of preparing knowledgeable employees. Learning is also a source of self-actualization and self-fulfillment. It makes the employees feel more valuable, reduces turnover, and increases productivity, which ultimately influences the organization's bottom line. Organizational learning is one of the major sources of achieving competitive advantage.

Valuing Employees and Partners: Performance excellence, especially in high-technology organizations, depends increasingly on the knowledge, skills, creativity, motivation, and dedication of employees. Valuing employees means committing to their satisfaction and development. Baldrige's principles suggest that challenges in the area of valuing employees include:

> "...(1) demonstrating your leader's commitment to your employees' success, (2) recognition that goes beyond the regular compensation system, (3) development and progression within your organization, (4) sharing your organization's knowledge so your employees can better serve your customers and contribute to achieving your strategic objectives, and (5) creating an environment that encourages risk taking."[15]

In any enterprise, especially high-technology organizations, success depends greatly on the partnership these organizations have with their customers, suppliers, subcontractors, and others. This is especially true now that the percentage of outsourcing work is growing rapidly. Building the right relationship with partners is very important. Valuing employees and creating partnerships with all constituencies of the organization is an excellent source of reducing turnover, improving the loyalty of employees, customers, and suppliers, and creating an environment for continuous progress.

There is one more area that is growing rapidly in the high-technology industry and requires continuous partnership improvement, and this is the area of strategic alliances. In recent years, we have observed a strong tendency toward mergers and acquisitions. This requires managers to learn how to work as partners with organizations from different countries and with different cultures.

Agility: Agility is the capacity for rapid change and flexibility. Businesses, especially high-technology organizations, have less and less time to introduce new products. In other words, the lifetime of a product is continuously shortening. This, in turn, requires a higher speed of innovation, design, and production. Today, customers require the best products, for a lower price, and they want them *now*. This requires an organizational capability of providing fast, more flexible responses to customers. Major process changes should be performed in significantly shorter time. Organizations require more and more cross-trained and empowered employees to work in a global environment. Focus on time has become a very important component of an organization's success, and design-to-introduction cycle time has become a critical measurement of performance excellence. To cope with this, organizations need to seek new forms of management, by reducing the organizational bureaucracy, introducing such concepts as concurrent engineering, and taking other measures to meet the demands of rapidly changing global markets.

Focus on the Future: The continuously growing competition in the marketplace requires organizations to focus on the future and be prepared for continuous change. This requires a clear understanding of the short- and long-term factors that affect the organization's business in the marketplace. It also requires a long-term commitment to the major stakeholders—customers, employees, suppliers, stockholders, the community, and other partners. Long-term planning requires the anticipation of many factors, such as customers' expectations, mergers and acquisitions, globalization, new technology, new customers and market segments, strategic moves by competitors, etc. All these should be considered when developing long-term strategic plans and objectives. A focus on the future also requires improving the long-term relationship with suppliers and subcontractors, employee development, and creating an innovative environment in which all constituents can participate in creating a high-performance organization.

Managing Innovation: In this context, innovation means making meaningful changes to improve an organization's products, services, and processes, which will allow it to create new value for the organization's stakeholders. Innovation should raise the level of quality and performance. Not only is it related to technology, new products, and new methods, but it is also important for all aspects of the organization and all business processes. This is especially important for high-technology enterprises, where products and processes have a short life cycle. Developing a culture in which risk taking is encouraged by management and innovation is motivated is an important requirement for a contemporary high-technology organization.

Management by Fact: To achieve performance excellence, an organization should continuously measure and analyze its performance. Measurements should include customers, products, and services; benchmarking and comparisons of operational market and competitive performance; and suppliers, employees, cost, and financial performance.

Managing by fact is probably one of the most difficult, yet important, core values to follow. With this core value, organizations are expected to collect data continuously in a systemic way. Management must ensure that data are gathered on the right variables and collected randomly and that the sample size allows people to come to meaningful conclusions and take the right corrective actions. Once a meaningful and balanced set of matrices have been introduced in an organization, research must be conducted that will include comparisons and relationships between hard and soft data, for example, a comparison between the scores of customer satisfaction and future business strategy or the relationship between scores of customer satisfaction and organizational profit. All this will provide the opportunity to use data as feedback to improve the system.

Public Responsibility and Citizenship: An organization is responsible for operating according to the requirements of business ethics and the protection of health, safety, and the environment. This includes the operations of the organization as well as the life cycle of its products and services. An organization should also emphasize resource conservation and waste reduction at the source. All of this should be considered at the earliest stages, starting from the idea and design. Practicing good citizenship refers to leadership and the support of publicly important purposes. Such purposes may include improving education and health care in the community where the organization is located, environmental excellence, resource conservation, and other practices.

Focus on Results and Creating Value: Focusing on results is an important component of performance excellence. Results should be used to create and balance value for all key stakeholders. Creating value for key stakeholders will allow you to build loyalty and obtain greater contributions from customers, suppliers, employees, and other partners.

232 SOME FUNDAMENTAL CONCEPTS OF MANAGING QUALITY

The Baldrige criteria require an emphasis on balance between short-and long-term performance, ensuring long-term survival and success. It is important to note here that one-half of the points in the Baldrige criteria are based on actual results and organizational achievements. This means that the Baldrige Award is concerned with not only doing the right things but also getting the right results.

Systems Perspective: The Baldrige criteria provide leaders with a systems perspective for managing an organization. The core values described above and the seven Baldrige criteria (described in the next section) form an integrated mechanism for the Baldrige criteria system. The application of the system allows leaders to manage the whole organization, as well as its components, to achieve organizational excellence.

Summary

The core values described above demonstrate that the Baldrige criteria cover almost all aspects of organizational performance, starting with visionary leadership and finishing with the focus on results and creating value for all stakeholders. The description of core values and principles ends with "systems perspective." It shows that the Baldrige criteria use a systemic approach in measuring and managing the process of achieving excellent performance. This

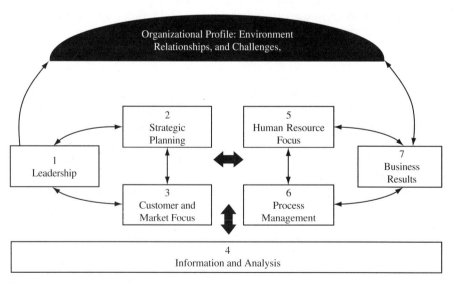

Figure 12-7 Baldrige Categories for Performance Excellence Framework: A Systems Perspective

Source: Reprinted with permission from the 2001 Criteria for Performance Excellence of The Baldrige National Quality Program, p. 5.[16]

is why we think that the Baldrige criteria provide an excellent framework that organizations can use to create their own management system of excellence.

12.8.2.3 The Baldrige Criteria for Performance Excellence Framework: A Systems Perspective
The Baldrige criteria consist of seven integrated categories that are shown in Figure 12-7. In addition, the figure includes the organizational profile, which sets the context for the way your organization operates: internal and external environments, key working relationships, strategic challenges, and other components that serve as an overreaching guide for your organizational performance measurement system.

As Figure 12-7 suggests, the seven categories are interconnected and function together as a system. Although every category is evaluated separately, the overall result of the system depends on the relationship between the seven elements.

Even though the Baldrige criteria are upgraded slightly every year, the main principle remains the same. The examiners measure excellence by analyzing an organization's performance in seven categories. Table 12-3 shows the Baldrige Award categories and point values for 2001. The National Institute of Standards and Technology publishes an annual booklet concerning these award criteria (individual copies can be ordered free of charge from NIST; FAX 1-301-948-3716).

The criteria and the scoring guidelines represent a two-part assessment system. The 2001 criteria are a set of 24 basic, interrelated, result-oriented requirements. The scoring guidelines (see Table 12-4) include three assessment dimensions—approach, deployment, and results—and the key factors used in assessment relative to each dimension. The Baldrige criteria allow management to receive a profile of strengths and areas for improvement relative to the 24 requirements, which measures a large variety of processes and actions that will contribute to performance excellence.

12.8.2.4 What Baldrige Criteria Score is Required to Become a World-Class Organization?
Table 12-4 is a scoring guideline that you can find on page 46 of the Baldrige *2001 Criteria For Performance Excellence*. This scoring scale can be used to assess your organization's performance. As Harry S. Hertz, Director of the Baldrige National Quality Program, suggested, "In the most competitive business sectors, organizations with world class business results are able to achieve a score above 700 on the 1,000 point Baldrige scale."[19] In his book *Baldrige Award Winning Quality*, Mark Graham Brown wrote, "According to a major quality consultant's data base from a survey they conduct on the Baldrige criteria, 'corporate America' rates 560 points out of the 1000 on the Baldrige scale. An organization I consulted with scored themselves at 700/1000 on a Baldrige assessment survey, and was shocked when they got knocked out of the first round upon actually applying for the Baldrige Award. The truth is that most companies think they rate much higher on the

Table 12-3 Baldrige Award Categories and Point Values

	Examinations Categories/Items	Maximum Points
Preface	*Organizational Profile*	
	P.1 Organizational Description	
	P.2 Organizational Challenges	
	1.0 Leadership	*(120 points)*
1	1.1 Organizational Leadership	80
2	1.2 Public Responsibility and Citizenship	40
	2.0 Strategic Planning	*(85 points)*
3	2.1 Strategy Development	40
4	2.2 Strategy Deployment	45
	3.0 Customer and Market Focus	*(85 points)*
5	3.1 Customer and Market Knowledge	40
6	3.2 Customer Relationships and Satisfaction	45
	4.0 Information and Analysis	*(90 points)*
7	4.1 Measurement and Analysis of Organizational Performance	50
8	4.2 Information Management	40
	5.0 Human Resource Focus	*(85 points)*
9	5.1 Work Systems	35
10	5.2 Employee Education, Training, and Development	25
11	5.3 Employee Well-Being and Satisfaction	25
	6.0 Process Management	*(85 points)*
12	6.1 Product and Service Processes	45
13	6.2 Business Processes	25
14	6.3 Support Processes	15
	7.0 Business Results	*(450 points)*
15	7.1 Customer-Focused Results	125
16	7.2 Financial and Market Results	125
17	7.3 Human Resource Results	80
18	7.4 Organizational Effectiveness Results	120
	Total Points	*1,000*

Source: Based on the 2001 Criteria for Performance Excellence of the Baldrige National Quality Program, p. 9[17]

Table 12-4 The Baldrige Award Scoring Guidelines

	Approach/Deployment	Results
0%	• No systemic approach evident; information is anecdotal	• There are no results or poor results in areas reported
10% to 20%	• The beginning of a systemic approach to the basic purposes of the Item is evident • Major gaps exist in deployment that would inhibit progress in achieving the basic purposes of the Item • Early stages of a transition from reacting to problems to a general improvement orientation are evident	• There are some improvement and/or early good performance levels in a few areas • Results are not reported for many to most areas of importance to your organization's key business requirements
30% to 40%	• An effective systematic approach responsible to the basic purposes of the Item is evident • The approach is deployed, although some areas or work units are in early stages of deployment • The beginning of a systematic approach to evaluation and improvement of basic Item processes is evident	• Improvements and/or good performance levels are reported in many areas of importance to your organization's key business requirements • Early stages of developing trends and obtaining comparative information are evident • Results are reported for many to most areas of importance to your organization's key business requirements
50% to 60%	• An effective, systematic approach, responsive to the overall purposes of the Item, is evident • The approach is well deployed, although deployment may vary is some areas of work units • An act-based, systematic evaluation and improvement process is in place for improving the efficiency and effectiveness of key processes • The approach is aligned with basic organizational needs identified in other Criteria Categories	• Improvement trends and/or good performance levels are reported for most areas of importance to your organization's key business requirements • No pattern of adverse trends and no poor performance levels are evident in areas of importance to your organization's key business requirements • Some trends and/or current performance levels—evaluated against relevant comparisons and/or benchmarks—show areas of strengths and/or good to very good relative performance levels • Business results address most key customer, market, and process requirements

Table 12-4 (*contd.*)

	Approach/Deployment	Results
70% to 80%	• An effective, systematic approach responsive to the multiple requirements of the Item and your current and changing business needs is evident • Approach is well-deployed with no significant gaps • A fact-based, systematic evaluation and improvement process and organizational learning/sharing are key management tools; there is clear evidence of refinement and improved integration as a result of organizational-level analysis and sharing • The approach is well-integrated with organizational needs identified in the other Criteria Categories	• Current performance is good to excellent in areas of importance to your organization's key business requirements • Most improvement trends and/or current performance levels are sustained • Many to most trends and/or current performance levels— evaluated against relevant comparisons and/or benchmarks— show areas of leadership and very good relative performance results • Business results address most key customer, market, process, and action plan requirements
90% to 100%	• An effective, systematic approach fully responsive to all the requirements of the Item and all your current and changing business needs, is evident • The approach is fully deployed without significant weaknesses or gaps in any areas or work units • A very strong, fact-based, systematic evaluation and improvement process and extensive organizational learning/sharing are key management tools; strong refinement and integration, backed by excellent organizational-level analysis and sharing, are evident • The approach is fully integrated with organizational needs identified in other Criteria Categories	• Current performance is excellent in most areas of importance to your organization's key business requirements • Excellent improvement trends and/ or sustained excellent performance levels are reported in most areas • Evidence of industry and benchmark leadership is demonstrated in many areas • Business results fully address key customer, market, process, and action plan requirements

Source: Based on the 2001 Criteria for Performance Excellence of the Baldrige National Quality Program, p. 46[18]

Baldrige scale than they really merit. If corporate America were really at an average level of 560 on the Baldrige scale, U.S. products and services would be beating everyone else's in quality."[20] According to Mr. Brown, around 80% of the Baldrige Award applicants score less than 600 points. Usually an organization needs a score of 600 points or greater to receive a site visit, but this is not a hard rule. What is important is for organizations to start performing a self-assessment against the Baldrige criteria to determine how far they are from world-class organizations. As Mr. Hertz wrote in the cover letter to the *2001 Criteria of Performance Excellence*, "While we make no promises for the future, on average, publicly traded Baldrige Award recipient companies have outperformed the Standard & Poor's 500 by four to one."[21] Even if an organization applies the Baldrige criteria only for internal use, the chance that it will become a world-class enterprise is high.

12.8.3 Global Criteria for Measuring Quality

When we started to use the Baldrige criteria as a base for developing our $TCPI^2$ measurement system, some managers were concerned about how this would fit within the various cultures of our organization, which are located in different parts of the world. In this regard, it was interesting to see that the Baldrige criteria gradually became global criteria for measuring quality. Table 12-5 is a comparison of the core values of the Malcolm Baldrige Award and those of the European Quality Award (EQA) System.

As you can see, the values are very similar. In his book *Insights To Performance Excellence 2001*, Mark L. Balzey suggests that one of three basic models are based, in whole or in part, all around the world: the Baldrige model, the Deming model, and the EQA model. He further provides 10 core values that are included in over 50% of the worldwide awards (see Table 12-6). Note that the Baldrige values given in parentheses show that the Baldrige Award system, which started in 1988, has been gradually modified to reflect the global needs of organizations.

12.8.4 Concluding Thoughts on the Baldrige Criteria

The Baldrige criteria are not only a comprehensive system for measuring the quality levels of an organization but, we believe, also the best set of guidelines on how to run an effective organization. The core value system on which the criteria were built is itself an excellent guideline that provides direction toward performance excellence for organizations. On the basis of these values, an organization can develop a strategic plan to achieve performance excellence. The seven criteria can be a foundation for a measurement system tailored to the particular needs of the organization. Managing for quality requires a system that allows the measurement of progress. The Baldrige criteria constitute such a system.

Table 12-5 A Comparison of the EQA System and Baldrige Award Core Values

EQA Value	Malcolm Baldrige Award
Customer focus	Customer driven
Supplier partnerships	Valuing employees and partners
People development and involvement	Organizational and personal learning
Process and facts	Management by fact
Continuous improvement and innovation	Managing for innovation
Leadership and consistency of purpose	Visionary leadership
Public responsibility	Public responsibility and citizenship
Results oriented	Focus on results and creating value

Source: Reprinted with permission from *Insights To Performance Excellence 2001: An Inside Look at the 2001 Baldrige Award Criteria*, by Mark L. Blazey, p. 396. Copyright © 2000, ASQ Quality Press, Milwaukee, WI.[22]

Table 12-6 Ten Common Core Values of the Worldwide Quality Awards

- Customer Orientation (Customer Driven)
- Continuous Improvement (Organizational and Personal Learning but was formerly called Continuous Improvement and Learning)
- Participation by Everyone (Valuing Employees and Partners)
- Committed Leadership (Visionary Leadership)
- Process Orientation (Managing for Innovation)
- Long-Range Perspective (Focus on the Future)
- Public Responsibility (Public Responsibility and Citizenship)
- Management by Facts (Management by Fact)
- Prevention (No clear corresponding value, but Agility corresponds in part)
- Learn from Others (Organizational and Personal Learning)

Source: Reprinted with permission from *Insights To Performance Excellence 2001: An Inside Look at the 2001 Baldrige Award Criteria*, by Mark L. Blazey, p. 396. Copyright © 2000, ASQ Quality Press, Milwaukee, WI.[23]

References

1   Peter F. Drucker, *The Essential Drucker*, Harper Collins Publishers, Inc., New York, NY, 2001, p. 20
2   J.M. Juran, *A History Of Managing For Quality*, ASQ Quality Press, Milwaukee, WI, 1995, p. 599
3   Kano et al., "Attractive Quality and Must-Be Quality," presented at Nippon QC Gakka: 12th Annual Meeting, 1984
4   J.M. Juran, *Planning For Quality*, 2nd edition, Juran Institute, Inc., Wilton, CT, 1990, pp. 1–10
5   Tom Peters, *Liberation Management*, Alfred A. Knopf, a division of Random House, Inc., New York, NY, 1992, p. 691

6 "Doctors' Letters Are Linked To Patients' Satisfaction," *The Wall Street Journal*, New York, NY, October 11, 1991, p. B4

7 Douglas F. Aldrich, *Mastering The Digital Marketplace*, John Wiley & Sons, Inc., New York, NY, 1999, pp. 268 269

8 J.M. Juran and Frank M. Gryna, *Quality Planning and Analysis*, 3rd edition, McGraw-Hill Companies, Inc., New York, NY, 1993, p. 158

9 Dorothy Leonard-Barton, *Wellsprings Of Knowledge*, Harvard Business School Press, Boston, MA, 1995, p. 52

10 J.M. Juran and Frank M. Gryna, *Quality Planning and Analysis*, 3rd edition, McGraw-Hill Companies, Inc., New York, NY, 1993, p. 159

11 J.M. Juran, *A History Of Managing For Quality*, ASQ Quality Press, Milwaukee, WI, 1995, p. 624

12 J.M. Juran, *Juran On Quality By Design*, The Free Press, a division of Simon and Schuster, Inc., New York, NY, 1992, p. 12

13 Bradley T. Gale, *Managing Customer Value*, The Free Press, a division of Simon and Schuster, Inc., New York, NY, 1994, p. 9

14 Philip B. Crosby, *Quality Is Free*, McGraw-Hill Book Company, New York, NY, 1979, p. 116

15 Baldrige National Quality Programs, *2001 Criteria for Performance Excellence*, National Institute of Standards and Technology, Gaithersburg, MD, 2001, p.2, http://www.quality.nist.gov/Business_Criteria.htm

16 Ibid., *2001 Criteria for Performance Excellence*, National Institute of Standards and Technology, Gaithersburg, MD, 2001, p.2, http://www.quality.nist.gov/Business_Criteria.htm p.5

17 Ibid., *2001 Criteria for Performance Excellence*, National Institute of Standards and Technology, Gaithersburg, MD, 2001, p.2, http://www.quality.nist.gov/Business_Criteria.htm p. 9

18 Ibid., *2001 Criteria for Performance Excellence*, National Institute of Standards and Technology, Gaithersburg, MD, 2001, p.2, http://www.quality.nist.gov/Business_Criteria.htm p. 46

19 Ibid., *2001 Criteria for Performance Excellence*, National Institute of Standards and Technology, Gaithersburg, MD, 2001, p.2, http://www.quality.nist.gov/Business_Criteria.htm p. I

20 Mark Graham Brown, *Baldridge Award Winning Quality*, 11th edition, Productivity, Inc., Portland, OR, 2001, p. 65

21 Baldrige National Quality Programs, *2001 Criteria for Performance Excellence*, National Institute of Standards and Technology, Gaithersburg, MD, 2001, p.i, http://www.quality.nist.gov/Business_Criteria.htm

22 Mark L. Blazey, *Insights To Performance Excellence 2001: An Inside Look at the 2001 Baldrige Award Criteria*, ASQ Quality Press, Milwaukee, WI, 2001, p. 396

23 Ibid.

Chapter 13

Managing Variation: A Requisite for Quality

When an organization has designed a competitive product, success in the market depends on the product's high quality, attractive cost, and short delivery time. Reducing process variation plays a major role in achieving these three requirements.

From the time Dr. Deming gave his popular courses on quality and statistical principles, you probably could not find a manager who, at one time or another, had not taken a course on statistical principles in which the concept of process variation was discussed. However, in their routine work managers don't always use the concept of process variation as a factor to improve quality or reduce cost. Today there is no shortage of different types of books on the subject of managing variation. Because of this, in this chapter we will simply provide an overview of this subject with an emphasis on managing for quality in high-technology enterprises just to refresh your knowledge and perhaps trigger new interest in statistical thinking.

13.1 MANAGING PROCESS VARIATION

We want to start the discussion on variation with a classic story. Even though you may be familiar with it already, we want to repeat it because of its power to illustrate the importance of managing variation.

Ford and Mazda were both producing an identical automatic transmission with identical plants and machinery. However, the warranty costs for Ford's transmission were many times higher than the warranty costs for Mazda-made transmissions. To find out the reason for this discrepancy, Ford decided to take two sets of new transmissions, one from each plant, dismantle them, and measure every part under the same conditions for both sets. Figures 13-1 and 13-2 show the results of measuring two groups of gears—one group from Ford's transmission and one from Mazda's.

As you can see, there is a significant difference in the process variation for the gears of the transmissions from the two companies. As the story goes, when the technician took the measurements of the gears made by Ford and then moved to measure the gears produced by Mazda, he thought that something was wrong with the gauge he was using for measurement because Mazda's product demonstrated a very narrow process variation. However, after checking the gauge, it was clear that the low variation of Mazda's gear was a result of a very high-quality manufacturing process. The results of the investigation clearly indicated the reason for the large difference in warranty cost.

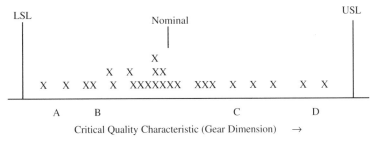

Figure 13-1 Ford's Gear Pattern of Variability

Source: Reprinted with permission from *American Samurai* by William Lareau, New Win Publishers, 1991, p. 194.[1]

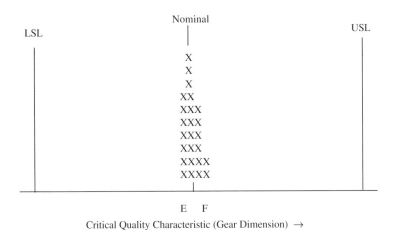

Figure 13-2 Mazda's Gear Pattern of Variability

Source: Reprinted with permission from *American Samurai* by William Lareau, New Win Publishers, 1991, p. 195.[2]

242 MANAGING VARIATION: A REQUISITE FOR QUALITY

As we know, product quality is always inversely proportional to variability and directly proportional to cost. In other words, when we reduce the variability, we increase the quality and reduce the cost. Reducing process variation also has an influence on reducing delivery time because of less rework, inspection, and other time-consuming operations. This is why managing variation should be considered one of the most important components of management and leadership. In this chapter, you will find some concepts that may help you manage variation.

13.1.1 The Peculiarities of Managing Variation in a High-Technology Enterprise

Because of the complexity of their products, high-technology enterprises must have processes that produce products and components at a low defective parts per million (ppm) level. This is why managing variation in high-technology enterprises has a high priority.

Our experience shows that the first requirement for a process to have a low ppm quality level is that its spread (process width) must consume no more than 62.5% of the specification width. In this case, we may expect a quality level of 1,000 ppm (see Fig. 13-3, curve A; this includes the assumption that the process average may deviate from the center ±1.5 sigma.)

The ultimate goal in a low ppm environment is to reduce the process variation to the extent where its spread only consumes 50% of the specification width, to achieve a quality level of 3.4 ppm or better (see Fig. 13-3, curve B).

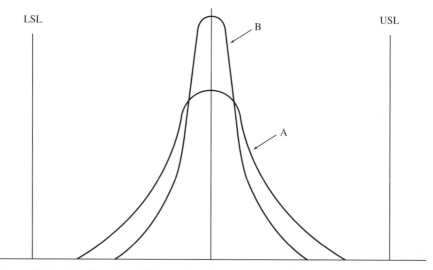

Figure 13-3 The Entry Level (Curve A) and Ultimate Goal (Curve B) of a Low Process

Assuming that in a low ppm environment, all processes are already in a state of statistical control and that the specification limits are optimized, the only way to achieve the goal mentioned above is through a significant reduction of the process variation.

This cannot be accomplished by only performing small incremental process improvements. A climate must be introduced in which breakthrough improvements can be achieved. This requires the application of Design of Experiments (DOE), redesigning the existing processes, investing in automation and new technology, reeducating the engineering and operations personnel, and other major rearrangements. Creating this kind of climate is not just a technological issue; it also requires motivating people to challenge the organization's assumptions and create a culture of innovation. This is why creating a low ppm environment must come from the top management of the organization.

13.1.2 Creating a Sense of Urgency

As Dr. Juran pointed out in his book *Managerial Breakthrough: A New Concept of the Manager's Job*,[3] often we may see situations in which management is too preoccupied with routine variances and they have no time left to recognize the urgent need for a breakthrough. It is human nature to get used to a particular environment and not react to gradual changes or chronic problems. We usually react much faster to sudden events that create a major change to what we are used to. For example, a machine may run for years with a particular amount of loss created from nonconforming parts. We take this as its "normal" capability. But when the machine starts to produce a significantly higher reject rate, the signal gets through and management immediately pays attention to the problem. As soon as the problem has been fixed and the machine is brought back to "normal," the tension that management feels drops, even though the machine continues to produce nonconforming parts. In the long run, these losses are much more significant than the accidental losses that management paid attention to earlier. Somehow, management is more used to reacting to strong and sudden effects than abnormalities that happen gradually and create chronic problems. We need to learn how to create a sense of urgency even when the problem is not so obvious. This is the art of management.

Below is an analogical example to illustrate more clearly the issue of how important it is to learn to create a sense of urgency.

The Yield Drop Problem

After a lot of engineering effort, the yield of the manufacturing process was brought up to 95%. When all ideas of incremental improvement had been exhausted, the management looked at a 5% loss as a normal state for the process.

The Planning Department scheduled an extra 5% production in the manufacturing schedule to cover the yield loss, and this did not create any problems for the organization until the day the yield suddenly dropped to 88%. This was where the problem began (see Fig. 13-4). Top management expressed concern and held meetings, and special teams were formed to resolve the problem. Different ideas were considered, including the suppliers' quality. Two days later after many actions had been taken, the process was back to normal and the average yield was back to 95% (see Fig. 13-5). It was not obvious what had caused the problem, but management was happy with the team's efforts.

Now let's compare the two charts in Figures 13-4 and 13-5. The loss from the two days of low yield is graphically represented in the striped triangle in Figure 13-4. The long-term loss from running the process at 95% yield is represented in Figure 13-5, which indicates that this yield was constant and the loss was significantly higher than the yield loss from the 2-day episode. Despite this higher loss, there were no urgent meetings or special teams formed to bring the average yield to a higher level. To be fair, we should say that the Engineering Department had a project for yield improvement and cost reduction, but it did not get as much attention as the 2-day yield loss. Why? The 2-day yield loss created a sense of urgency because it infringed on the shipment schedule and

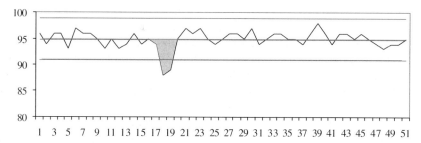

Figure 13-4   Loss Through Lack of Control

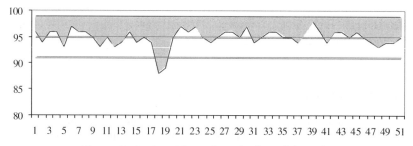

Figure 13-5   Loss Through Lack of Breakthrough

management was not prepared for this loss. The other loss—the constant 5% yield loss—was a chronic problem that the manufacturing team was used to and one they were prepared for. It was not even considered a problem. It is human nature that we react much faster to sudden events. If a problem stays unresolved for a long time, it becomes chronic and we get used to it, so it is not seen as a problem.

In a low ppm environment, we need to create a sense of urgency; a need to bring our process close to perfection. When AMD was working on the manufacturing processes to create a low ppm environment, special teams were formed to work on processes that had earlier been considered normal. Any process that has a Cpk (see Section 18.8) of ≤ 1.5 was considered a problem—an area that required urgent work. Creating a sense of urgency is an important component of the TCPI^2 movement.

13.2 TAGUCHI'S QUALITY PHILOSOPHY

Taguchi's view on quality deserves close attention. The purpose of this section is to familiarize you with his concept. To read more on this subject, please refer to our recommendations at the end of the chapter.

13.2.1 Beyond Conforming to Specifications

Reducing the variability of the process is a more progressive way to continuously improve the process than by just making the parts fit the specification requirements. This is in line with Taguchi's Loss Function philosophy. Below is a brief description of the Taguchi concept, followed by an illustrative example.

For many years, manufacturers have been used to the traditional way of thinking about quality, which is based on conformance to specification. This means that if we measure a part for a particular parameter and it shows that the part is within specification limits, this part is considered to be of good quality. In a similar vein, if a part doesn't conform to the specification requirements, it is considered to be of bad quality (see Fig. 13-6). This is a simple, black-and-white approach. It is how we measure our supplier's incoming material and how we measure our outgoing product. How good is this mental model? As Dr. Deming wrote, "There is obviously something wrong when a measured characteristic barely inside a specification is declared to be conforming; outside it is declared to be nonconforming. The supposition that everything is all right inside the specifications and all wrong outside does not correspond to this world."[4]

As Figure 13-6 suggests, in the traditional way of thinking, part A, which is barely inside of the upper specification limits, is considered good quality but part B, which is slightly outside of the upper specification, is considered bad quality. In reality, the difference between these two parts may be related to

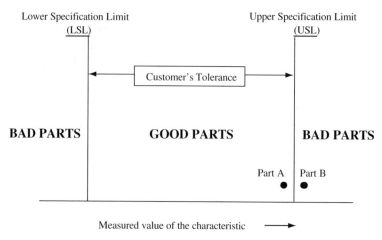

Figure 13-6 The Traditional Way of Assessing Quality

inspection accuracy or tester error. If they are measured a second time, their places may be reversed. Not much trust can be accorded to the quality of part A, and the rejected part B could possibly be as good as part A. This is why 100% of the parts falling within the tolerance does not guarantee that the customer received 100% conforming products.

13.2.2 What a Manager Should Know about Taguchi's Loss Function Concept

The new organizational environment requires a more reliable way to define quality. One alternative is Taguchi's Loss Function. This is closer to the Japanese way of thinking. Supporting this alternative, Dr. Deming wrote, "A better description of the world is the Taguchi Loss Function, in which there is minimum loss at the nominal value, and an ever-increasing loss with departure either way from the nominal value."[5] Taguchi's concept of quality suggests that any time a particular measurement of a product deviates from the target value, we may face customer dissatisfaction and have losses to the society. From an organizational point of view, this means loss of profit.

As Figure 13-7 suggests, the loss is zero when the product quality meets the target. As soon as the quality starts to deviate from the target, the product generates losses to the society at an increasing rate. To elaborate this point, we will use a simple example adapted from Phillip J. Ross' book *Taguchi Techniques for Quality Engineering*, "Batteries supply a voltage to a light bulb in a flashlight. There is some nominal voltage, let us say 3 volts, that will provide the brightest light but will not burn out the bulb permanently. Customers want the voltage to be as close to the nominal voltage as possible, but battery manufacturers may be using a wider tolerance than allowed by the battery specification.

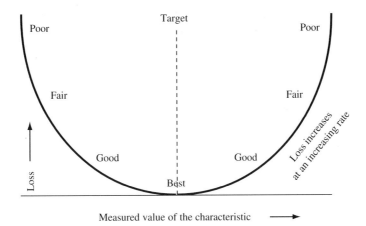

Figure 13-7 The Taguchi Loss Function for Examining a Manufacturing Process
Source: Reprinted with permission from "The Value of Continuing Improvement" by P.T. Jessup, Proceedings of the 1985 International Communications Conference. Copyright © 1985 IEEE.[6]

So, as a result, some flashlights burn dimly and others burn out the bulbs permanently."[7] This example emphasizes that to satisfy the customer, the manager needs to put efforts into continuously producing his product not just to the specifications, but as close as possible to the nominal. To do this, constant activity toward reducing the process variability around the target should be put in place. And this can be accomplished only by introducing a statistical method that will help to bring the process into a state of control and enhance the process capability.

13.2.3 Concluding Thoughts on Taguchi's Quality Philosophy

To conclude this section, we want to use a story written by Andrea Gabor in which he described Larry Sullivan's reaction when he visited Japan with other Ford employees. Gabor wrote, "Visiting a maker of cigar lighters, the Ford team ran across a group of Japanese workers who were trying to improve the process that makes detents, one of the lighter's mechanical parts. In one room, eight workers hovered around an eleven-foot-long control chart spread across the table. 'I figured they must be having problems in the field or with customer complaints,' observed Sullivan. When he looked more closely at the control chart, he realized that the process was not only within specification limits, but that, on average, the process performed within less than one third of the range of the specification limits."[8] Even though this story takes us back to the 1980s, when Ford was trying to learn the secrets of Japanese quality, the moral of the story still deserves attention today. The story shows that Taguchi's methods are not just a methodology but a philosophy, and to implement it, we need a culture, a mind-set that high quality and low cost will not be achieved just by making parts to specifications or tightening specifications.

This story was just about a simple process for making cigar lighters, which have a limited number of parts, and how the Japanese team was unhappy with a process spread that consumed only 33% of the specification width. What about a more complicated product that is made from a large number of parts? Here the importance of reducing the parts' variability becomes even greater because of the tolerance *stack-up risk*. If Taguchi's loss function concept is considered as just one part of the engineering methodology, it will never be implemented. It is up to management to build up a mind-set and an understanding of its importance. Switching from the old way of thinking (that making parts to the specification requirements is good enough) to a more progressive mind-set (that high quality can only be achieved through reducing process and product variation) is a psychological issue and takes a lot of management's time and effort to make happen. The application of Taguchi's Loss Function is an excellent way to convince management to release financial resources for projects related to process improvement. This is because it allows us to talk in dollars and cents—to talk about actual cost savings.

William Lareau recalls, "While working for Ford [by using Taguchi's Loss Function], I was able to successfully support a proposal for a million dollar per year investment in quality improvements because I was able to show that 'we' could reduce loss to society by over 30 million dollars over a 5-year period. Bingo—project approved."[9] For a start, just changing the organization's mind-set, thinking and acting in the direction of setting the process average on the target, and continuously reducing the variability around the target will bring significant economic and qualitative results. It is just a matter of changing a mental model, and everything will proceed from there. In this context, Douglas Montgomery's definitions of quality and quality improvement fit perfectly: "*quality* is inversely proportional to variability;"[10] and "*quality improvement* is the reduction of variability in processes and products."[11] Following these definitions, organizations will be able to continuously reduce all kinds of waste related to scrap, rework, field services, etc. and, as an end result, improve customer satisfaction by reducing the cost and increasing the quality of products and services.

13.3 CREATING A LOW PPM ENVIRONMENT

Contemporary high-technology enterprises usually have processes and products whose quality is measured in defective ppm. In this section, we will share AMD Manufacturing Services Group's experiences in creating and managing quality in a low ppm environment.

Creating an environment in which the process generates high-quality products with defective rates measured in defective ppm is not easy. It takes a lot of time, knowledge, effort, and money to achieve this goal. It usually requires a change in technology, metrology, and the organizational structure. In addition, it also requires a psychological change—a change in people's perception.

In his book, *TQC Wisdom of Japan: Managing For Total Quality Control*, Hajime Karatsu, the winner of the Deming Prize, wrote, "Some people believed that while defects might be reduced significantly, they could not be eliminated completely as long as the processes were controlled only statistically. But our actual performance proved this perception wrong. To measure defects, we relied first on percentages, in other words, so many defects per 100. Today, however, many plants measure in terms of ppm, that is, so many defects per million units."[12] This statement clearly illustrates the influence of perception on actual results. We cannot make a significant change if we continue to think in familiar ways. We cannot improve something if we continue to measure results using out-of-date methods. Not only is achieving a low ppm environment related to technology, but it is also related to a mind-set. We must believe it is possible to develop processes that produce only conforming products. Only then will we improve from percent defective to low ppm quality levels.

At AMD we have defined a low ppm environment as one that has all its processes running at a 1,000 ppm or lower quality level. When we started the journey toward achieving this quality level, there was a real need for employee commitment. In the beginning, it was very difficult to have all employees buy into this mode when a large number of critical processes were running with a Cpk ≤ 1, which is roughly equivalent to or worse than a process quality level of 67,000 ppm (with an adjustment of a 1.5 sigma shift). (We will describe the meaning of Cpk and a 1.5 sigma adjustment in section 13.8.) We needed some activities to help the employees overcome this psychological barrier and thereby make it easier for people to perceive the reality of the journey to a low ppm environment. The first step in this direction was necessarily the launch of a comprehensive statistical principles educational program for all employees. The second step was the development of a long-term plan with a set of measurable goals. To make the plan easier to understand, we divided the whole program into three major parts that reflected different zones of activities (see Table 13-1).

Table 13-1 The Process Quality Zones[3]

	Preparation for the Low ppm Environment		The Low ppm Environment
	Zone 1	Zone 2	Zone 3
Quality Level	Cpk < 1.0 (Higher than 67,000 ppm)	$1 \leq$ Cpk ≤ 1.5 (67,000 to 1,000 ppm)	1.5 60 Cpk (Lower than 1,000 ppm)
Goal	Cpk = 1.33 (6,200 ppm)	Cpk = 1.5 (1,000 ppm)	Cpk = 2 (3.4 ppm)

Note: This table includes a 1.5σ shift for all listed values of ppm.

Zone 1

Zone 1 is related to the processes that run with a Cpk level of less than 1 (Cpk $\leq$ 1), which is equivalent to a quality worse than 67,000 ppm. At the time we started the program, most of the processes had this quality level. To be able to deliver high-quality products to our customers, many extra activities were required, such as testing, inspections, and screenings. The losses that were related to these activities in Zone 1 were high. To improve the situation, a goal was set for Zone 1 to achieve a Cpk = 1.33, which is equivalent to a process average of 6,200 ppm process quality. Project teams were formed to conduct process capability studies, eliminate unnatural process variations, and bring the process into a state of statistical control. This allowed us to achieve a Cpk = 1.33 for the majority of the individual processes. Almost no capital investment was needed.

In hindsight, moving from Zone 1 to Zone 2 was an easy task compared with the goals achieved in other zones. It was a time of learning and encouragement, a time when people were starting to believe in the *magic* of statistical principles. Shewhart's control charts were established to control the most critical parameters, and a reporting system was developed to report the results.

Zone 2

Encouraged with the relatively fast results in achieving the goal for Zone 1, a new goal was set for the critical process parameters, a Cpk = 1.5, which is approximately equivalent to 1,000 ppm. This task required more innovation and capital investment. It also required time to perform different DOE and reduce some process variations. The goal was achieved, which created more employee confidence and enthusiasm. Reducing the process spread around the target became a source of quality improvement, cost reduction, and knowledge, and statistical thinking gradually became a part of the organizational culture.

Zone 3

Accomplishments in Zones 1 and 2 provided us with an entry ticket into the low ppm environment, where the task was to bring all processes to a level below 1,000 ppm and achieve a goal of Cpk = 2 or better (3.4 ppm or lower). This was the most difficult part of the program. The closer we came to the ultimate quality goal of practically zero nonconformities, the more difficult the task became. In some areas, the traditional control charts became inconvenient to use and new statistical tools were required.

This chapter is dedicated to the description of some new techniques and management approaches required for Zone 3—the low ppm Zone.

Forming three zones of process improvement, setting reasonable goals, and enjoying the step-by-step results made the 67-fold improvement of process quality levels a reality. It allowed us not only to have processes capable of satisfying customers but also to receive significant savings by reducing process variations. Roughly speaking, every time we moved from one zone to another, we achieved an approximate 10% reduction in cost by reducing the amount of inspection, retesting, scrap, and other losses related to process imperfection. Today, the major task is to stay in the low ppm zone and continue our efforts to reduce ppm levels.

13.4 CHALLENGING AND CHANGING OUR ASSUMPTIONS ABOUT QUALITY

Usually when we talk about quality improvement, we challenge our process and seek ways to reduce the nonconforming quality levels by improving the process capability. But we very rarely challenge the way we think about quality improvement. We almost never challenge our assumptions. Why? Probably because we can't see them or touch them—they are transparent. For example, it's easy to see the extra variation in a process, the losses from low yield, rework, scrap, etc., but how much loss is accountable to wrong assumptions? Do we operate with assumptions that create losses? Do we know all of them? Here we will use only one example to illustrate the point that challenging assumptions is a part of continuous improvement.

What is an assumption? In its simplest form, an assumption is something that we think we know, and because we know it, we rarely challenge it. Sometimes our assumptions isolate us from reality and prevent us from making progress. To illustrate this, we will use a true story that shows the importance of constantly checking and challenging our assumptions.

Figure 13-8 is an old design of a lead frame. It is a part of an electronic device that had been used for years. The assumption was that a plastic ring was required during assembly to keep the leads straight by holding them together. Later, when the device was assembled, the ring was cut off and scraped. Although this caused a loss of extra material and additional processing, the assumption was that it was worth it to keep the leads aligned. This assumption was held for years, even though there was no real evidence for its support.

Later, this topic was brought up in a meeting related to continuous improvement. An engineering manager explained, "We could probably move to lead frames designed without a ring, but if we do this we will need to rework our process and teach our employees to handle the parts more carefully. We are not ready now." This again was just an assumption. How can the employees learn to handle the parts differently while the ring is still holding the leads? It was not necessary for them to be more careful because the ring was at work. The ring was not just holding the leads of the device, it was also holding back process improvement and trapping the engineering mind in a "ring" that kept the wrong assumption alive. Later, under the pressure of competition, it was finally decided to move to a new technology, which was making lead frames without a ring. After the implementation of the new technology, we initially observed an

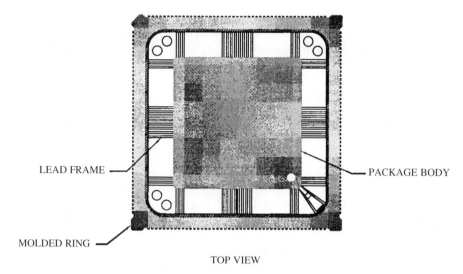

Figure 13-8 Old Lead Frame Design

increase in the reject rate because of the learning period needed to get used to the new assembly process, but people learned fast because they were free of the ring that held up progress. The reject rate moved very rapidly into a low ppm level, and millions of dollars were saved by reducing the need for extra materials and eliminating some assembly process steps.

This is just one example to demonstrate the fact that bringing our assumptions to the surface and challenging them is a way to reduce costs and improve quality—a way to reduce the barriers to progress. Challenging our assumptions requires knowledge—the more we know, the more we can challenge our operating assumptions. However, it is not easy to see our old assumptions because they are part of our identity, built into our thinking and perception. If we don't bring them to the surface and challenge them, how will we know what assumptions don't work anymore? To achieve and maintain a low ppm environment, we need to utilize all sources of improvement, one of which is challenging our assumptions.

13.5 DEFINING THE BOUNDARIES BETWEEN LOW PPM AND PERFECTIONISM

In a low ppm environment there is a tendency to prevent the occurrence of any nonconforming units. Many organizations have developed a set of specifications that regulate all quality issues, including the size of scratches on the product surface and other cosmetic errors that have nothing to do with the product function or performance. So, how much further can we go without entering the zone of perfectionism?

Human beings have an instinctive drive for perfection. How do we prevent ourselves from overdoing the effort to achieve a perfect product? We are not questioning the functional issue of whether to allow or not allow some nonconforming units of product to ship to the customer. Rather, we are talking about cosmetic or other related attributes that don't have an impact on the product's performance but on which the customer or supplier is imposing tight requirements that can create loss to both the supplier and the customer.

In a low ppm environment, when a high sensitivity to perfection is built up, we also need to remember that perfection can be antagonistic because of the need for extra material, processing, testing, and inspecting. This creates a loss of time, material, and money that does not add any value to the product. The following is a simple example to illustrate the point we want to make. AMD produces a microprocessor module. There is a very tight specification that regulates the cosmetics of this device, so if there is a visible scratch on its surface, the unit could be rejected. In this example, we're not talking about poor surface roughness or other surface conditions that could affect heat transfer properties, but only a noncritical, minor scratch. It could be explained by saying that customers are paying good money for this module and don't want a scratch on it. This is understandable. However, if this part is not going to a retail store, but to a customer who will install it in a computer, the ultimate customer will never see it. Should the part be rejected for a nonfunctional scratch? Another more serious example is design tolerance. Influenced by the drive for perfection, we also sometimes design tight tolerances for features that have nothing to do with improving performance. This is also a waste of money that ultimately impacts the consumer's pocket or the producer's margin.

We all have, more or less, a drive for perfection. Sometimes we act like perfectionists without even realizing that the managerial decisions that we are making will not add value to the user. Top management will more easily support your overreacting proposal than allow you to loosen a too-tight specification. Why? It is safer. The hidden cost losses are not as visible as your intention to *protect* the customer. It would be easier to get a proposal for an overdesigned bridge accepted than an initiative to reduce the amount of material in the bridge without jeopardizing its optimal safety. We don't always know what "optimal" is because it is human nature to be on the safe side of the optimum.

In the process of introducing a low ppm environment through quality-conscious design of products and processes, we also need to be concerned about overdesign. This can be achieved by forming special cross-functional teams that include representatives from design, manufacturing, maintenance, marketing, and customer service departments. Such teams will review the product and process in the design stage to determine whether there are elements of overdesign, oversafety, or other design issues that will negatively affect product cost or do not add value to the usability of the product. In other words, the major goal of the design review team is to eliminate the nonessential characteristics at the design stage before the product or process is built.

13.6 THE IMPORTANCE OF DOUBLE-LOOP LEARNING IN QUALITY IMPROVEMENT

Single-loop and double-loop learning is a separate topic by itself, which has been described in some detail in Part II, *Managing in a Knowledge-Based Organization*. In this section, we will just touch upon the surface of this subject and give an example of an application related to the subject of working in a low ppm environment.

One of the important conditions to achieving and maintaining low ppm quality levels is practicing double-loop learning (see Fig. 13-9). It took years for our employees to fully appreciate the concept of double-loop learning, which together with other means allowed us to build a low ppm environment. In his book *Overcoming Organizational Defenses: Facilitating Organizational Learning*, Chris Argyris describes the difference between single-loop and double-loop learning. He writes, "Single-loop learning solves the presenting problems. It does not solve the more basic problem of *why* these problems existed in the first place."[13]

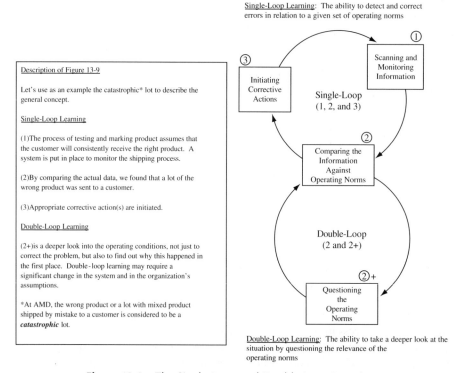

Figure 13-9 The Single-Loop and Double-Loop Learning

For example, at AMD we had a problem with what we called *catastrophic lots*. These were rare cases in which the customer received a lot of the wrong product or the product somehow was mismarked. This problem was mainly related to human error, but in a low ppm environment we could not tolerate this type of failure. Giving this issue a high priority, the management developed measures that have helped reduce the ppm level related to it. The process has been improved, which allows the supervisor to track the error and find out who made the mistake. Special meetings were held to bring this issue to everyone's attention. A separate reporting system was developed to monitor the improvement and record the actions taken. This helped to reduce the occurrence of catastrophic lots, but the problem still wasn't completely resolved. This is all related to single-loop learning.

Even though all the measures just mentioned resulted in an improvement, they did not solve the basic problem of *why* these problems existed in the first place. One of the authors of this book had the opportunity to participate in a dialogue related to this problem in one of our offshore plants. The purpose of the dialogue was to respond to the question of whether we could have a system in which catastrophic lots do not happen. Employees from different departments, including operations, participated in this dialogue session. The dialogue helped to challenge most people's assumption that it was impossible to have zero catastrophic lots over a long period of time. After sharing thoughts during this dialogue, the participants still did not have concrete answers to the *how* and *why*, but it got the people thinking about the problem and there was a tacit understanding that something should be changed. Later the plant $TCPI^2$ Board received a lot of recommendations that suggested a much deeper approach to fixing the problem—an approach that changed the system. One of the suggestions inspired a major decision to totally reorganize the way we form and process lots. Special prevention software was developed. This was a radical change in the procedure, and the results can be seen in Figure 13-10. Only one year later, the goal of zero catastrophic lots was achieved, and the assumption that it was impossible to have zero bad shipments collapsed. This is a less complicated example of double-loop learning.

In all our $TCPI^2$ activities, in addition to continuous incremental improvement, we moved to a more innovative approach—double-loop learning, which has become a part of our organizational culture. We have started to learn how to find the answer to *why* a particular problem exists. In other words, we have learned how to learn. It is interesting to note here that after the introduction of double-loop learning, not only were better and more stable results achieved, but also some assumptions, sayings, and even vocabulary were changed. For example, in situations related to catastrophic lots, sayings such as "Human beings make mistakes," "Nothing is perfect," and "We learn from our mistakes" were given a different interpretation. Employees started to differentiate between a mistake related to risky decision-making and a mistake made because we did not pay enough attention to our responsibility. In the first case, the risk was taken to achieve better results. In this case, sometimes a failure can provide

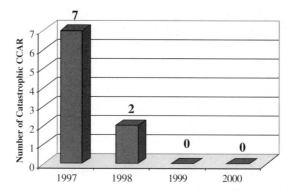

Figure 13-10 The Results of Eliminating Catastrophic Lots

a lot of "meat" for learning, which is beneficial in the end. However, in the second case, the mistake does not arise from good intentions, but is just a pitfall. We don't need to crash into a wall to learn how to drive carefully. This helps sometimes, but most of the time it results in a condition in which you don't need to learn anymore. All these expressions have no meaning in the case of catastrophic lots. The management knew that there was no excuse for sending catastrophic lots to customers. For example, how do we prevent sending a lot of the wrong product to a customer? Second-loop learning should occur here.

13.7 DO WE NEED STATISTICAL PROCESS CONTROL FOR AUTOMATED EQUIPMENT?

Automation was the technological enhancement that accelerated the movement to a low ppm environment. The electronics industry now has smart machines, automated equipment, and robots that will stop producing parts if something is wrong with the process. In other words, they either produce conforming parts or they produce nothing, but they don't produce nonconforming units. For example, a wire-bonding machine will stop any time a part is not bonded, the device is not properly set, or for any other condition that might create nonconformities. This is an excellent attribute for a low ppm environment.

Many machines are equipped with computers that are capable of doing continuous measurements of the units they produce during the manufacturing process and calculating statistical parameters such as averages, standard deviation, etc. Some of the automated equipment is also capable of comparing the process results with preset numerical standards. So, should we use statistical techniques for automated processes? Is there any value in using sampling techniques when we already have measurements from every unit produced? The simple answer to both questions is yes, and we'll try to explain why. Some people have a misunderstanding about the relationship between a process and the output of a process. This is probably due to a lack of appreciation for the intrinsic variability of the manufacturing and measurement processes themselves. We need to recognize that even in situations in which we have a complete data record of the measurement of every part produced, this is still only a sample of the output of the process. It is very important to recognize that the process is *future oriented* in time, whereas the record of measurement is *past oriented*. This means that unless statistical control is attained, the data from the past production cannot be used to predict the variability of the process in the future. So, without the use of statistical techniques to monitor the process performance, there is no evidence of process stability. Our experience shows that maintaining computerized control charts on automated processes is not only appropriate, but a necessity.

13.8 MOTOROLA'S SIX SIGMA METHODOLOGY

As we already mentioned, Motorola's Six Sigma philosophy has multiple meanings. It can be considered as a corporate initiative strategy, vision, goal, benchmark, etc. However, in this section we will only describe some elements related to the methodological aspect of assessing six-sigma quality.

The Motorola Six Sigma approach to measuring a process quality level is based on two major assumptions: that the measurements taken from the process follow a normal distribution, and that the process mean (μ) may shift from the nominal or target value by ± 1.5 sigma (σ). To explain Motorola's Six Sigma concept, and in particular the meaning of the 1.5 sigma shift, let's first assume a situation in which the process is centered and always stays at the nominal, or target value, and the upper and lower specification limits are respectively at ± 3 sigma of the mean (see Fig. 13-11). We know that the process average usually varies, but let's assume that the process average is stable and stays on the target. In this case, we would have a process quality level of 2,700 ppm. As Figure 13-11 suggests, if the specification limits were set at ± 6 sigma (not ± 3 sigma) apart from the nominal, the process quality level would be 0.002.

Table 13-2 shows different ppm process quality levels that depend on how far the specification limits are from the center. For example, if we had a situation in which the upper and lower specification limits were at ± 4 sigma, the process quality level would be considered 63 ppm. All this is true only under the assumption that there is no variation of the process average. But, as we know, in real life there is always a variation in the process average. It is only a matter of how much actual variation occurs. This variation comes from a number of sources, such as tool wear, setup deviations, different materials, process imperfections, etc. As we can see in Table 13-2, if the assumption is that the process average always stays at its setting, without shifting, a process of five-sigma level would be an excellent process (0.57 ppm).

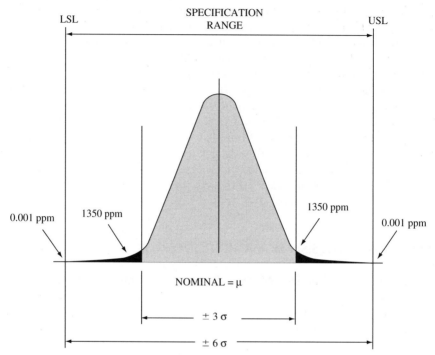

Figure 13-11 Six Sigma Process with Mean Centered at Nominal

Source: Reprinted with permission by Motorola University from Mikel J. Harry, *The Nature Of Six Sigma Quality*, Motorola University Press, 1997, p. 11. Copyright © 1988 Motorola, Inc.[14]

Table 13-2 Key Six Sigma Parameters

Sigma Level	Centered Process C_p	PPM	Shifted (±1.5) Process C_pK	ppm
3	1	2,700	0.5	66,803
4	1.33	63	0.833	6,200
5	1.67	0.57	1.167	233
6	2	0.002	1.5	3.4

Source: Reprinted with permission from Fred R. McFadden, "Six-Sigma Quality Programs," *Quality Progress*, June 1993, p. 39. Copyright © 1994 American Society for Quality.[15]

Now let's take a look at what would happen with the process quality if we consider a process shift of 1.5 sigma, which Motorola suggests as the average compensation factor for different kinds of shifts. The result from a 1.5 sigma shift in a six-sigma process is shown in Figure 13-12. As you can see, if

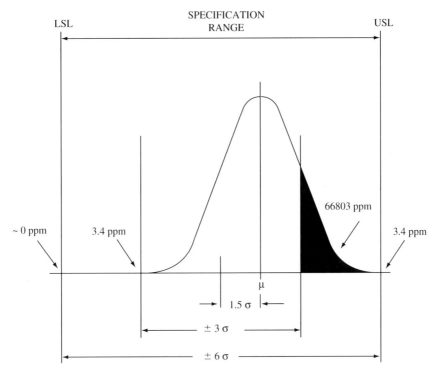

Figure 13-12 Six Sigma Process with Mean Shifted from Nominal by 1.5 Sigma

Source: Reprinted with permission by Motorola University from Mikel J. Harry, *The Nature Of Six Sigma Quality*, Motorola University Press, 1997, p. 12. Copyright © 1988 Motorola, Inc.[16]

considering a 1.5 sigma shift, the distance between the upper specification limit (USL) and the process mean (μ) is 6 sigma minus 1.5 sigma, or 4.5 sigma. The area outside the USL represents a 3.4 ppm nonconformity rate. Motorola's concept suggests that if we have a situation in which the process follows a normal distribution and the specifications are set six sigma apart from the nominal or target, the process will be capable of having only 3.4 ppm, even if the process mean deviates ±1.5 sigma from its nominal setting.

The selection of 1.5 sigma as a compensation for process shift is an arbitrary one, and it is based on a lot of observations. In some organizations, a shift compensation factor of 1 sigma or less may be sufficient (see Table 13-3). However, the main value of the Motorola methodology is that it suggests a way to achieve and measure close to perfect processes.

Conclusion

Motorola's approach requires that the process variation be very small, so that the specification limits will be ±6 sigma away from the process average. In

Table 13-3 The Number of Defectives (Parts Per Million) for Specified Off-Centering of the Process and Quality Levels

Shift Compensation Factor	Distance between Process Average and Specification Limits				
	3 sigma	4 sigma	5 sigma	5.5 sigma	6 sigma
0	2,700	63	0.57	0.034	0.002
0.5 sigma	6,440	236	3.4	0.71	0.019
1 sigma	22,832	1,350	32	3.4	0.39
1.5 sigma	66,803	6,200	233	32	3.4
2 sigma	158,700	22,800	1,300	233	32

Source: Adapted with permission from Pandu R. Tadikamalla, "The Confusion Over Six-Sigma Quality," *Quality Progress*, American Society for Quality, November 1994, p. 85[17]

other words, the specification tolerance will be twice the natural spread of the process. This will allow us to have a process quality level of 3.4 ppm, even if the process average shifts 1.5 sigma from the nominal. This approach to process capability fits within the requirements of a low ppm environment.

References

1. William Lareau, *American Samurai: A Warrior for the Coming Dark Ages of American Business*, New Win Publishers, Clinton, NJ, 1991, pp. 194–195.
2. Ibid.
3. J.M. Juran, *Managerial Breakthrough: A New Concept of the Manager's Job*, McGraw-Hill Book Co., 1964, New York, NY, p. 24.
4. W. Edwards Deming, *Out of the Crisis: Quality, Productivity and Competitive Position*, Cambridge University Press, Cambridge, MA, 1982, 1986, p. 141.
5. Ibid.
6. Peter T. Jessup, "The Value of Continuing Improvement", *Proceedings of the 1985 International Communications Conference*, Institution of Electrical and Electronics Engineers, Inc., 1985, p. 90
7. Phillip J. Ross, *Taguchi Techniques for Quality Engineering: Loss Function, Orthogonal Experiments, Parameter and Tolerance Design*, McGraw-Hill Book Co., New York, NY, 1988, p. 3.
8. Andrea Gabor, *The Man Who Discovered Quality: How W. Edwards Deming Brought the Quality Revolution to American—The Stories of Ford, Xerox, and GM*, Penguin Books, New York, NY, 1990, p. 50.
9. Lareau, *American Samurai: A Warrior for the Coming Dark Ages of American Business*, p. 201.
10. Douglas C. Montgomery, *Introduction to Statistical Quality Control*, John Wiley & Sons, 3rd ed., New York, NY, 1997, p. 4.
11. Ibid., p. 5

12 Hajime Karatsu, *TQC Wisdom Of Japan: Managing For Total Quality Control*, JUSE Press, Ltd., Tokyo, 1981, English translation by Productivity Press, Inc., Portland, OR, 1988, p. 8
13 Chris Argyris, *Overcoming Organizational Defenses: Facilitating Organizational Learning*, Allyn and Bacon, Needham Heights, MA, 1990, p. 92
14 Mikel J. Harry, *The Nature of Six Sigma Quality*, Motorola University Press, Schaumburg, IL, 1997, p. 11
15 Fred R. McFadden, "Six-Sigma Quality Programs," *Quality Progress*, American Society for Quality, Milwaukee, WI, June 1993, p. 38
16 Mikel J. Harry, *The Nature of Six Sigma Quality*, p. 12
17 Pandu R. Tadikamalla (1994), "The Confusion Over Six-Sigma Quality," *Quality Progress*, American Society for Quality, Milwaukee, WI, November 1994, p. 85

Chapter 14

Some Major Quality Initiatives

Let's assume that you are assigned to lead a new organization and you want to introduce an initiative that will help you achieve the most competitive quality level. What approach would you select: Kaizen, Statistical Quality Control (SPC), Total Quality Control (TQC), Total Quality Management (TQM), Reengineering, Six Sigma Quality, or another concept? You would probably use the most modern approach. If the above question was asked when Dr. Deming was preaching statistical methodology, you would have probably picked SPC. If this was asked when Reengineering was used in many organizations, you would probably have picked that approach. The question is, why do we select organizational initiatives and improvement approaches as though we were choosing a new model of car?

Do you remember when every organization was looking for a statistician to help introduce statistical principles into their process? How many statisticians are there in your organization now? If you have been in business long enough, you have probably witnessed several quality improvement crusades. This is normal. What is not normal is that when we have introduced a new initiative, all the earlier initiatives have quietly died and been buried without an autopsy. This is a great loss in the form of knowledge, experience, and skills.

Who says that TQM is less important when Reengineering is introduced? Who says that the concept of SPC is not needed; regardless of whatever quality movement is currently in your organization? Why, when focusing on a new initiative, do we forget about the "good stuff" of the old initiative? Look at a symphony orchestra. When introducing new instruments, they don't throw out all of the old ones. They preserve those that have value to the whole musical composition. Getting quickly disappointed with one quality initiative and continuously seeking for new ones without preserving the good parts of the previous concept is one of the major reasons why management in many organizations doesn't get the results they expect from new initiatives.

The purpose of this chapter is to provide the reader with short descriptions of some quality initiatives that we believe deserve attention, regardless of how old they are. This may trigger your interest in deeper study of these programs and rating the importance of their application when creating your own quality improvement system. Our experiences in introducing the TCPI^2 macro system (see Chapter 3) show that management for every organization needs to develop their own quality initiative to progress; whether that initiative is based on TQM, Reengineering, Six Sigma Quality, or any of the new or existing concepts. What is important here is that the initiatives fit the needs of your organization and your plans for the future. Below are descriptions and comparisons of some of the more popular quality concepts.

14.1 TWO TYPES OF IMPROVEMENT

Before describing and comparing different quality initiatives, it is very important to establish an understanding of the meaning of the term "improvement." There are two types of improvement. One type is related to continuous control, and the other is related to breakthrough-type improvement.

Continuous process control is directed toward maximizing the utilization of the existing process capabilities and producing prevention measures and gradual incremental improvements to maintain existing standards. It does not consist merely of maintaining the status quo. Here we are not seeking out problems, we just take care of them when they occur. This is a continuous type of improvement that usually doesn't require capital investments and involves employees who work on the process.

The second type is a breakthrough-type of improvement that requires innovation for its introduction. The management of the organization sets priorities and plans capital investments in R&D, technologies, and new equipment. As you can see from Figure 14-1, improvement has a broad meaning that includes both incremental and breakthrough activities toward process enhancement.

According to Juran, "Breakthrough means change, a dynamic, decisive movement to new, higher levels of performance."[1] Breakthroughs also can be seen as two types: (1) breakthroughs that take the existing process up to a significantly higher level, and (2) breakthroughs that totally replace the existing process (see Case Story 14-1). Both types require innovation and creativity, and both types belong to the concept of improvement.

Having elaborated the meaning of the term "improvement," we are ready to compare different quality initiatives. This will help the management of organizations better understand the concept of these initiatives and establish applicability in their organizations.

264　SOME MAJOR QUALITY INITIATIVES

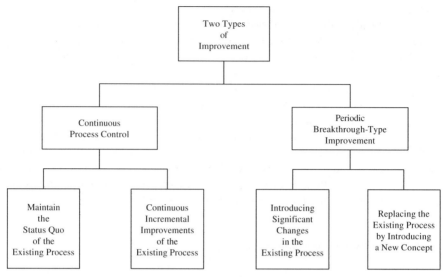

Figure 14-1　Control and Improvement Work Together

Case Story 14-1　Illustrating the Difference Between Incremental Improvement and Breakthrough-Type Improvement

For many years, AMD was struggling to reduce (or totally eliminate) shipping the wrong product to the customers (called wrong shipments). From year to year, the Distribution Manager was reporting a gradual reduction of wrong shipments. This improvement was achieved by better control, but going beyond a certain level of improvement was impossible. The wrong lots were occasionally sent to customers, which is an embarrassing fact. "We are human, and humans make mistakes" could be heard from the employees working in distribution.

In 1998 a new system was introduced that was based on computerization and bar coding, and since then there have been no wrong shipments. As one of the workers from the distribution area said, "Now you need to make an effort to ship the wrong lot." This is considered a breakthrough improvement—an innovation.

14.2　THE KAIZEN APPROACH TO QUALITY IMPROVEMENT

In its literal translation from Japanese, Kaizen [pronounced *kai* (as in "eye")-*zen*] means continuous improvement. In his book *Kaizen* (1986), Masaaki Imai wrote, "... When applied to the workplace, Kaizen means continuing improvement involving everyone—managers and workers alike."[2] As you can see from this definition, as a quality initiative Kaizen is related to the movement of employee involvement through teamwork, quality circles, and other organizational forms that empower people in the quality movement. Kaizen by itself is a strategy that clearly delineates responsibilities for maintaining standards to the worker, with the management's role being the improvement of existing standards. It is very important to recognize that in Japan Kaizen is not just a strategy, but also a process-oriented philosophy of working life. In Japan Kaizen and innovation work together, where the first—Kaizen—is directed toward maintaining and improving the existing working standards through small, gradual, incremental improvements, and require almost no capital investments. The second—innovation—calls forth radical breakthroughs, and improvements that require large capital investments for enhancing the existing or developing new processes, technologies, and equipment. However, as Figure 14-1 suggests, Kaizen and innovation are two forms of activities that create improvements. As you will see later in this chapter, in the long run the Japanese achieve greater results with less capital because they use Kaizen together with innovation.

Kaizen and Innovation Work Together

To better understand the importance of utilizing these two concepts together for process improvement and transformation, it would be meaningful to compare some general characteristics of these approaches. As you can see from Insert 14-1, there is a critical distinction between Kaizen and Innovation. The right column reflects how some American companies work, and the left column is a more Japanese philosophy of management.

In recent years, we have observed a tendency in many American companies toward recognizing the importance of using continuous incremental improvements in conjunction with innovations. If American companies continue to maintain superiority in Innovation and at the same time use Kaizen as a complementary component for organization enhancement, they will be unbeatable.

14.3 TOTAL QUALITY CONTROL (TQC)

Dr. Armand V. Feigenbaum is widely known for his significant contribution to the worldwide quality movement by developing an approach that extends the concept of quality well beyond the manufacturing domain. He argued that quality in manufacturing could not be achieved if the products were improperly designed, poorly distributed, insufficiently marketed, and improperly supported in the customer's site. He also argued that every function within the organization is responsible for quality. Dr. Feigenbaum's philosophy is now known as Total Quality Control (TQC), which is defined as "an effective system for integrating the quality-development, quality-maintenance, and quality-improvement efforts of the various groups in an organization so as to enable

> **Insert 14-1**
>
> **A Comparison of Kaizen and Innovation**
>
> | KAIZEN | INNOVATION |
> |---|---|
> | *(Continuous Incremental Improvement)* | *(Breakthrough-Type Improvement)* |
> | Conventional technology | New technology |
> | Small improvements | Huge leaps |
> | Everybody involved | Mainly a few group's efforts |
> | Maximize use of conventional | Look for new technology knowledge |
> | Improve the existing processes | Start over with new stuff |
> | Focus on adaptability | Focus on creativity |
> | Attention to small details | Attention to big picture |
> | Widely shared information | Information is restricted |
> | System is adapted to the people | People adapt to the system |
> | Little cost to practice | Expensive to practice |
> | Small, gradual changes | Huge, infrequent leaps |
> | Everyone is a key player | Selected key players |
> | Can always be profitable | Requires growth industry to make money |
> | Management patient in pursuing changes and improvements | Management wants quick results |
> | No risk involved | High risk involved |
>
> *Source*: Based on *American Samurai* by William Lareau, New Win Publishers, 1991, pp. 156–157[3]

marketing, engineering, production, and service at the most economic levels which allow for full customer satisfaction."[4]

As this definition suggests, TQC is not just a function of the Quality Department but requires the efforts of many other departments to achieve the desired results in customer satisfaction. According to Dr. Feigenbaum, the main principle of TQC, and its basic difference from many other quality initiatives, is that to provide genuine effectiveness, control must start with identification of the customer's quality requirements. Only when the product has been placed in the hands of a customer who is completely satisfied by it does TQC achieve its purpose.

To achieve this goal, TQC works as a system that guides the coordinated actions of people, machines, and information. As Dr. Feigenbaum suggests, any product is affected at many stages of the industrial cycle (see Fig. 14-2): "Marketing evaluates the level of quality which customers want and for which they are willing to pay; Engineering reduces this marketing evaluation to exact specifications; Purchasing chooses, contracts with, and retains vendors for

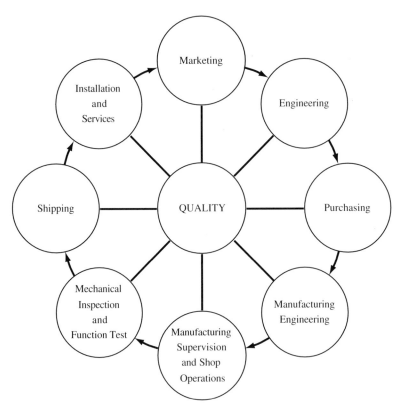

Figure 14-2 The Scope of TQC

Source: Based on *Total Quality Control* by A.V. Feigenbaum, © McGraw-Hill, 1991, p. 11. Used with permission of the McGraw-Hill Companies[6]

parts and materials; Manufacturing Engineering selects the jigs, tools, and processes for production; Marketing Supervision and shop operators exert a major quality influence during parts marking, subassembly, and final assembly; Mechanical inspection and functional test check conformance to specifications; Shipping influences the caliber of the packaging and transportation; Installation and Product Service help ensure proper operation by installing the product according to proper instructions and maintaining it through service."[5]

As TQC has come to have a major impact on management and engineering practices, it has provided a foundation for the development of Total Quality Management (TQM). It should be noted, however, that even though TQM as a concept is considered to be a higher-level, more developed system, TQC still remains an important concept that is applicable in many areas of the industry.

14.4 UNDERSTANDING THE CONCEPT OF TQM

Total Quality Management (TQM) was and still remains one of the best quality initiatives for continuous improvement and innovation. As we read in a special section in *Fortune* magazine, under the title "Quality 2000: The Next Decade of Progress,"

> "The future is not what it used to be. In the decade since 1985, thousands of organizations have adopted and applied the principles of total quality management (TQM). As a result, the worldwide standards for the quality of products and services have never been higher. TQM has proven to be a highly adaptive management tool that is here to stay. It has passed the fad test. TQM has delivered on its promise, but has not reached its potential. The challenge for the next decade is renewal. The global business community is in the midst of a revolution that is shaping how people and organizations will operate and work–well into the next century."[7]

TQM is a system capable of delivering significant results in improvement. However, this depends on how management executes the implementation and maintenance of this concept. TQM is not just a technological concept. It cannot be implemented as we install a machine. It requires a change in the organization's culture, new forms of measuring results, and new forms of motivation. It should be supported from the top, but it cannot be imposed upon employees. It needs employees to recognize its importance and requires employees' commitment to make it work. If you compared TQM of the 1980s with TQM today, you would observe a significant improvement of this concept. Management's role is not only to apply TQM as it is, but also to find ways for its modification and adjusting it to the needs of their organizations.

Despite the disappointment of some organizations when TQM did not meet their expectations, this quality initiative is receiving recognition from many organizations worldwide. In 1997, JUSE (the union of Japanese scientists and engineers) announced a formal change from the term TQC (Total Quality Control) to TQM (Total Quality Management).[8] As we mentioned above, almost all concepts (such as Kaizen, TQC, TQM, and others) are a collection of good thoughts, methods, and tools that should be in the quality "toolbox" and used properly when needed. As time passes, different philosophies arise, such as six-sigma quality, Reengineering, etc. Some of them are popular longer, and some of them die quickly. TQM today is probably the most frequently used term in the United States, whereas TQC was until recently most often used in Japan. But things change over time, and organizations and countries change their way of thinking.

The purpose of this part of the section is to give a short description of the TQM philosophy, which we think will continue to be one of the most contemporary approaches of organizational transformation.

What Is TQM?

Total Quality Management (TQM) is a systemic way of achieving and maintaining total quality. To be successful, the TQM system must be customer-focused, people-centered, and sensitive to the external environment. Figure 14-3 is a pictorial representation of the meaning of TQM, and represents the thoughts of the TQM Committee of JUSE. The pictogram reflects a complex system that consists of ten interrelated blocks.

The wavy line on the bottom of the figure symbolizes that TQM is operating in a continuously changing business environment. It suggests that a good TQM system needs to be able to adapt itself to the continuously changing external and internal environments. This is a requirement for any open system.

TQM starts with the implementation of a vision (block 1) of how the leadership sees quality in their organization from a long-term perspective. A strategy is developed to materialize the vision (block 2). This suggests that TQM requires a quality culture whose values and beliefs influence the people's actions and behaviors. Without a quality culture, there is no TQM. On the basis of the quality culture (see Section 12.5), the organization develops the concerns and policies necessary to achieve excellent performance.

The TQM system should be based on the implementation of scientific methods (block 3). In management, organization, technology, and people development, Human Resources (block 4) is probably one of the central parts of the TQM system. This aspect should be considered at all levels of the organization hierarchy; it should start at the top and go right down to the worker who actually produces the products or services for the customer. Emphasizing the importance of the human factor in the TQM system is probably one of the major differences between TQC and TQM.

These days it is actually impossible to satisfy customers without having a strong information system (block 5) that allows interaction and communication among the customers, suppliers, and other parts of the enterprise. This is especially important in the period of globalization, when our customers and suppliers are all around the world. TQM should be viewed as a macrosystem that consists of a number of subsystems, which includes a management system (block 6) that is mainly focused on process management rather than people management; a quality assurance system (block 7) that is related to the assumption that the product going to the customer will probably perform well all its life; and cross-functional management systems (block 8) that go beyond the four walls of a particular department or division and even beyond the boundaries of a single organization. All these components are embedded in one function—management—that is responsible for continuous improvement and periodic innovation (see the shaded "M" in Fig. 14-3). To succeed in TQM, every organization must have a set of technological core capabilities and core competencies (block 9a) that will not only allow the organization to design and manufacture good products, but will also allow it to have the necessary speed to produce the products. The customers want the highest quality for the lowest

270 SOME MAJOR QUALITY INITIATIVES

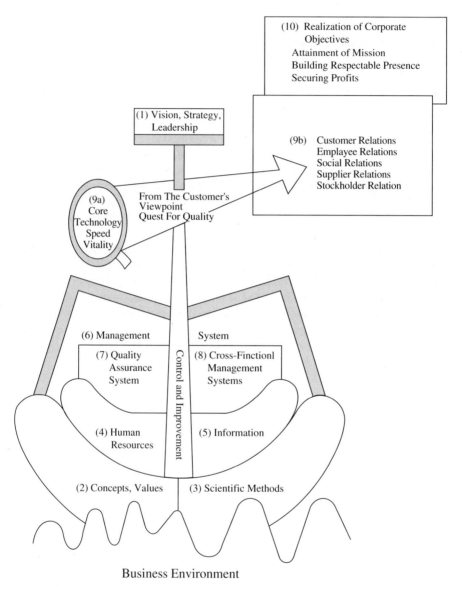

Figure 14-3 The Overall Picture of TQM.

Source: Reprinted with permission from "A Manifesto of TQM-Quest for a Respectable Organisation Presence," by the TQM Committee, Societas Qualitus, Vol. 10. No. 6, Jan/Feb 1997, p. 4[9]

price, and they want it now. It is the function of TQM to provide all this to the customers.

As the arrow directed from block 9a in Figure 14-3 suggests, quality is not making products to the specification (though this is certainly important); it is not what the CEO thinks it is; quality is what the customers think it is. The customer's perspective and viewpoint determines our success in performance excellence.

As in any system, and especially in a comprehensive system such as TQM, relationships between its elements are the foundation of good system performance. Block 9b reflects all kinds of relationships: with customers, employees, social groups, suppliers, stockholders, and others. And finally, the top block in Figure 14-3 (block 10) suggests that if all the blocks were properly instrumented, the corporate objective would be realized and the mission would be accomplished. As you can see, the overall picture of TQM embraces issues from speed through the entire organization. This is why TQM is an activity that belongs to all the people in the organization and, if it is properly performed, will benefit all the employees and the society.

14.5 COMPARING TQC AND TQM

Total Quality Control (TQC) and Total Quality Management (TQM) are two popular quality improvement initiatives that have been widely adopted by many organizations. Here we will briefly discuss their similarities and differences.

Figure 14-4 represents TQC, which is mainly the Japanese approach, and TQM, which is the Western approach. The main difference between these

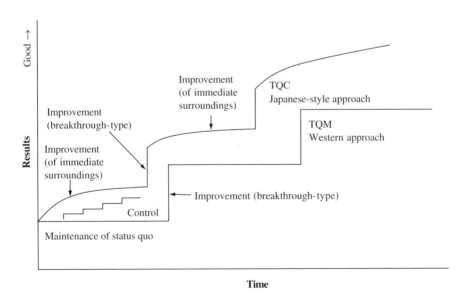

Figure 14-4 Comparing the Concepts of TQC and TQM

Source: Adapted with permission from *Introduction To Quality Control* by Kaoru Ishikawa, 3A Corporation, Japan, 1990, p. 70.[10]

approaches is that many years ago the Japanese recognized the importance of performing continuous incremental improvements in conjunction with occasional breakthrough-type improvements. Incrementalism became a part of their organizational culture. In contrast, the U.S organizations usually maintained the status quo of their processes and periodically introduced quantum leap innovative improvements.

Figure 14-4 demonstrates that even though the breakthrough-type improvements (the vertical lines) are smaller in the Japanese curve than in the Western curve, in the long run the overall improvement results are greater with the Japanese style. This is due to the application of continuous incremental improvements (the horizontal lines) that add up to additional improvements, which usually don't require capital investment. In the period between the breakthroughs, the Japanese companies use Quality Control Circles to perform small incremental improvements. In contrast, the Western companies usually use the period between breakthroughs just to maintain the status quo. We should note, however, that since globalization started to evolve rapidly, a lot of American and European companies have started to practice the introduction of incremental improvements in their organizations, especially in their plants located in Asia. For example, AMD uses a process improvement concept in their overseas plants that is similar to the Japanese philosophy. Quality Control Circles and other types of small group activities are continuously working on incremental improvements. In this regard, TQC and TQM philosophies gradually become similar approaches. However, as a concept, TQM remains a much broader philosophy, which allows people to consider the next step of achieving performance excellence.

We believe that those companies that do not utilize the power of small group activities to introduce incremental improvements in the periods between breakthroughs will spend more capital on innovation and, in the long run, will fall further behind in overall process improvements, whereas those organizations that emphasize both innovation and incremental improvements will have great success.

14.6 THE SIX SIGMA PHILOSOPHY

In 1988 Motorola introduced its Six Sigma quality initiative and achieved tremendous results. In 1995 General Electric adapted and enhanced Motorola's concept, which also generated great results and saved the company a lot of money. Today, six sigma has become a symbol of high quality, and the concept is applied in many organizations.

High-technology industries require processes that are capable of producing error free products. Motorola's six sigma concept is the solution. Since the time the six sigma philosophy was first introduced, many book and articles have been written, so you probably could not find an executive, manager, or engineer who doesn't know what six sigma means. However, many practitioners see this concept more as a toolbox with many techniques that can be used to improve the process. Others see this as a program, but by definition a program has a beginning and an end, whereas six sigma doesn't have an end. As soon as you achieve this level of quality, more sophisticated, new processes and products are implemented and the efforts toward six sigma quality start again. In our opinion the concept of six sigma is more a management philosophy that can stand by itself as a quality initiative or can be incorporated as a major block of an even larger quality improvement system. We think that, independent of what name the organization gives to its transformation initiative, it must include the concept of six sigma quality. For example, 10 years ago the TQM concept did not strongly emphasize the importance of having close to zero nonconformities in the products. This reduced its importance as a quality initiative, even though all other elements of the TQM system were very impressive. In his 1992 book, *Managing For The Future*, Peter Drucker wrote, "Finally, the leading Japanese companies are moving from Total Quality Management (TQM) to Zero Defects Management.... Dr. Deming is still a folk hero. But what the leading companies increasingly practice was expressed in a recent statement by one of the top manufacturing people at Toyota. 'We can't use TQM.' he said. 'At its very best—and no one has reached that yet—it cuts defects to ten percent. But we turn out four million cars; and ten percent defects rate means that 400,000 Toyota buyers get a 100 percent defective car.'"[11] We can say the same thing about any global high-technology industry. But this was 10 years ago. Today, the TQM philosophy includes techniques and requirements similar to the six sigma concept to achieve high-level quality processes, and because of this, TQM became more attractive. The main point we want to emphasize here is that the six sigma quality concept should be the "main dish" to any quality initiative. Without having error-free products, low prices, and on-time delivery, other quality components will not help.

In high-technology industries it is impossible to talk about, for example, reengineering as a concept without having six sigma quality under the same umbrella. As we mentioned earlier, at AMD we introduced our TCPI^2 quality initiative. This included elements of TQC, TQM, Reengineering, and other popular concepts, but six-sigma quality was one of the core concepts that made the TCPI^2 initiative successful.

Some people may think of the six sigma concept as a mechanism of improving existing processes and bringing them to a level close to perfection, that six sigma is closer to the concept of incrementalism than to innovation. This is incorrect. Motorola, General Electric, and other organizations use the term "six sigma" more as a general name of a larger quality initiative that includes many other concepts. When we work on reducing process variation in all kinds of business processes, this triggers our organizations to introduce radical changes that will allow them to achieve performance excellence in all kinds of business processes, including service.

Radical changes are needed for further improvement, especially when organizations have already achieved 4.5, or 5 sigma process quality levels. These radical changes may relate to the process in question or to the introduction of new processes. The need for projects grows as a chain reaction. In his book, *The Complete Guide To Six Sigma*, Thomas Pyzdek wrote, "Welch launched the effort in late 1995 with 200 projects and intensive training programs, moved to 3,000 projects and more training in 1996, and undertook 6,000 projects and still more training in 1997. According to *Business Week*, the initiative has been a stunning success, delivering far more benefits than first envisioned by Welch. In 1997, Six Sigma delivered $320 million in productivity gains and profits, more than double Welch's original goal of $150 million."[12] Certainly you need innovative efforts to achieve such results.

The Losses Incurred By Not Having Six Sigma

How much revenue is your company losing by not having processes that run at a six sigma quality level or close to it? If you made a rough calculation to respond to this question, you will probably find that your organization's losses fall between 15 and 25% of the total sales. It depends on the variation of your process. In their book *Six Sigma*, Mikel Harry and Richard Schroeder wrote, "In 1995, when General Electric calculated its overall sigma levels to be 3.5, they discovered they were wasting $5 billion each year in the cost of quality. United Technologies Corporation Chairman and CEO George David says that poor quality costs his company more than $2 billion a year."[13] This is money that the organization and its shareholders can have just by reducing the process variation. Having processes with high variation creates not only monetary losses, but also losses related to the company's reputation. The introduction of six sigma quality concept will not only return to your organization the 15–25 percent of revenue to which it is entitled but will also increase the organization's reputation as an enterprise that can be trusted; an enterprise with products that don't need to be tested or inspected at the customer's facility. The cost of product nonconformity and failure will gradually decrease as the process approaches six sigma (see Fig. 14-5).

As Harry and Schroeder wrote, "Companies operating at a three sigma level that marshal all their resources around six sigma can expect to make one sigma shift improvement each year. These companies will experience:

- A 20 percent margin improvement
- A 12 to 18 percent increase in capacity
- A 12 percent reduction in the number of employees
- A 10 to 30 percent capital reduction"[15]

As you can see, the implementation of the six sigma quality concept will allow your organization not only to produce high quality products but also to have a positive influence on other economic characteristics such as capacity improvement and capital reduction.

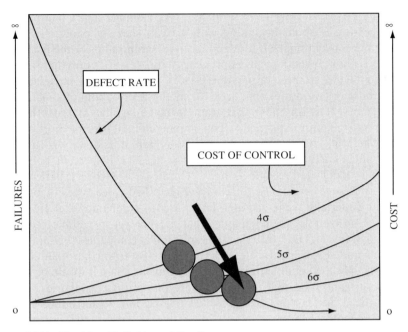

Figure 14-5 The New Definition of Quality

Source: From *Six Sigma: The Breakthrough Management Strategy Revolutionizing the World's Top Corporations* by Mikel Harry and Richard Schroeder, copyright © 1999 by Mikel Harry and Richard Schroeder. Used by permission of Doubleday, a division of Random House, Inc.[14]

14.7 REENGINEERING AND BEYOND

In 1993, Michael Hammer and James Champy published their book *Reengineering The Corporation*, which now is widely known as a bestseller that captured the minds of many executives and managers throughout the world. Their book came out at a time when organizations were trying different quality initiatives such as SPC, TQC, TQM, and other programs, but they were not getting all the results they were expecting. At the same time, the competitive environment was forcing organizations to improve the quality and reduce the price of their existing products, and come up with new products and services to satisfy and foresee the customers' needs; all within a limited time frame. Because of this, organizations were open to accepting something radically new that might help them satisfy their needs. The Reengineering concept filled this role.

14.7.1 The Concept of Reengineering

The main idea behind reengineering is that organizations that want to stay competitive must introduce radical changes in their business processes by "starting all over, starting from scratch."[16] As Hammer and Champy wrote, "Reengineering is about beginning with a clean sheet of paper. It is about rejecting the conventional wisdom and received assumptions of the past. Reengineering is about inventing new approaches to process structure that bear little or no resemblance to those of previous eras."[17] The concept of reengineering is not for those organizations that seek for methods of continuous incremental improvements. It is for those that want to make significant breakthroughs. Hammer and Champy explained, "If a company falls 10 percent short of where it should be, if its costs come in 10 percent too high, if its quality is 10 percent too low, if its customer service performance needs a 10 percent boost, that company does not need reengineering."[18] On the basis of their experiences, the authors of *Reengineering The Corporation* identified three types of organizations that undertake reengineering. These are: a) companies that find themselves in deep trouble, (b) companies that are not yet in trouble but whose management foresees that trouble is coming, and c) companies that are in peak conditions and see reengineering as an opportunity to further maintain their superiority over their competitors. In these turbulent times, it would be difficult to find an organization that doesn't fall into one of these categories. There is always a need, at least in part of your organization, to introduce a breakthrough-type improvement.

We presume that most of you are already familiar with the reengineering philosophy, which already has been described in several books by various authors. The main purpose of this section is to explain how reengineering as a concept is related to other quality initiatives such as TQC and TQM. This will help the manager better understand when and where to incorporate the reengineering approach when developing quality systems for organizational transformation.

14.7.2 Comparing Reengineering with Other Quality Programs

How much does reengineering differ from other existing quality initiatives? In *Reengineering The Corporation*, Hammer and Champy argued, "... nor is reengineering the same as quality improvement, total quality management (TQM), or any other manifestation of the contemporary quality movement."[19] Elaborating the difference between reengineering and other quality programs, they used Kaizen as a general term of all quality improvement programs. Hammer and Champy wrote, "Quality programs work within the framework of a company's existing process and seek to enhance them by means of what the Japanese call Kaizen, or continuous incremental improvement. The aim is to do what we already do, only do it better. Quality improvement seeks steady incremental improvement to process performance. Reengineering...seeks breakthroughs not by enhancing existing processes, but by discarding them and replacing them with entirely new ones."[20] One may conclude from their statement that all quality programs are related only to continuous incremental improvement and have nothing to do with innovation, that all quality initiatives are equivalent to Kaizen. As we have described in Section 14.2, Kaizen is a mind-set inextricably linked to maintaining and improving standards. But, as we demonstrated earlier, in a broader sense improvement can also be described as Kaizen and innovation (see Fig. 14-1).

Graphically, the reengineering concept, which is based only on periodic breakthroughs, can be expressed as in Figure 14-6. The vertical lines represent breakthrough results, and the horizontal lines show that new advancements always have a tendency to slip back to the old way of doing things if special mechanisms are not in place to hold the achieved level. The continuous incremental improvement activities are mechanisms that will maintain the status quo and allow the management to achieve small improvements.

TQM is an excellent example of a quality system that includes incremental improvement and innovation (see Fig. 14-4). This is not to say that reengineering is just another name for TQM. Reengineering as a concept certainly contributed to the quality movement by focusing management's attention onv the need to get detached from the old way of doing things if they no longer work. The ability to bring existing assumptions to the surface, to analyze andchallenge them to see which of them need to be discarded, and to decide which new assumptions need to be brought in is an art of management. The Reengineering concept helps do this.

As often happens to new initiatives, as soon as they become popular, organizations jump on them, expecting an easy way out of the problems they face. However, some organizations become disappointed by not achieving fast results, and they start to blame the new initiative. The same thing happened to Reengineering. Regardless of whether they needed it or not, or whether they

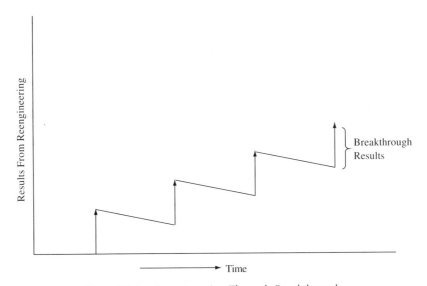

Figure 14-6   Reengineering Through Breakthroughs

were ready or not, many organizations started using Reengineering in their-processes and became disappointed from not achieving the expected results. In an article by Gene Hall et al. published in *Harvard Business Review*, we read, "In all too many companies, Reengineering has been not only a great success but also a great failure. After months, even years, of careful redesign, these companies achieve dramatic improvements in individual processes only to watch overall results decline."[21] On the basis of their research into Reengineering projects in more than 100 companies, the authors of this article came up with five factors common to the successful application of the Reengineering concept (see Insert 14-2).

Even though these factors reflect the specifics of reengineering, they also can be considered when implementing other quality initiatives such as TQM, Six Sigma Quality, etc. Hammer and Champy also recognized that not all organizations that were engaged in Reengineering achieved dramatic results. They wrote, "Our unscientific estimate is that as many as 50 percent to 70 percent of the organizations that undertake a reengineering effort do not achieve the dramatic

---

Insert 14-2

Five Keys to a Successful Redesign

The following five factors are common to successful reengineering efforts.

1. Set an aggressive reengineering performance target. The target must span the entire business unit to ensure sufficient breadth. For example, aim for a $250 million pretax profit increase to result from a 15% cost reduction and a 5% revenue increase measured across the business unit as a whole.

2. Commit 20% to 50% of the chief executive's time to the project. The time commitment may begin at 20% and grow to 50% during implementation stage. For example, schedule weekly meetings that inform the top manager of the project's status.

3. Conduct a comprehensive review of customer needs, economic leverage points, and market trends. For example, customer interviews and visit, Acompetitor benchmarking analysis of best practices in other industries, and economic modeling of the business.

4. Assign an additional senior executive to be responsible for implementation. The manager should spend at least 50% of his or her time on the project during the critical implementation stage.

5. Conduct a comprehensive pilot of the new design. The pilot should test the design's overall impact as well as the implementation process, while at the same time building enthusiasm for full implementation.

Source: Reprinted with permission from Gere Hall et al, "How to Make Reengineering *Really* Work," *Harvard Business Review*, November-December 1993, p. 128.[22]

> **Insert 14-3**
>
> **The Top Ten Ways to Fail at Reengineering**
>
> 1. Don't reengineer but say that you are.
> 2. Don't focus on processes.
> 3. Spend a lot of time analyzing the current situation.
> 4. Proceed without strong executive leadership.
> 5. Be timid in redesign.
> 6. Go directly from conceptual design to implementation.
> 7. Reengineer slowly.
> 8. Place some aspects of the business off-limits.
> 9. Adopt a conventional implementation style.
> 10. Ignore the concerns of your people.
>
> *Source*: From *The Reengineering Revolution*, by Michael Hammer and Steven A. Stanton, p. 33. Copyright © 1995 by Hammer and Company. Reprinted by permission of HarperCollins Publishers, Inc.[24]

results they intended."[23] Recognizing the importance of further helping organizations understand the philosophy of reengineering, in 1995 Michael Hammer and Steven A. Stanton came out with *The Reengineering Revolution*. In this book, they described the elements of how to make reengineering successful and gave new illustrative examples of reengineering. By analyzing the mistakes organizations usually make when trying to apply reengineering, they came up with "Ten Top Ways to Fail at Reengineering" (see Insert 14-3). By avoiding those mistakes, an organization can succeed in their reengineering efforts.

Later in the same book, Hammer and Stanton wrote, "Management must clarify the nature of customer-focused quality and explain how TQM is related to reengineering. It may be wise to reinforce this relationship by integrating the units responsible for the two approaches."[25] This statement infers that managers must have a clear understanding of what the two concepts in question can offer. We believe that elements of reengineering, TQM, SPC, and other existing initiates that have been polished by time can be integrated into one macro-system that would fit the needs of the organization.

14.7.3 Beyond Reengineering

In 1996, Michael Hammer came out with a new book, *Beyond Reengineering*, in which he wrote, "I have now come to realize that I was wrong, that the radical character of reengineering, however important and exciting, is not its most significant aspect. The key word in the definition of reengineering is 'process': a

280 SOME MAJOR QUALITY INITIATIVES

complete end-to-end set of activities that together create value for a customer."[26] This is a significant change in the way of thinking about reengineering. "Process," as the term itself suggests, is continuous and not episodic. An organization consists of multiple processes that have their own aims and objectives; processes are holding people accountable for what they are doing; process is how organizations compensate and motivate people for their performance; process is how organizations forecast, plan, and grow. Process-centered management is not about forgetting the past and starting with a clean sheet of paper. Almost all initiatives such as TQC, TQM, Six Sigma Quality, and TCPI^2 are based on systems and processes that embrace incremental improvement, radical improvement, breakthroughs, innovation, creativity, and other forms of process enhancement.

Figure 14-7 is reprinted from Hammer's book *Beyond Reengineering*. As you can see, it resembles Figure 14-4, which represents TQC and TQM. The different is only in the interpretation. Hammer continues to suggest that TQM is related only to incremental enhancements and that reengineering is where breakthroughs occur (see Fig. 14-7). He writes, "...TQM and reengineering fit together over time in the life story of a process. First, the process is enhanced until its useful lifetime is over, at which point it is reengineered. Then, enhancement is resumed and the entire cycle starts again."[27] As we see it, TQC and contemporary TQM, which recognize the importance of incremental

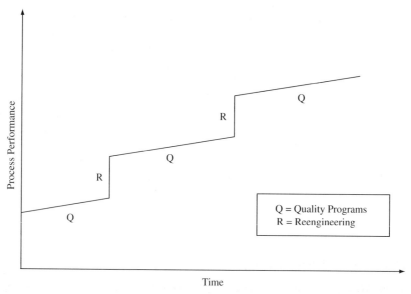

Figure 14-7 An Interpretation of Reengineering

Source: From *Beyond Reengineering* by Michael Hammer, p. 83. Copyright © 1996 by Michael Hammer. Reprinted by permission of HarperCollins Publishers Inc.[28]

improvement instead of only maintenance of the status quo (see Fig. 14-4), resemble Michael Hammer's new interpretation of process-centered reengineering. In the TQC and TQM curves (see Fig. 14-4), the vertical lines reflect breakthrough-type improvements, whereas in Hammer's chart the vertical line represents "reengineering." This certainly does not suggest that all the above-mentioned initiatives are the same. *Reengineering The Corporation* and *Beyond Reengineering* contain a lot of valuable concepts that you will not find in other publications on continuous improvement of business processes. The task of contemporary management is to select from all the existing concepts on improvement those that will best fit their needs. This is what we did at AMD when we designed our Total Continuous Process Improvement and Innovation (TCPI^2) program. Having designed it in 1988, we continue to enhance it by using new concepts and adapting experiences from other organizations such as Motorola, General Electric, and Intel.

The discussion of reengineering would not be complete if we did not mention the development of another branch of thought on the same subject. The following section will introduce you to a holonic approach to reengineering, which is a natural extension of business process reengineering.

14.7.4 Creating Holonic Business Systems

The approaches to business improvement that we have described so far in this chapter were related to activities that take place mainly within the four walls of the enterprise. Here we want to familiarize you with an approach that goes beyond the traditional way of thinking: seeing a company as a network of many interconnected and interdependent organizations who bring their unique capabilities together to form a process needed to serve the customer. In 1995, an international team (U.S., U.K., and Italy) of coauthors, Patrick McHugh, William A. Wheeler, III, and Giorgio Merlin, published the book *Beyond Business Process Reengineering*, which includes a description of a holonic approach to doing business. Their approach does not replace any of the existing initiatives of process transformation. What they suggest can be considered an approach that challenges the conventional way of thinking about an enterprise. And as the holonic concept requires us to rethink what an organization is, it also requires us to rethink what a process is and what quality improvement means. This is why we decided to include a discussion about creating holonic enterprises in this chapter. To do this, we will adopt a metaphor and a story from the book we mentioned above. We think that this will give you a feeling for what a holonic business process is. We hope that this will be enough to trigger your interest in this subject, but before we do this, we will familiarize you with some terminology that is used to describe the holonic approach.

14.7.4.1 Holon and Holonic Networks Arthur Koestler first coined the word "holon" in his book, *The Ghost of the Machine* (1969).[29] The term came from the Greek *holos*, meaning "whole," and *on*, inferring "part." A holon can be interpreted as Janus-faced, which means it has two faces, one inward and the other outward. A holon has the characteristic of being both a part and a whole. A company is a holon in that it is integrative. It is a whole and at the same time it is a part of a larger whole. In this context, the term holon is used to interpret a process within a process.

Considering a company or an organization as a "holon," McHugh et al. define a holonic network as "a set of companies that acts integratedly and organically; it is constantly re-configured to manage each business opportunity a customer presents.... Each configuration of process capability within the holonic network is called a virtual company."[30] Now that you have become familiar with the main terms related to this subject, we will introduce you to a case study that will allow you to grasp the concept of a holonic network (see Case Story 14-2).

A process that can produce results as described in this story can be achieved only in a holonic enterprise. The effectiveness of such a holonic enterprise can

---

Case Story 14-2   The Sikisui Holonic Network

The customers, a wife and husband enter the [Sikisui] sales office to buy a home. There they find several computer terminals, each linked to the company's design office. First, the couple designs the house's basic layout, restricted only by building codes and the size of the lot the house will be built on. Then they begin to personalize it.

They discuss options, with or without the help of the sales agent. They try out "what ifs" on the screen. Eventually, they devise a plan that meets their exact specifications. The sales agent offers them tea while they wait for their quotation to be ready, including the cost of building the house. The couple agreed to the quotation, and come to terms on a payment schedule. The contract is signed.

Within minutes, the company has sent orders to all of its suppliers to prepare the materials for this custom designed house. If a supplier of one item cannot fulfill its order within three days, it is responsible for finding another vendor in the network of suppliers who can.

Two days before the house is to be built on the customer's site, the components of the house are at the building company's factory. Here they are assembled into modules. Sikisui, with its sales staff and design centers, acts as the integrator in this holonic network that includes dozens of suppliers. Each home—designed by the prosumer—is supplied by a different group of suppliers—in effect a different virtual company for each product. Fifty percent of the parts in the typical home are ordered on a kanban system (i.e., they are repetitive orders), while the other 50 percent are special orders for each home.

Source: Reprinted with permission from *Beyond Business Process Reengineering* by Patrick McHugh, et al, John Wiley & Sons, Limited, 1995, p. 16–17. Copyright © 1995 P. McHugh, G. Merli, W.A. Wheeler.

be attested to by the fact that "annual revenues per Sikisui employee are almost $1 million each."[32] After reading this story, you can probably see the applicability of the holonic network concept to any industry. But to provide you with a deeper understanding of this subject, we will use the Portuguese man-of-war as a metaphor.

14.7.4.2 The Portuguese Man-Of-War as a Metaphor of a Holonic Network For many years we have had a mechanistic view of an enterprise. We were used to using a machine as a metaphor for it. But a holonic enterprise can best be described by using an organic metaphor. McHugh et al. use the Portuguese man-of-war as an excellent metaphor for a holonic business network. As they describe it, a holonic network resembles organization characteristics in that it is continuously evolving through interaction with the environment. Any organization is an open system that is continuously evolving with the environment; but in a holonic network the speed and response to the external environment are much greater and more effective. The holonic network is also more flexible than a single organization. It is capable of moving to new markets without feeling that it needs to protect its old position. Finally, the holonic network has greater capacity and flexibility. Instead of buying new equipment and hiring and training new employees, it can take on new partners with their available capacity and competencies. In a holonic system, each holon has some unique capabilities that allow the holon-organization to be valuable to the network. This uniqueness makes the node capable of surviving in the network, although it is important to note that this uniqueness is more powerful when the holon is within the system than when it stands alone.

As we mentioned in Chapter 1, a complicated concept is easier to explain and easier to understand when using a good metaphor. We will use the Portuguese man-of-war as a metaphor to describe the life and work of a holonic system. As Art Spikol describes in his murder mystery novel *The Physalia Incident*, the man-of-war is really a colony of animals—"a variety of creatures called hydrozoans, each group of which handles a particular life function and gradually becomes modified so that it can better handle that function.... As time goes by, the coordination between the individuals becomes better and better.... The nervous system becomes ever more integrated as these individuals become more specialized. And the more the individuals become coordinated, the more the colony acts like a single organism."[33] This is such a wonderful match to the topic in question. It is almost a definition of a holonic enterprise. No wonder the authors of *Beyond Business Process Reengineering* selected the man-of-war as a metaphor for their topic. The Portuguese man-of-war, a distant cousin of jellyfish, is really a colony of hundreds of hydrozoans, which in our case can be called holons. Each holon-hydrozoan has its own job to do. Some of them catch food, and others digest it, etc. The man-of-war, which is considered here as a holonic network, communicates between its various holons through the blood supply and nervous system, just as a commercial holonic system would use computer information technology. Each animal is born into the colony and lives its life within its structure and rules. This is analogous to how companies join, adapt, and live in a holonic network. More than one holon exists for each task in the Portuguese man-of-war, and although they compete with other holons, they are also dependent on each other's success. In the same way, holons in a commercial holonic network both compete and are interdependent. The success of the holonic enterprise totally depends on the interaction of its holons. As McHugh et al. define it, "A holonic network in the commercial world is a group of business that, acting in an integrated and organic manner, is able to configure itself to manage each business opportunity that its customer presents. The Portuguese man-of-war can configure its holons to deal with each new fish trapped in its tentacles."[34]

In Case Story 14-2, as soon as the customer signed the contract, the company configured itself to manage a particular business opportunity. Continuing to use the Portuguese man-of-war metaphor, we can find more analogies that can help us better understand the concept of the holonic business network. For example, the holonic network is not an individual business just as the man-of-war is not a single animal; the man-of-war reacts to the capture of a fish or to a change in the environment, and not to the activity of a single holon. In a holonic business network, the managers have to learn to manage by processes and not by functions. By studying holonic networks more deeply, you will observe that they are process based and live by the logic of processing. There is no hierarchy in holonic processes. There is only a minimal management overhead, which is mirrored in each holon.

There are certainly many other peculiarities that differentiate the holonic approach of doing business from other traditional ways; however, we think that the story of building a home in 9 days (Case Story 14-2), and the metaphorical interpretation of the holonic concept we have described will help you grasp the meaning of the holonic way of doing business. Our intention in this section was to spark your interest in this subject, and we hope we achieved that.

14.8 CONCLUDING THOUGHTS ON QUALITY INITIATIVES

In this chapter, we briefly introduced some old and some relatively new quality initiatives. This was to remind you about the great old stuff and at the same time introduce some new quality concepts that have become available to you in the last decade. When will the next new quality initiative come out, and what will it look like? As the history of quality development shows us, previous major changes in the approaches to quality improvement have occurred approximately every 20 years (see Fig. 14-8).

The approaches to quality improvement can be briefly described as follows:
The first initiative in the quality field was *Operator Quality Control*, which was inherent in the manufacturing job up to the end of the nineteenth century.

In the early 1900s, the industry progressed to *Foreman Quality Control*. This period saw the large-scale advent of the modern factory concept, in which many

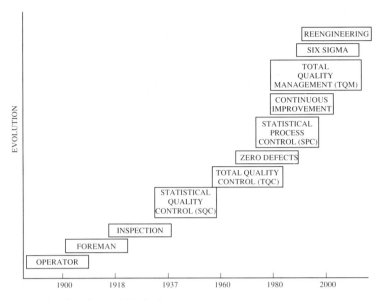

Figure 14-8 Quality Control Evolution

Source: Extended from the chart described in *Total Quality Control* by A.V. Feiganbaum, © The McGraw-Hill Companies, 1991, p. 16[35]

individuals performing a single task were grouped so that they could be directed by a foreman, who then assumed responsibility for the quality of the work.

Then in the 1920s and 1930s, large *Inspection* organizations came on board, which were separately organized from production. This quality program remained active until tremendous mass production required a more effective approach.

In the 1940s, *Statistical Quality Control* became popular. At the time, it mainly replaced 100% inspection by the introduction of sampling plans.

In 1979, Philip B. Crosby's book *Quality Is Free*, became a bestseller. His name is associated with the quality initiative Zero Defects, which was popular in the military and in industrial organizations in the 1970s. Crosby has advocated the notion that Zero Defects can and should be the target in the quality movement.

In 1951, Feigenbaum came out with his *Total Quality Control* book, which formulated a new concept of managing quality.

In the 1980s, Dr. Deming influenced American corporations with his lectures on statistical principles. This was when organizations in the U.S. became interested in SPC and started introducing this concept in their organizations. Continuous incremental improvement was recognized more by the Japanese organizations. Influenced by Dr. Deming's activities in Japan, a number of Japanese authors further developed and implemented different quality initiatives in their country.

In 1986, the book *Kaizen*, written by Masaaki Imai, reflected a movement of continuous improvement in Japan. Kaobu Isakawa is also famous for his contribution to quality control. Dr. Taguchi is another contributor to the quality movement. He is most known for his approach to Design of Experiments and the concept of Quality Loss Function.

Total Quality Management can be considered a science that was based on the theories and practical experience of many quality gurus such as Deming, Juran, Crosby, and Feigenbaum. TQM makes it clear that quality is not determined by the workers on the manufacturing floor, nor is it determined by a particular department. The TQM concept shows that quality is determined by the senior management of the organization, who by virtue of their position and obligation, are responsible to all stakeholders for the success of the business.

In 1988, Motorola's *Six Sigma Quality* initiative came on board, and received great recognition in the U.S. and abroad.

In 1993, Hammer and Champy published their concept of *Reengineering*, which became another movement toward organizational transformation.

As you can see, in the latter part of the twentieth century, initiatives for quality transformation developed much faster than earlier in the century. The management today has the opportunity to use the experience gained in quality development and, at the same time, participate in the introduction of new approaches and concepts that may more closely match the requirements of the continuously changing environment.

References

1. J.M. Juran, *Managerial Breakthrough*, McGraw-Hill, Inc., New York, NY, 1964, p. 2
2. Masaaki Imai, *Kaizen*, McGraw-Hill, Inc., New York, NY, 1986, p. xx
3. William Lareau, *American Samurai: A Warrior for the Coming Dark Ages of American Business*, New Win Publishers, Clinton, NJ, 1991, pp. 156–157
4. Armand Vallin Feigenbaum, *Total Quality Control*, 3rd ed., McGraw-Hill, Inc., New York, NY, 1991, p. 835
5. Ibid., pp. 11–12
6. Ibid., p. 11
7. "Quality 2000: The Next Decade of Progress", *Fortune*, 19 September 1994, p. 158
8. The TQM Committee, "A Manifesto of TQM(1)—Quest for a Respectable Organization Presence", *Societas Qualitas*, vol. 10, no. 6, Jan/Feb 1997, p. 1
9. The TQM Committee, "A Manifesto of TQM(1)—Quest for a Respectable Organization Presence," *Societas Qualitas*, vol. 10, no. 6, Jan/Feb 1997, p. 4
10. Kaoru Ishikawa, *Introduction To Quality Control*, 3A Corporation, Tokyo, Japan, 1990, p. 70
11. Peter F. Drucker, *Managing for the Future*, Truman Talley Books/Dutton, New York, NY, 1992, p. 182

12 Thomas Pyzdek, *The Complete Guide to Six Sigma*, Quality Publishing, L.L.C., Tucson, AZ, co-published with McGraw-Hill Companies, New York, NY, 1999, p. 5
13 Mikel Harry and Richard Schroeder, *Six Sigma*, Doubleday, New York, NY, 1999, pp. 29–30
14 Ibid., p. 31
15 Mikel Harry and Richard Schroeder, *Six Sigma*, Doubleday, New York, NY, 1999, p. 2
16 Michael Hammer and James Champy, *Reengineering The Corporation*, HarperCollins Publishers, Inc., New York, NY, 1993, p. 2
17 Ibid., p. 49
18 Ibid., p. 33
19 Ibid., p. 49
20 Ibid. p. 49
21 Gene Hall et al., "How To Make Reengineering *Really* Work," *Harvard Business Review*, Boston, MA, Nov-Dec 1993, p. 119
22 Ibid., p. 120
23 Michael Hammer and James Champy, *Reengineering The Corporation*, p. 200
24 Michael Hammer and Steven A. Stanton, *The Reengineering Revolution*, HarperCollins Publishers, Inc., New York, NY, 1995, p. 33
25 Ibid., p. 97
26 Michael Hammer, *Beyond Reengineering*, HarperCollins Publishers, Inc., New York, NY, 1996, p. xii
27 Ibid. 1996, p. 82
28 Ibid., p. 83
29 Arthur Koestler, *The Ghost In The Machine*, Pan Books Ltd, London, 1970.
30 Patrick McHugh, Giorgio Merli, and William A. Wheeler, III, *Beyond Business Process Reengineering*, John Wiley and Sons, Inc., New York, NY, 1995, p. 4
31 Ibid., pp. 16–17
32 Ibid., p. 17
33 Art Spikol, *The Physalia Incident*, Penguin Books, New York, NY, 1989, p. 51
34 Patrick McHugh, Giorgio Merli, and William A. Wheeler, III, *Beyond Business Process Reengineering*, John Wiley & Sons, Inc., New York, NY, 1995, p. 82
35 Armand Vallin Feigenbaum, *Total Quality Control*, 3rd edition, McGraw-Hill Companies, New York, NY, 1991, p. 16

Chapter 15

Achieving High Quality Through Transformational Changes

"It is not the strongest of the species that survive, nor the most intelligent, but the ones most responsive to change."

—*Charles Darwin*

We have all come to expect quality in the products we buy. In the past, we had different opinions about the influence of quality on the product cost. Today, it is pretty much universally understood that it is less expensive to make high-quality products than low-quality products because of the cost associated with inspection, screening, scraping, and performing other non-value-added activities to prevent shipments of nonconforming products. In other words, quality has become more of a given factor and less of a competitive factor that differentiates one company from another. This is because every company must have a certain quality level for their processes and products to stay in business. Even though this is the correct way of thinking, it may, however, become a distracting factor, and organizations may lose their focus on quality. Because of this, it is important to note that quality, to be a ticket of entry to the marketplace, requires a process of continuous improvement and innovation. You cannot use the same ticket forever. To gain the necessary market share, an organization needs to create customers. This involves creativity, innovation, and learning how to deal with changes.

In this chapter we will focus on managing changes. Much has been written about this topic. Researchers and practitioners have come up with different concepts and methods that allow us to understand the process of organizing change. Because of this, we will limit ourselves to giving brief descriptions of some of the concepts that we feel a manager must know.

15.1 UNDERSTANDING THE CONCEPT OF CHANGE

If you had a conversation with your grandfather, complaining about the pressure you have from the change you are facing today, he would probably tell you that he said the same thing to his grandfather. People from all generations have had to confront the repercussions of change; that is nothing new. So why do we feel there is a difference in the pressure of change when we compare it to change in the past? Why is it so difficult today to constantly adjust to changes?

The answer to these questions probably lies in the magnitude and speed of change, in the increased seriousness of its implication, and in the diminishing shelf life of the effectiveness of our response to change. For example, from year to year we can observe a tendency toward a continuous reduction in manufacturing cycle time and development time. As Tom Peters wrote in his book *The Tom Peters Seminar*, "Or consider IBM's Austin, Texas, personal computer operation. In the last three years, what was already a thoroughly modern operation has, according to *The Economist*, cut its average manufacturing cycle time from 7.5 to 1.5 days and new product development time from 24 to 8 months, increased its product portfolio from 19 to 85, and simultaneously shrunk the payroll from 1,100 to 423 people."[1] We could not have observed this magnitude in speed of change earlier. It has occurred only in our time. In his book *Ackoff's Best*, Russell L. Ackoff observed, "The speed with which we can travel has increased more in our lifetimes than it has over all the time before our births. The same is true for the speed with which we can calculate, communicate, produce, and consume."[2]

Today the speed of change is so fast that it takes great effort to catch up with it. We need to learn how to learn faster and to make new products with speeds we could not imagine before. Moore's Law states that the number of transistors on a semiconductor chip doubles approximately every 18—24 months. This statement is still operational, even though the complexity of microprocessors has gotten significantly greater. Catching up with the speed of change is not an option—you either do it or go out of business.

This is why organizations have enthusiastically accepted the concept of Reengineering. It encourages management to starts with "a clean sheet of paper,"[3] which means seeking new solutions, achieving radical breakthroughs by innovation, and giving credit to the importance of continuous incremental process improvement. Rapid change requires innovation. Because of this, management needs to learn how to manage creativity and innovation and how to build an organizational environment that will motivate people to innovate. As Figure 15-1 suggests, there is a conflict between the nature of people and the environment in which we live.

By their nature, human beings and their organizations seek stability. But the world in which we live and work is unstable and is getting more and more dynamic over time. Globalization is influencing the size of the organization's external environment. The growth of technology, which allows us to have breakfast in one part of the country and dinner in another part, and the ability

290   ACHIEVING HIGH QUALITY THROUGH TRANSFORMATIONAL CHANGES

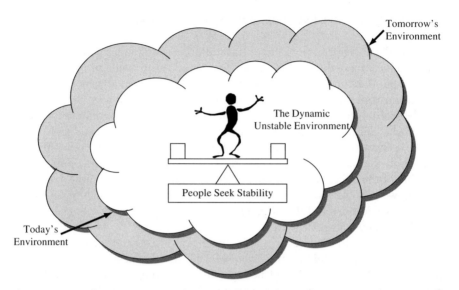

Figure 15-1   Balancing Between the Desire for Stability and Dynamics of an Unstable Environment

to see and communicate with other people in any part of the world without leaving our offices, is making the external environment larger and larger. This, in turn, is making the environment more turbulent and unpredictable, which then makes it more difficult to manage and lead.

There is a strong correlation between the speed of change and management style. This can be illustrated by an analogy of flying aircraft. At a high-enough altitude, when the weather is good, a pilot can enjoy the stars or the blue sky, keeping his hand on the yoke (steering control) with little or no change in course. He can even put the aircraft on autopilot. But if there is turbulent weather, the worse it gets, the more difficult it is for the pilot to maintain the course. He needs to introduce frequent changes in the direction, speed, and altitude; and whatever he does, he cannot avoid the ups and downs that disturb the pleasure of his passengers. When the environment gets really bad, the pilot needs take more actions. He needs to think and act fast. In his book *Beyond The Stable State*[4], Donald A. Schön shows that as the change increases, the complexity of the problems that we face also increase. The more complex the problems are, the more time it takes to solve them. The more the rate of change increases, the more the problems that we face are changed and the shorter is the life of the solutions we find to solve the problems. Therefore, by the time we find solutions to many of the problems that we face, usually the most important ones, the problems have so changed that our solutions to them are no longer relevant or effective; they are stillborn. In other words, many of our solutions are solutions to problems that no longer exist in the form in which they were solved. As a result, we are falling further and further behind our time.

In our analogy, if the pilot does not take action quickly, he may face a situation in which the problem disappears or takes another form. In this case, his solution will be to the wrong problem. You are probably familiar with situations like this; for example, when the demand for a particular product goes up and we decide to react by adding more capacity. This certainly takes time. When the capacity becomes available, the demand for this particular product falls, and now you have extra capacity, which creates a new problem. The point here is that the increased rate of change requires a change in the way we do business. Change requires a revision of our assumptions, new values, relearning, and rethinking a lot of things. This is why organizational transformation should become a way of life.

15.2 DO WE LOVE OR HATE CHANGE?

While writing this book, we were involved in implementing a new business information system that will create a significant change in the way we do things here in AMD. To make sure that employees will support this change, we conducted seminars on the topic "Managing Change." In one of these seminars, we asked the audience to divide into teams and discuss a series of topics; such as the reasons people resist change, solutions of how to reduce the resistance to change, etc. The intent was to generate some ideas for a dialogue on change.

Surprisingly, without our instruction, one team came up with a list of reasons for why people *like* change. Their intention was to prove that if people were treated properly, and if the motives for change were right, they would not only *not* resist change but would be happy to participate in implementing the changes. The team challenged the existing assumption that people will always resist change.

This accident triggered our interest in doing some deeper studies in this direction. We found that most people in our organization have a positive attitude toward change. Some of them even love change and feel bad being in a frozen, unchanging environment. Our exit interview with some employees showed that not having the opportunity to experience change was their main reason for leaving the organization. For example, when AMD became a leader in producing microprocessors, which required a lot of changes, the turnover of engineers was significantly reduced. How much do we know about what exactly people think about work, and in particular about change? The times have changed, the professionalism of people has changed, their mental models continuously change, but you as a manager sometimes continue to operate with old assumptions and old categories.

In his book, *The Pursuit Of WOW!*, Tom Peters tells a story (see Insert 15-1) that he concludes with a message, "Don't treat change with kid gloves. Don't assume that 'they' (or you) can't cope with it. Do realize change is as normal as breathing, and that we want *lots of it* (along with *lots* of constancy)."[5]

> **Insert 15-1**
>
> **A Thirst for Change**
>
> Change is a pain. As I write, the powers that be are fixing a road between my home and office. That means I have to detour, and I've been taking the same route for 10 years. The detour really upsets me.
>
> That detour is peanuts compared to the introduction of self-managing teams or a new performance-evaluation scheme. Still, it illustrates how the smallest disruptions can irritate in a big league way.
>
> **Lesson:** We all need constancy in our lives. Wise managers spend a lot of their time helping employees find new constancy in the midst of perpetually stormy commercial seas. (Which describes most every commercial sea, these days.)
>
> But that's not the end of my little screed. Most people would stop here: (1) Change is a pain. (2) Help people cope.
>
> That is *exactly* one-half of the story. A ton (literally, probably of solid psychological research says that we humans need stability... and equally we need stimulation. The change-is-a-pain activists ("They don't' want it," "They can't handle it," etc., etc.) ignore the other half.
>
> Did you return to the same vacation spot this summer as last? Did you go to the same restaurant last Friday as the Friday before? Did you re-read John Grisham's *The Firm* instead of getting a new book for your last plane ride? See my point? In tiny ways (those I just listed) and much bigger ones (new project, new hobby, new night-school course, new job searches), we seek stimulation while we also value stability.
>
> Source: From *The Pursuit Of WOW!* by Tom Peters, p. 81–82. Copyright © 1994 by Excel/A California Partnership. Used by permission of Vintage Books, a division of Random House, Inc.[6]

If we wanted to reduce the resistance in an electrical circuit, we would increase the thickness of the wires, increase the voltage in the circuit, change the materials of the wires, etc. In short, we design a condition where the resistance is not a problem. It is difficult to think digitally: people like or people don't like change. It depends on many variables. What management needs to do is to design a system with less resistance. Resistance produces unnecessary heat.

There are a large variety of reasons that determine people's attitudes toward change. To manage change effectively, we need to understand these reasons and act accordingly. In his book, *Management Masterclass*, Neil M. Glass presented an interesting model for analyzing the different reasons for resistance. He recommended that we group people by how well they understand the reasons for change and how emotional or politically involved they are in the change itself (see Fig. 15-2).

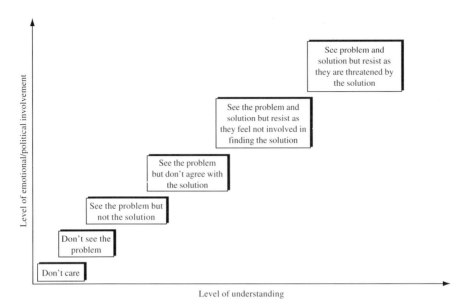

Figure 15-2 Some of the key reasons for resisting change

Source: Reprinted with permission of the publisher, Nicholas Brealey Publishing, from *Management Masterclass* by Neil M. Glass, 1996, p. 121[7]

Even though it would probably be difficult to actually group people in this way, Figure 15-2 can help the manager better understand how different states of mind and feelings can influence people's readiness to participate in a change and to take action accordingly to achieve a higher level of emotional and political involvement of people in the organization. According to Neil M. Glass, people respond to change on three levels:

> "*Emotional*—What will this change mean for me personally? How will my life change? Will my position in other people's eyes be different? Will I be able to cope with changes to my role? and so on.
>
> *Political*—Will I lose control over resources/people/decisions? Will I still be part of some key groups? Will I still be able to influence the decisions that affect me and my area?
>
> *Rational*—Is this change right for the organization?"[8]

Helping people to attain the correct answers to these questions is the management's responsibility. This will strongly influence the overall level of change experience.

15.3 RECOGNIZING THE CYCLES OF CHANGE

To manage transformation successfully, we need to understand and recognize the cyclical character of change. People's reactions to change depend on their perceptions, feelings, and aspirations, which are triggered by external events. Organizations also have a cycle of change as they go through their different stages of development. For example, if an organization is in its start-up period, it requires an entrepreneurial approach to change. When it is in a period of growth, it needs an incremental approach to change. When its organizational pattern is starting to deteriorate, it requires two types of change management simultaneously: one that will maintain the status quo, and the other will encourage innovation. Having an understanding of the cyclical characteristics of change will help management facilitate the change process more effectively.

In this section, we will describe three interrelated theories that will allow you to see the cycle of change from different perspectives.

15.3.1 Janssen's Four-Room Apartment

While describing the importance of implementing a program that would require a lot of change, an AMD executive shared his thoughts, "Everybody probably understands the importance of the program. Most of the people support it. But why aren't some employees, or even departments, fully participating in its implementation?" In response, someone in the audience said, "Because some people are still in the *Contentment* room." After pausing, the executive continued, "We need to help them to open the door and move more quickly towards *Renewal*." The audience understood his metaphor and it made sense to them.

This terminology came from the Four Rooms Of Change theory that was developed by a Swedish social psychologist, Claes Janssen. Colloquially, his theory is known as the "Four-Room Apartment." Its description first appeared in Dr. Janssen's book, *Personal Dialectics* (1975). At AMD, Janssen's four-room model is well known and recognized as a useful and easy to understand concept. What makes this concept unique and has kept it alive for more than a quarter of a century?

Remember Shewhart's Plan/Do/Study/Act concept, which Deming popularized in Japan, and later became known as Deming's Cycle? Dr. Shewhart wrote about it in 1986, but due to its deep meaning and simplicity, it is still popular today. We believe that Janssen's model of change is another theory that will live a long time because of its unique, simple interpretation of a complicated phenomenon. Following is a short description of the theory.

As Janssen suggests, each person, team, department, or organization lives in a "four-room apartment" (see Fig. 15-3).

While living our lives as individuals or organizations, we move from room to room, depending on our feelings, perceptions, and aspirations, which are influenced by the internal and external environments. As the organizations (or individuals) grow, they change rooms. The rooms represent different cyclical phases of development. Every person or organization needs continuous *Renewal*, but the only way to get there is through *Denial* and *Confusion*. These are the two rooms where we mobilize and force our energy to renew. Let's make a cycle by going through all four rooms and seeing what is going on in each of them.

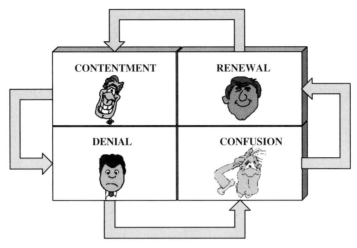

Figure 15-3 Janssen's Four-Room Apartment
Source: Based on Claes Janssen's Theory

What Is Going on in Each Room?

In *Contentment*, we are in a state of satisfaction. We like things as they are. From an organizational perspective, in this stage we are not as concerned about innovation or reengineering as we are about maintaining the status quo. From an individual perspective, we like our job, our income, and the working conditions. In other words, we feel calm and able to do the work and live the life we have chosen. People in an organization that are in this phase usually want to be left alone. Attention is focused on the "here and now," and the "I'm okay, you're okay" attitude is at work.

But suddenly, something changes the external environment: a competitor comes up with a new product, and your organization's market share is at risk; your company announces a merger with another organization and you don't know what will happen; or a large reengineering effort takes place in your organization that may affect your job. Many things may change the status quo, and you (or the management of your organization) don't feel good anymore. You (or your organization) move into a crisis—that is, a turning point. It is important here to note that even though a crisis may seem to be caused by a sudden effect, actually the change that causes a crisis is usually slow and cumulative—it does not occur overnight. And because it develops gradually, we don't feel the crisis approaching. There are plenty of signals, but they don't get through to us.

In a crisis, we move to *Denial*, where we are perceived as unaware and afraid of change. It is difficult, or maybe even impossible, to realize how long we stay in *Denial* because as soon as we notice we are there, we move into *Confusion*. We fear change when we are in *Denial*. It is a place where we have no clear feelings, and what is happening every day feels empty and irrational. This

Contentment	Renewal
• Satisfied with the status quo • "We like it just as it is" • "Leave me alone" syndrome	• Getting things together • Self confidence • Desire to make things happen
Denial	Confusion
• No clear feelings • A sense of emptiness • Fear to change	• A sense of unreality • Something feels wrong • Not knowing what to do

Figure 15-4 What is Going on in the Four Rooms of Change

Source: Adapted with permission from Dr. Claes Janssen's website, *Quaternity*, http://www.move.to/quaternity[10]

forces us to move into the next apartment—*Confusion*. In *Confusion*, we feel that we don't know the right things to do or how to make things right. We have a sense of unreality. And while it is certainly no fun to be in *Confusion*, this is the time and place where we begin to analyze what is going on, make comparisons and do benchmarking. This causes us to open the door to another room and go into *Renewal*. Here we have a sense of "getting it all together;" we will be more confident, energetic, and creative, which creates a desire to make things happen. How long will we stay in the *Renewal* room? As Dr. Janssen wrote, "From there [*Renewal*], if one takes the consequences of whatever truths one has seen in this inspired frame of mind, making them real in action, one moves to another contentment, which is frequently richer, more effective, more fun than the old one. And happiness—earthly happiness, circumscribed by compromises, but nonetheless real, or personal effectiveness, if you prefer to call it that—is simply to have the doors open, so that one can move freely through the rooms."[9] Figure 15-4 is a condensed representation of what is going on in each room.

How to Keep the Doors Open

As we mentioned earlier, every time we start a new cycle through the four rooms, we come once again to *Contentment*, which makes us more knowledgeable, more effective, and richer. Today, as never before, the external environment forces organizations and individuals to move faster through the apartments. To do this, management needs to keep the doors open so that people can move freely from one room to another. This means making sure that people know the current state of the organization, where it wants to go, and how it is planning to get there. This also means that employees know their role in the organizational transformation; they know how the organizational change will affect them. All this will keep the doors open for greater opportunities.

In this regard, as Figure 15-5 suggests, three major elements are at work here: an understanding of the current reality, an inspired vision that will take us to the desired state, and an effective strategy to get there. These elements together will create a field between the current and desired states, which will pull towards the desired state. Experience has shown that today's knowledge workers don't like to be pushed. They feel more comfortable when they are in a field that motivates them to move freely toward the future state. And, as soon as the vision becomes a reality, they need help to develop a new desired state, which will maintain a constant creative tension.

The idea of seeing vision as a field came from Margaret J. Wheatley. In her book, *Leadership and the New Science*, she wrote, "If vision is a field, think about what we could do differently to create one. We would do our best to get it permeating through the entire organization so that we could take advantage of its formative properties. All employees, in any part of the company, who bumped up against that field, would be influenced by it. Their behavior could be shaped as a result of 'field meetings,' where their energy would link with the field's form to create behavior congruent with the organization's goals."[11]

Just as the strings of a musical instrument need tension to produce the right sounds, people need tension to be able to create great things. This, together with the proper "tuning," will keep the doors of the apartment open, and will help people and organizations move freely from *Contentment* through *Denial* and *Confusion* to *Renewal* (see Fig. 15-6).

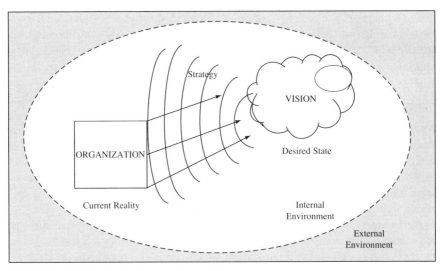

Figure 15-5 An Attractive Vision of What We Want to Achieve Will Create a Pull Towards Its Realization

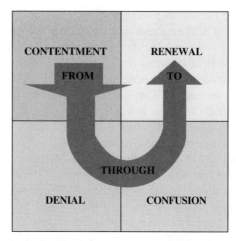

Figure 15-6 The Only Way to Renewal is Through Denial and Confusion

An organization or an individual, as any other living system, must grow to survive. As we grow, we must continuously move through the four rooms.

15.3.2 Three Phases of Change Management

As an organization grows and develops, it requires different approaches to change management. To describe these differences, we will use the S-shape curve that was used earlier in Chapter 6 when we talked about innovation. For this purpose, the S-shape curve will be divided into three parts, which will form three distinct phases, each with its own unique qualities and characteristics. Because of their peculiarities, each phase requires different management styles.

Phase 1: Entrepreneurial Management

When an organization is starting up, all the management efforts are concentrated on forming a structural pattern that will suit the organization's needs to survive and grow. This is an unpredictable period that requires a trial and error approach—a period when management applies their entrepreneurial knowledge and skills. At this stage, the management team firmly believes that they have a good idea that will fill a gap in the marketplace.

As Figure 15-7 suggests, this is a period of ups and downs, tension and frustration, and temporary successes that generate hope. This turbulent process goes on until the organization finally finds the repeatable pattern for success. This is the beginning of building a foundation for organizational growth.

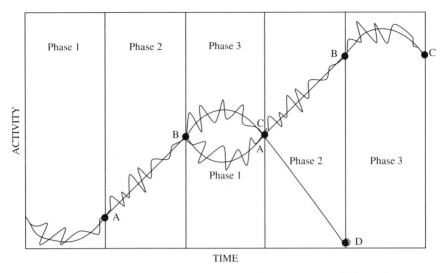

Figure 15-7 The Representation of Organizational Growth Through the S-Shape Curve

The superimposed thin line on the S-curve reflects the process of small and large changes within every phase. Studies show that these oscillations are not random, and with time they form a natural pattern.

There are many qualities that are imperative in Phase 1 that management needs to possess, some of which are reflected in Insert 15-2.

Without most of the qualities reflected in Insert 15-2, it would be difficult, if not impossible, for management to create a new organizational pattern.

In this start-up period, not all leaders of a start-up organization have long-term vision and are capable of seeing how their organization will look in ten years. Their main objective is to survive at least long enough to prove that the new organization and their products have something to offer. The long-term vision sometimes comes later. What is important, though, is to have a clear goal and the exceptional determination to achieve the goal. The organization must believe that the customer will value its products or services because they are either better or preferably better and different.

As we know, the number of start-up organizations that succeed in finding their organizational pattern of success is significantly smaller than the number of those who fail. There are different reasons for this—perhaps the new organization counted on a product that wasn't needed in the marketplace, or the management had a good invention, but didn't have the capability of developing a demand for it, or the management's entrepreneurial skills were not good enough. It is probably too much to expect every start-up organization to succeed, but the failure rate could be significantly reduced if these organizations would spend more time studying the market to discover the real needs for products or services. As studies show, this is one of the most frequent reasons for failure. Another

> **Insert 15-2**
>
> **Managerial Qualities Required in Phase 1**
>
> 1. Enterpreneurship and inventive creativity
> 2. A strong belief in what you are doing
> 3. A positive and winning attitude
> 4. The ability to create an environment of trust and sharing
> 5. The ability to take responsibility
> 6. A sense of the market's needs
> 7. The ability to cultivate good market contacts
> 8. The ability to motivate people
> 9. The ability to resolve conflicts
> 10. The ability to listen for signals from the external environment
> 11. The ability to experiment to find what succeeds and what will fail
> 12. Exceptional determination and commitment, as well as ideas and resources
> 13. Great flexibility and adaptability to meet the unpredictable circumstances that inevitably appear
> 14. The ability to take risks, act quickly, make effective decisions, and solve problems
> 15. The ability to recognize the importance of celebrating every small and great success
> 16. The ability to focus on success and not lose ground when some failure occurs

reason is that enterprises that actually have a unique product to offer fail because they give up too soon, for whatever the reason may be.

Finally, we want to emphasize one more time that management in Phase 1 is completely different from the classic forms of management. Here management has more to do with inventions, flexibility, commitment and quick decisions. These management qualities are imperative to succeed in the start-up period.

Phase 2: Incremental Improvement Management

If an organization survives in Phase 1 and finds its pattern, it moves into a new phase that has its own managerial requirements. Now, when the foundation has been laid and the main focus is on stabilizing the organization by repeating the pattern, trial-and-error and other managerial efforts that were essential earlier are replaced by managerial qualities that are needed to maintain stability and focus on organizational growth. At this stage, if the leaders continue to apply their entrepreneurial forms of management by inertia, it can lead the new organization into disaster. The management qualities required for in Phase 2 are those that will allow them to improve those things that work and screen out those things that don't fit the pattern that was developed earlier. Now the focus is on making quality products and shipping them to the customer on time. Customer satisfaction and expanding the market are now the major priorities. The opportunity for growth is absolutely different from that in the entrepreneurial period. Experimentation is replaced by policies that allow management to maintain organizational stability and order.

Just as a tree in its early period of development concentrates its energy mainly on growing roots and only later focuses on its environment, so does an organization in Phase 2 concentrate on its suppliers, product lines, marketing and sales. At this stage, management focuses on process and product improvement, customer-supplier relations, and other aspects required to achieve product uniformity, efficiency, and effectiveness. In short, this is a period for developing systems that will maintain uniformity and facilitate pattern repetition and expansion.

In Phase 2, the management's role is to establish limits that will allow the organization to concentrate its activities on the methods and forms that were already proven to be successful. This is a period of *growth through limitation*. By focusing the organization's energy on replication of the proven successful pattern, we can observe rapid growth (see Fig. 15-7).

Phase 2 is also a period of cultural formation. Gradually the organizational culture becomes strong enough to create barriers to anything that does not fit the already established pattern. "We don't do this in our organization" becomes a rule that is supported by policies and procedures.

All this, however, does not mean that creativity has no place in Phase 2. Here creativity is focused on a continuous incremental improvement of the existing processes and products. For example, introducing Statistical Process Control (SPC), Design Of Experiment (DOE), or Total Cycle Time (TCT) are activities that require a lot of creativity to improve the existing processes, products, and systems.

This phase also includes activities related to achieving quantum leap improvements; but again, these breakthrough improvements are related to enhancing the existing organizational pattern. For example, a significant change in technology can improve the yield of a process from, let's say, 80% to 99.9%. This could be considered a breakthrough that could bring great savings, but it is still within the limits of the established organizational pattern.

So, how long can a company stay in Phase 2? As we mentioned in Section 15.3.1 when describing the "Four-Room Apartment," if an organization is in the "Contentment Room," it is in a state of status quo. The organization enjoys its established state, which was achieved by repeating its systems pattern again and again. This is the same state as in Phase 2. If there were no need for risk taking, there would be no need to get out of the Comfort Zone. However, as the organization becomes larger, it reaches a point where its ability to grow just by using the established practices becomes exhausted. A need for change arises.

The achieved success by continually repeating the organization's established pattern introduces a change in the external environment that, in turn, also forces the organization to get out of the Comfort Zone. For example, a computer maker starts producing a high-speed computer. For awhile, the Organization enjoys a growth in market share, but this changes the environment. Competition and new customer demands forces the computer maker to come up with a new product that differs significantly from the previous one. In other words, in order to continue to grow, organizations must leave their Comfort Zone (Phase 2) when the lifetime of their products reaches the natural limit.

Phase 3: Breakthrough Management

When an organization moves from Phase 1 to Phase 2, it is an expected and desired breakpoint. It is a sign that the organization has passed the infant mortality stage. But a move from Phase 2 to Phase 3 is more difficult, and the organization may face much resistance. Who wants to get out of the Comfort Zone when a threat to survival is not obvious? If management observes enough symptoms that suggest that a radical change is needed, it is better to make changes at an earlier stage (see Fig. 15-7, point B) where there will be more time to introduce a change. If management waits until reaching the slope of the S-curve (see Fig. 15-7, point C), it may be too late, and the organization may go out of business. In change management, timing is a very important component of success. Insert 15-3 shows some symptoms that indicate the organization has entered Phase 3 and needs to prepare for change.

Insert 15-3

Some Symptoms that Suggest the Organization Is in Phase 3

- The competition is formally reporting about its readiness to introduce a more advanced product or technology
- The sales organization has a hard time holding market share and/or its market share is declining
- The company must reduce the price of its product in a healthy or normal market because of competitive pressure
- It becomes difficult to continue to reply on the same cost reduction programs
- New competitors have arisen in the global market
- There has been a decrease in productivity gains
- The overall economic health of the organization is declining

---

To overcome the resistance to change in Phase 3, management needs to create a strong sense of urgency, to make people see and feel the need for change. This can be done by sharing information with employees on what is going on in the internal and external environments; by involving employees in planning and materializing these plans; by demonstrating to employees the consequences of what may happen if a breakthrough change is not made. For example, when Jerry Sanders, the CEO of AMD, visualized that for the organization to survive and grow it needed to become a microprocessor company, he mobilized himself and his staff's energy to not only make it happen, but also to create a sense of urgency and convince employees that this was the right way to go. This allowed him to mobilize employees to move in a new direction. Today AMD is recognized worldwide as a competitive producer of microprocessors. AMD succeeded in introducing a breakthrough change because it started the change activities early in Point B, when there was still time to survive with an established organizational pattern.

It is important to note here that when a radical change is introduced, it creates a domino effect, which requires a lot of creative changes in Phase 3 (see Fig. 15-7). For example, the development and implementation of a microprocessor technology required a significant change in AMD's packaging process. Without the implementation of Flip-Chip packaging technology, producing high-speed microprocessors would be impossible. When a breakthrough change is produced, it automatically triggers the need for multiple incremental and creative changes to achieve growth in Phase 2.

As Figure 15-7 suggests, in Phase 3 we have a special situation where activities occur simultaneously in two directions: one is to maintain the existing business growth as long as possible. This is not an easy task because of the pending obsolescence of current products. More efforts are needed from the marketing and sales side to maintain the existing market share; more creativity is needed from the manufacturing and engineering side to reduce cost and create competitive, profitable products to attract customers; more organizational efforts are needed to offer attractive service and shorter delivery response. All this together can allow the organization to stay in business longer, but these are temporary measures.

The second simultaneous activity at this stage is to start the introduction of a new system or product that will take off just at a time when the old pattern is beginning to decline. This dual activity is often called *bifurcation*. This is why we see in Phase 3 that one of two lines is declining, which represents the old organizational pattern; the other is the renewal line and shows the rise of a new invention—a new breakthrough. Even though the two processes of change occur concurrently, they require completely different approaches in management. Here we need managers who know how to maintain the status quo, and managers with entrepreneurial skills. In other words, in Phase 3 we must perform partially as in Phase 1, and at the same time, partially as in Phase 2. At this stage, there is no one style of management. In different parts of the organization, different forms of management may be needed. It depends on the situation and the objectives of that particular division or department.

Final Thoughts

We mentioned earlier that the failure rate in Phase 1 is very high, and that this is understandable, but why do initiatives such as Total Quality Management (TQM) or Reengineering, which are usually undertaken by mature organizations, also have a high failure rate? Why does the same organizational transformation approach succeed for one organization and fail for another? There are many different reasons why major change initiatives fail, but we will limit ourselves to some of the reasons that we feel deserve special attention because of their general character.

One of the reasons is related to timing. Organizations that recognize the need for change at Point B (see Fig. 15-7) and act accordingly have more time and a greater chance to succeed. At this point, organizations are exposed to many signals from the environment, but somehow the signals don't always get through. Many organizations wait until the need for a breakthrough is very obvious—when hard data shows a reduction in market share, when there is a significant loss of revenue or profitability, etc. When an organization is undertaking a radical change at Point C, the probability of success is significantly less and the time for experimentation is limited. Making changes while on the downward slope of the S curve is dangerous. There is a high probability of going out of business (see Point D).

A second reason why organizations fail with the implementation of transformation change is related to culture. An established organization that is at the end of Phase 2 or in Phase 3 usually has a strong cultural pattern that had been formed for many years and served as a source of success. When a new change initiative is underway, management needs to make sure that the existing culture—the organization's mindset and its values and assumptions—match with the requirements of the new initiatives (see Fig. 15-8, a). If not, actions must be taken to reshape the organization's culture (see Fig. 15-8, b and c). The older an organization is, the thicker is its culture, and the more difficult it is to change.

To introduce a radical change, we often need significant capital investment, but we also need a culture that will support the new undertaking. Unfortunately, in reality, management does not pay a lot of attention to the cultural part of change, which is as important as having enough money to make the physical change.

Another reason for the failure in organizational transformation is related to the human side of change. This reason includes the cultural aspect we just described, but has a broader scope (see Fig. 15-9). For example, Enterprise Resource Planning (ERP) is a relatively new change initiative as compared with, let's say, TQM, but ist has become quite popular lately. AMD is in the process of implementing ERP throughout the company. Management

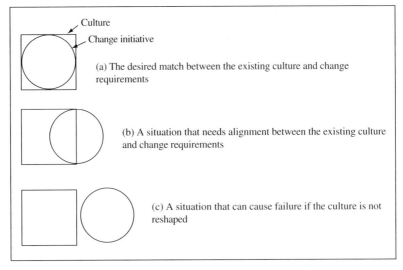

Figure 15-8 The Alignment Between Change and Culture

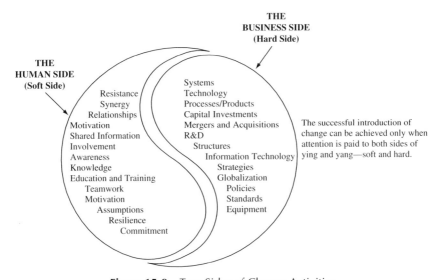

Figure 15-9 Two Sides of Change Activities

recognizes that the success of ERP implementation is strongly related to the issue of how seriously we take the soft side of the project. As Figure 15-10 suggests, problems that occur during ERP implementation are twice as likely to be on the human side than on the technical side. ERP, as is true of any other

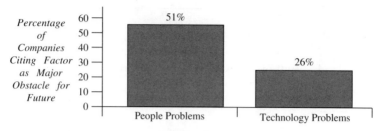

Figure 15-10 Obstacles for ERP Implementation
Source: Reprinted with permission from Deloitte Consulting[12]

great transformational initiative, has many potential capabilities, but to utilize them we need to pay more attention to the human factor.

Here we can use a violin as an analogy. A good violin costs a lot of money because of its great capability to produce music, but to materialize this capability we need a player who knows how to play it. The player creates the music, and the violin is just a tool. Management often expresses their disappointment when an implemented initiative does not meet their expectations. For example, to extract all the "music" from ERP that it can produce, we need knowledgeable people who are motivated to fully utilize its capability. This frequently means changing ingrained business processes.

Conclusion

To succeed with introducing radical change:

1. We need to find the right initiative—one that matches the organization's needs;
2. We need to maintain momentum and find the right time to begin undertaking the change;
3. We need to make sure that we have the right match between the organizational culture and the change initiative. If necessary, a change in culture may be needed;
4. We need to remember that change is not made by technology or by a new sophisticated machine, but by people. Recognizing the human side of change is very important.

15.3.3 Second-Order Change

First-order change—incremental change—deals with routine organizational activities and problems to maintain quality levels. For example, the application of statistical process control allows us to utilize the full capability of the process, but it will not serve as a tool to introduce significant changes. Second-order change—transformational change—is related to restructuring the process, challenging the organization's basic assumptions, and dealing with new and unknown elements in the environment. Second-order change requires invention and innovation, risk taking, and overcoming people's resistance to change. In Amir Levy's article "Second-Order Planned Change: Definition and Conceptualization," he suggests that second-order change involves the following four components.

1. The organizational paradigm, which is the propositions or underlying assumptions that unnoticeably shape perceptions, procedures, and behaviors. Levy calls these "the metarules—the rules of the rules."
2. The organizational mission and purpose.
3. The organizational culture, including beliefs, norms, and values.
4. The core processes, which include the organizational structure, management, throughput and decision-making processes, recognitions and rewards, and communication patterns.

Levy suggests that these four elements are interconnected and influence each other (see Fig. 15-11). A change in the organization's mental model will lead to a change in its mission, culture, and core processes.

This does not, however, mean that a change in the organization's core processes will necessarily influence and create a change in its culture, mission, and mental model. Levy's thought is very important because it helps us to better understand the relationship between the "hard" and "soft" stuff of an

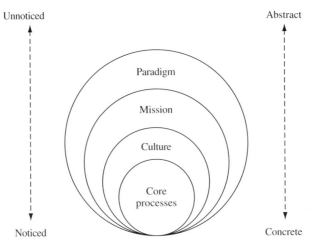

Figure 15-11 The Content of Second-Order Change

Reprinted from A. Levy, "Second-Order Planned Change: Definition and Conceptualization." *Organizational Dynamics* Summer 1986, p. 16 © 1986, with permission from Elsevier Science.[13]

organization. It suggests that a change in the organization's mind-set triggers a change in all other components of the organization.

The Stages of Change

Any time an organization goes through a transformational change, it goes through a life cycle that consists of four sequential stages. According to Levy (1986), these stages are:

1. *Decline or crisis* occurs when needs—either internal or external—are not appropriately met.
2. *Transformation* occurs when the need for change is accepted and commitment to the change is made.
3. *Transition* occurs when plans, ideas, and visions related to the change are translated into actions.
4. *Stabilization and development* occur when a second-order change has been "institutionalized, tuned up, maintained, and developed by first-order changes."

Levy further breaks these stages into smaller elements (see Fig. 15-12).

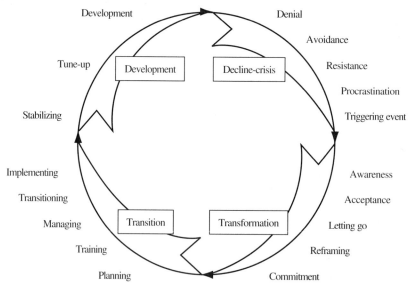

Figure 15-12 The Cycle of Second-Order Change

Source: Reprinted from A. Levy, "Second-Order Planned Change: Definition and Conceptualization," *Organizational Dynamics*, Summer 1986, p. 14, © 1986, with permission from Elsevier Science.[14]

You can see in Figure 15-12 that Levy considers the process of transformational change to be cyclical.

15.4 FORCE FIELD ANALYSIS

There are different methods related to change management. One well-known method that can be used to plan and manage change is Force Field Analysis. Dr. Lewin developed this special technique to help management manage change. He presented this concept for organizational analysis and change in his book, *Field Theory In Social Science* (1951). Since then, many authors and practitioners have used his concept.

At AMD, this method is frequently used when conducting workshops on change management. Participants are encouraged to analyze change by listing all forces inhibiting a desired change on one side of a line and all forces reinforcing the change on the other side. These forces are weighted according to their estimated strength of influence; and are indicated by assigning every force line a different length and thickness.

The progress of change depends on how much balance we have in the forces. Overcoming the areas of least resistance, rather than concentrating on the forces for change will make more progress toward change.

An Example of Applying the Force Field Analysis

To apply this technique, you begin by drawing a diagram similar to that in Figure 15-13. The dotted line represents the goal for the change effort. For example, you want to implement an integrated business system that can link all the processes together to provide services to the stakeholders of your enterprise (see Fig. 15-14).

The solid horizontal line represents the current state. Please note that this state needs to be described in as much detail as you can. The current situation is considered to be relatively static because the forces are in balance. For example, the current business system consists of small, separate modules. Even though it cannot totally satisfy the future needs of the growing business, people are used to it and feel comfortable with it.

The next step is to analyze the force field. The first set of forces are called ***driving forces***. These are forces that, if not opposed by the opposite forces, would press for change in the direction you desire. In other words, they are forces that operate in your favor. These forces exist in every situation, and to ignore them is to not take advantage of pressures already working to create the change you desire. These are forces that already exist in your system. What you need to do is just use them. For example, the driving forces of implementing an integrated business system in your company may be the growing complication of worldwide business, the demand of faster processing of business

310 ACHIEVING HIGH QUALITY THROUGH TRANSFORMATIONAL CHANGES

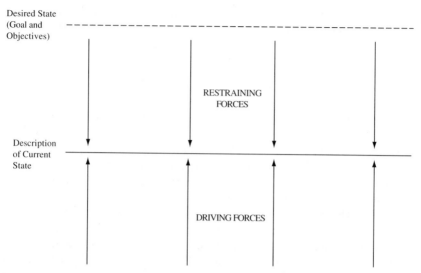

Figure 15-13 Outline From For Lewin's Force Field Analysis

Source: Based on *Field Theory In social Science* by Kurt Lewin, HarperCollins Publishers, Inc., 1951[15]

transactions, the need to have information to interact with customers, suppliers, and partners, etc. (see Fig. 15-14).

The opposing forces are usually named **restraining forces**. When you identify them you will find out that there may be just a few of them or there may be a large variety of them. This process is less helpful if the forces are listed in general terms; for example "the employees resist the new system." You can benefit from this process when the forces are stated in very specific terms, such as "employees are afraid of losing their jobs after the new system is implemented" (see Fig. 15-14). To come out with a specific set of restraining forces, we need to have knowledge of the process in question. We can use brainstorming techniques or conduct dialogues to develop a knowledge sharing process.

The analysis of these two sets of forces is by itself very useful, even if no other steps are undertaken. Once the current situation, the goal for change, driving forces, and restraining forces have been identified and labeled, three approaches to creating change can be considered.

1. We can consider increasing the driving forces (You must accept the new system if you want to work here). This is often the first thing that comes to mind, but this is usually the least effective. The reason for this is obvious: if you push people, they push back.
2. You can remove the restraining forces. For example, to overcome the restraining force of developing sophisticated software technology, you can carefully select a good software company that can work with you as a partner to implement the new business system.

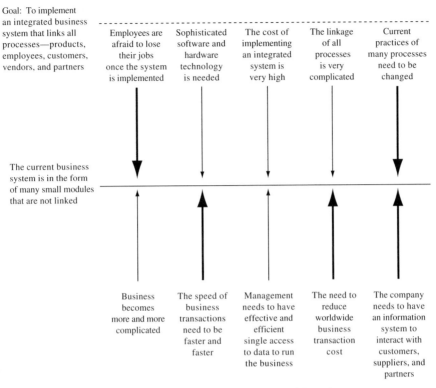

Figure 15-14 An Example of a Force Field Analysis

3. Another option is to not alter the existing forces, but reinterpret the situation in such a way that what was previously seen as restraining forces are now viewed (and work as) driving forces. To do this, you need more efforts, creativity, and innovation. But this action is more effective and can be successful. Here we need a lot of work related to elaborating why we need the change, how will it benefit the company and the individual, what will happen if we will maintain the status quo, etc.

There are some general principles that need to be considered when using this technique.

1. It is not recommended that this technique be used in changing individuals. It works best for groups and organizations.
2. Remember that the magnitude of all forces will not be the same. Forces are not balanced so much quantitatively as qualitatively. This should be individuated on the diagram by drawing more significant forces with longer or broader arrows (see Fig. 15-14).

3. The three change options we mentioned above do not have to be used one at a time. All three can be applied at once.

The purpose of describing this technique was just to give you a feeling for how the force field analysis can be applied to introduce change. To become more comfortable using these techniques, reading other material will help, but not much. What is needed is practice. As with any other technique, it can be properly understood in the process of application.

15.5 SOME STRATEGIES FOR CULTURAL CHANGE

All the quality initiatives that we described above more or less require a change in the organizational culture. At the same time, new strategies such as TQM, Reengineering, or Six Sigma, will themselves gradually shape or even totally change the existing culture. This is because people are influenced by strategic changes and will start to think, act, and behave differently, which is what culture is all about. Sometimes an organization may not even change the formally articulated set of core values, but the technological transformation gradually influences the existing culture and changes it. As time passes, the culture is transformed in such a way that it reflects the needs of the newly introduced strategy. In short, culture influences strategy and strategy influences culture. As we will show in this section, the implementation of a new strategy is also an action of shaping or changing the existing culture.

In writing this section, we were strongly influenced by the book *Strategies For Cultural Change*, written by Paul Bate, whose philosophy is very close to our way of thinking about culture.

15.5.1 The Meaning of Cultural Change

Before describing different types of cultural change strategies, it is important to have a common understanding of what we really mean by the term "cultural change." Some organizations have a tendency to think about cultural change mainly when something is going wrong in the organization. For example, when an enterprise starts to lose market share, we may say, "Something is wrong here—we need to change the culture." At this point, management starts to think about the organization's purpose, vision, mission and values, conducts surveys, and holds off-site meetings to think about cultural changes. However, when everything is running smoothly in the organization, management has a tendency to forget about the existence of culture and take it for granted—they want to preserve the culture. With this way of thinking, we treat culture as something that can be fixed—just as we can fix a machine by replacing a faulty component. From this perspective, culture is considered as a separate part of the organization that needs to be fixed once in a while so that the whole system will work.

The point we want to make is that a culture is not something that an organization *has*, but something that an organization *is*. Thinking in this way, we see the organization not as a machine, but as a society, and a society is a culture by itself. In other words, culture is not a component of the organization, but it *is* the organization. Culture is not a tool to improve the organization. Rather, culture is a way of thinking about the organization, a mental model for interpreting organizational life.

Somehow, managers used to think about organizational and cultural activities as separate issues. In practice, all organizational activities are cultural activities. Culture, in this context, is seeing the organization from a human perspective, from the perspective of their relationship. Paul Bate put it nicely in his book *Strategies For Cultural Change*, in which he wrote, "The culture perspective focuses on the 'human-ness' of organizations, regarding them as social rather than physical entities, made up of people talking, acting, interacting and transacting with each other. Hence the idea that culture exists not so much 'inside' or 'outside' people as 'between' people. This conception helps us to see a cultural change effort as a form of social intervention aimed at altering the quality of this 'between-ness' in some way or other."[16] If we as managers and leaders learn how to manage the between-ness of the organization more successfully, it would mean that we have learned how to work successfully with our customers, suppliers, subcontractors, and alliances. It would also mean that we have learned how to better manage the human side of the organization. All this should be considered as organizational change, and at the same time, cultural change. As Debra Meyerson and Joanne Martin put it, "Any change among and between individuals, among the patterns of connections and interpretations, *is* cultural change."[17]

15.5.2 The Substitutability of Strategy and Culture

In his article "The Significance of Corporate Culture," Karl E. Weick argues that culture and strategy are substitutable concepts. He offers an interesting way to prove this argument by using a quiz (see Fig. 15-15). The reader is asked to decide whether the first word in each statement should be considered as "strategy" or "culture." The trick here is that you will not be able to determine what is what. The answer is rather "either" (or "both") to all questions. Although all four statements were originally made on strategy, Weick remarks, "What I find striking is the plausibility of either term in each sentence. It is as if there were a common set of issues in organizations that some of us choose to call culture and others choose to call strategy."[18] Paul Bate supports the argument that culture and strategy are substitutable. He goes even further by saying, "To be clear here, I am not suggesting that culture is *like* strategy (and vice versa), nor am I saying that culture and strategy are closely *related* (as are Weick and others when they talk about the one being an 'outgrowth', 'offshoot' or 'dimension' of the other). What I am saying is that the one *is* the other: culture is a strategic phenomenon; strategy is a cultural phenomenon."[19]

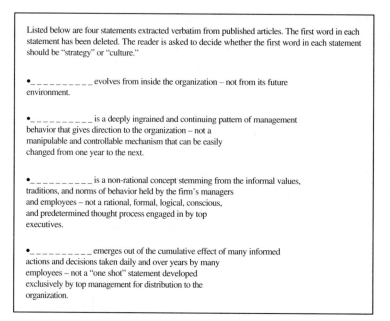

Figure 15-15 Weick's Culture Quiz

Source: From "The Significance of Corporate Culture," by Karl E. Weick in *Organizational Culture* (1985), Perter Frost, et. al., copyright © by Sage Publications Inc., p. 381. Reprinted by permission of Sage Publications, Inc.[20]

This broader definition makes culture a real source of a company's success; and has important implications for management. It means that cultural change programs cannot be successfully developed without considering change in other aspects of corporate life.

15.5.3 Methods and Forms for Cultural Change

If you are familiar with the general methodology of change, you will probably not see a big difference in the way we introduce cultural, strategic, technical, or any other organizational changes. However, there are some peculiarities that are captured in the book *Strategies For Culture Change*, by Paul Bate that deserve attention. Figure 15-16 shows four different approaches to cultural change that are based on Bate's methodology. In real life, none of these approaches is used separately. In different parts of the organization and in different situations, different approaches are used in combination. But there is one approach that we would single out for use in any situation, at any time or place. This is the *indoctrinate* approach, because it is based on familiarizing people with organization philosophy and other theories currently in use. We will give a short description of each of the four approaches but will pay more attention to the indoctrinate approach. This will allow us to provide a fuller picture of the methods and form of change that can be used concurrently with

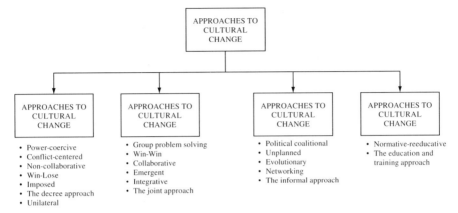

Figure 15-16 Four Generic Approaches to Cultural Change

Source: Based on *Strategies for Cultural Change* by Paul Bate, 1994, p. 168. Used by permission of Butterworth-Heinemann.[21]

the indoctrinate approach, if needed. All this may trigger thoughts on what approaches should be used for cultural change.

A point that should be emphasized here is that none of these approaches should be considered cultural-specific. In practice, there is little difference between the ways people try to change culture and the way they would typically go about changing structure, technology, operational systems, or any other aspects of their organization. However, the approaches described here may allow leaders to come up with a specific methodology that will be based on these four approaches and the specific needs of their organizations.

1. The Aggressive Approach

The aggressive approach might be described as cultural vandalism, a willful attack on the traditional values of the organization. This approach to cultural change fits in the general process of change, which is "unfreeze, "change," and "refreeze." To *unfreeze* is to destroy the conventional values. To achieve this, people are taught that their conventional values or ways of thinking are erroneous. Then to *change*, new correct values are popularized, and introduced by demotions, promotions, and firings.

Having dissolved the coherence of the existing culture, policies, memos, and letters are introduced to squeeze people into a new scheme with little regard for their feelings or preferences. In L. Helms' article "Tides Change at Seafirst," we read, "Management decides the firm must become market-driven so it sends out a memo to 6,000 employees saying 'starting tomorrow we are going to be market-driven.' And there it sits."[22] This form of cultural change certainly needs a lot of "policing" by requiring back-up information, auditing, etc. The application of the aggressive approach also creates a high turnover, which forms a large hole in the organization's "container" of knowledge and requires a lot of extra money to train and educate newcomers.

Considering all this, we could come to the clear and simple conclusion that this is not good. However, in reality, it is not as simple as that. Many examples demonstrate that despite its questionable efficiency, the *aggressive approach* is being used with increasing frequency in organizations. Today we can observe a lot of examples, which suggests a rejection of the concept of incremental and evolutionary change. The newly appointed chief executives don't have the time for incrementalism. They want it to be done now. How much of this approach is needed for cultural transformation depends on what leaders want to achieve and how much time they have to introduce the transformation. For example, organizations that apply the Reengineering concept, which is based on introducing radical changes in technology, also require a radical change in their value system. In their book *Business Process Reengineering*, Henry J. Johansson et al. wrote, "While Business Process Reengineering does not connote a specific set of values, the goals of BPR (breaking down functional barriers and thinking of business activities as processes) do force many companies to modify their values."[23] The effectiveness of such intrusion in the organization's culture is painful, but some organizations that introduce mergers and acquisitions, or any other radical transformations, are practicing the aggressive approach in cultural change.

2. The Conciliative Approach

The conciliative approach is appreciated by those who believe that cultural change can be successfully achieved by nondramatic, gradual, and routine means. The difference between this approach and the previous one is striking: Change is a quiet exercise; change occurs by degrees, as the new culture is slowly grafted on to the old, without any of the fanfare and razzmatazz of the previous approach. The conciliators possess a very practical theory for cultural change. One of its most interesting aspects is that it does not accept that there has to be a dialectical confrontation between different interests for change—even major change—to occur. Its belief is that the best way to achieve change is by a process of *convergence*, rather than divergence, by *conformity* rather than deviance, by *order* rather than rebellion. Mutuality is the key principle: You get your authority by not undermining other people's authority. Figure 15-17 is a summary of assumptions of the conciliative approach.

3. The Corrosive Approach

Leaders using the corrosive approach see cultural change as an essentially political process, whose main purpose is to effect a major change in the locus and distribution of power and authority within the corporate hierarchy. This approach recognizes the informal power more than the formal authority: It is not the visible organizational chart that is important, but the invisible network of the power structure.

> General Assumptions
>
> The best ways to achieve change is through:
> - Collaboration rather than confrontation.
> - Convergence rather than divergence.
> - Conformity rather than deviance.
> - Order rather than rebellion.
>
> Mutuality is the Key Principle: You get your authority by not undermining other people's authority.

Figure 15-17 A Summary of Assumptions of the Conciliative Approach

Source: Based on *Strategies For Cultural Change* by Paul Bate, 1994, p. 182. Used by permission of Butterworth-Heinemann.[24]

Corrosive and *conciliative* approaches differ in the way that pluralism is managed. Whereas conciliatives are prepared to be open and accommodating in outlook, corrosives tend to be covert and devious, skillfully manipulating relationships to achieve their ends. People who hold to this frame of reference see the changing world as booty to be grabbed and everyone within that world as a potential enemy rather than a brother. On the surface of this world, there is a peace and tranquility, but beneath the surface there is a good deal of commotion and turbulence. Despite its covert operations, there is in fact nothing secret or exclusive about the informal system in which the corrosives participate. What is important to note here is that membership of the informal system is not the privilege of the few, but a facility used by all. Employees involved in the informal network are not accidental. This is not to say that business and pleasure are not sometimes combined, but social activities are usually regarded primarily as a way of cementing and strengthening the work relationship. The business angle always imposes constraints on the social side of things. A degree of collaboration is possible, but full cooperation is not. Management needs to know in which parts of the organization this approach has a place, what the invisible network of power looks like, and how it influences the overall organization's culture.

4. The Indoctrinative Approach

From a relatively large variety of strategies for cultural change, we prefer the indoctrinative approach as a core strategy, with the use of some elements of other strategies if needed. This approach focuses on the concept of cultural change as a *learning process*. Learning, of course, is a very broad concept, and incidental learning can be said to occur wherever there are people and wherever there is interaction. However, what makes the idea of establishing a new culture through learning programs so different is that, as the term itself implies, learning is planned and programmed, not incidental. This strategy requires the design, development, and deployment of a comprehensive body of knowledge that is based on organizational psychology and other theories that reflect all parts of the culture. The educational and learning process is also based on the company's philosophy and needs to achieve a cultural change.

To shape or change the culture through a learning program, an initial group, let's call them cultural "explorers," may be formed. The main purpose of this group would be existential: to provide the participants with the opportunity to gain a deeper appreciation of the corporate culture and the manner in which it impacts them in their daily lives.

We think that a cultural program could be designed for 20 hours and spread over the year. Some of the disciplines could be delivered corporate wide, and the others could be given on a local basis. The program could be designed and led by corporate specialists, but lectures could be given from people within the local organization. The courses would contain two types of disciplines: one related to technical aspects derived from the organization's strategy and plans, the other to include the basics of the organization's new leadership cultural doctrine. It would be useful to select some important documents written by the organization's leaders in the last one or two years for the potential participants of the course to read in preparation for a discussion.

The leaders' participation in this course, with a delivery of some short case stories of "what is going on," will enhance the cultural program. This is to help the participants associate a face with the message. The main intent of the cultural education program is:

- To elaborate deeper on what culture is;
- To familiarize people with "what is going on" and "what is to be expected from the future";
- To give people the opportunity and help them to position themselves on the road to progress;
- To give people an opportunity to have a dialogue on what our values and principles are;
- To bring to the surface and challenge our assumptions;
- To challenge our professional ethics;
- To talk about changes;
- To think about loyalty, contribution, self-esteem, and other soft stuff.

The approaches to cultural change we have just described are not something from which you would pick the one you like the best. What type to use depends on what an organization is trying to achieve. Most of the time these approaches are used in different combination, and sometimes all of them are in operation in different areas of the organization. The main purpose of describing the four approaches of cultural change is to help you understand that there is no one way of shaping the culture.

References

1 Tom Peters, *The Tom Peters Seminar*, Vintage Books, a division of Random House, Inc., New York, NY, 1994, p. 7

SOME STRATEGIES FOR CULTURAL CHANGE   319

2   Russell L. Ackoff, *Ackoff's Best*, John Wiley & Sons, Inc., New York, NY, 1999, p. 3
3   Michael Hammer and James Champy, *Reengineering The Corporation*, HarperCollins Publishers, Inc., New York, NY, 1993, p. 49
4   Donald A. Schön, *Beyond The Stable State*, Random House, Inc., New York, NY, 1971
5   Tom Peters, *The Pursuit of WOW!*, Vintage Books, a division of Random House, Inc., New York, NY, 1994, p. 82
6   Ibid, pp. 81–82
7   Neil M. Glass, *Management Masterclass*, Nicholas Brealey Publishing, Ltd., Naperville, IL, 1996, p. 121
8   Ibid, p. 122
9   Claes Janssen, *The Four Rooms of* Change, Website Quaternity, http://www.move.to/quaternity
10 Ibid
11 Margret J. Wheatley, *Leadership and the New Science*, Berrett-Koehler Publishers, Inc., San Francisco, CA, 1992, p. 54
12 "ERP Second Wave," Deloitte Consulting
13 Amir Levy, "Second-Order Planned Change: Definition and Conceptualization," *Organizational Dynamics*, Summer 1986, American Management Association, New York, NY, pp. 15–16, with permission from Elsevier Science
14 Ibid, p. 14
15 Kurt Lewin, *Field Theory In Social Science*, HarperCollins Publishers Inc., New York, NY, 1951
16 Paul Bate, *Strategies for Cultural Change*, Butterworth-Heinemann, Ltd., Oxford, England, 1994, p. 15
17 Debra Meyerson and Joanne Martin, "Cultural Change: An Integration of Three Different Views," *Journal of Management Studies 24:6*, November 1987, p. 623
18 Karl E. Weick, "The Significance of Corporate Culture," in the book, *Organizational Culture*, edited by Frost, Moore, Louis, Lundberg and Martin, Sage Publications, Inc., Newbury Park, CA, 1985, pp. 381–382
19 Paul Bate, *Strategies for Cultural Change*, Butterworth-Heinemann, Ltd., Oxford, England, 1994, p. 19
20 Karl E. Weick, "The Significance of Corporate Culture," in the book, *Organizational Culture*, edited by Frost, Moore, Louis, Lundberg and Martin, Sage Publications, Inc., Newbury Park, CA, 1985, p. 381
21 Paul Bate, *Strategies for Cultural Change*, Butterworth-Heinemann, Ltd., Oxford, England, 1994, p. 168
22 L. Helms, "Tides Change at Seafirst," ABA Banking Journal, November 1988, pp. 69–74
23 Henry J. Johansson, Patrick McHugh, A. John Pendlebury, William A. Wheeler, III, *Business Process Reengineering*, John Wiley & Sons, New York, NY, 1993, p. 197
24 Paul Bate, *Strategies for Cultural Change*, Butterworth-Heinemann, Ltd., Oxford, England, 1994, p. 18

PART V

Reshaping the Organizational Culture

To introduce the subject of this section, we will start with an improvised story that could take place in any organization. Every year, usually in April, most companies have their "Career Day" where the employee's children are invited to visit their parent's workplace. This provides the children with the opportunity to know more about their parent's occupation and what an organization is all about.

The story we are about to tell took place in a fictional company called Mantron Corporation. This year the company was visited by significantly more kids than before, which is probably an indication that a new generation is growing up, and tomorrow they will take our place in the workforce. The kids moved in groups from one company building to another, where different surprises and new information were waiting for them. In one of these buildings, a dialogue between an executive and the visitors took place. At the end of his short speech to the visitors, the executive said, "Our company is a great place to work, and when you grow up, you are very welcome to come and work with us. Are there any questions?" After a minute or so, the silence was interrupted by a girl about 12 years old, "What do you mean by saying, 'Our company is a great place to work?' What makes the Mantron Corporation greater than any other company? What would entice me to come and work here other than the fact that my Mom has worked here for 22 years? In other words, what makes this company different?'" Again there was silence. All the kids looked at the executive, who had not been expecting a multiple question like this. "Well," the executive said, "our culture is different. We have a great culture here." "What is a great culture?" asked a boy sitting in the corner of the room. This time, the executive smiled and picked up a booklet that was on the podium. He said, "You will find our company purpose, vision, mission, and values in this brochure. We call this the PVM&V booklet. We describe our culture here. When you go to the lobby, please pay attention to the poster on the left-hand wall. It illustrates our five core values, which are the glue that keeps our company together." There were no more questions, and when the kids went through the lobby they all studied the poster on the wall, but did not seem very excited. One child turned to another and said, "Where is the glue he was talking about? I don't see it."

Really, where is the 'glue' that holds an organization together? How do you describe an organizational culture? What if the executive had read one of the best definitions of culture to the young visitors? Would that have excited them? Can we show the culture of our organization? Is this something you can see, hear, or touch? Where does culture come from? Can we change it? Should we change it? We will try to answer these questions in this part.

Chapter 16

The System of the Organizational Culture

16.1 ORGANIZATIONAL CULTURE: WHAT IS IT?

Usually when a hot discipline erupts into the business market, a number of definitions will arise; and the larger the number of definitions, the more confused we become. One example of this is the term "organizational culture." What is it? We all know it and can feel it, but when it comes to defining it, we have some difficulties.

AMD held a series of workshops on the subject of organizational culture to find out how much the employees knew about culture. We wanted to use this information to design an effective educational program. One of the exercises during these workshops was to define culture: What is it? The walls of the conference room were covered with paper, and people were asked to write down their opinions. Table 16-1 is an extract of what was written on the walls.

All these descriptions somehow reflect or are related to the meaning of culture, but we need a definition that will help us grasp the whole, broad meaning of culture. Edgar H. Schein gave one such definition that reflects our way of thinking. In his book *Organizational Culture and Leadership,* he wrote that the culture of a group could be defined as:

> "a pattern of shared basic assumptions that the group learned as it solved its problems of external adaptation and internal integration, that has worked well enough to be considered valid and, therefore, to be taught to new members as the correct way to perceive, think, and feel in relation to those problems."[1]

In our opinion, this definition is more attractive and complete because it connects culture to group learning and to problem solving. In short, it reflects the actual life of an organization. However, we felt that we needed to develop

323

Table 16-1 Definitions of Culture (From Dialogues)

1	Core values
2	A set of means to achieve the desired results
3	A value system
4	The way things get done here
5	A way to excellence
6	A set of behavioral boundaries
7	Norms of behavior
8	A road map for action
9	A way of thinking and doing
10	A way of life
11	A behavior that has a history
12	The way we communicate with each other
13	A collective way of working
14	A way of shaping the organization
15	A system of meanings
16	The way we work and live here
17	A philosophy
18	Guiding beliefs
19	Directions for performance
20	The way we and the whole organization work

our own working definition—one that would fit more closely with our way of cultural thinking. We came up with the following definition:

> An organizational culture can be described as an open system of shared values and behavior that come from the organization's experience and are proven to work well enough to be accepted as the correct way of doing business.

We used the term "open system" to emphasize that an organizational culture is connected to and influenced by the external environment and that, because of this, it is dynamic and subject to change. By using the term "shared values" we mean not only the core values, but also the operational values (this concept will be elaborated on in Chapter 18). Together with norms of behavior, they form the organizational culture. This definition contains the meaning that all shared values come from internal learning, not from outside. It also means that shared values must be proven by time. In the following section, you will find a description of the organizational culture system.

16.2 THE STRUCTURE OF THE ORGANIZATIONAL CULTURE SYSTEM

As Figure 16-1 suggests, an organizational culture can be represented as an open system, which is strongly influenced by its external business

Figure 16-1 The System of the Organizational Culture

environment.* At the center of the system are the purpose, vision, mission, and goals, which work together as an inspiring and focusing mechanism for the organization. The shared core values are the "glue" that holds the organization together. These values are tacit, invisible, and usually taken for granted by the people who work in this organization. They are taught to newcomers as the correct way of cultural thinking. In an organization with a strong culture, if employees do not act according to the core values, they are considered to have deviant behavior and have only two options, either adapt themselves to the internal culture, or leave the organization. The management's obligation is to create a fast-learning environment for newcomers and help them adapt to the organizational culture.

The most robust part of the whole system is *shared core values*, which do not change easily under the influence of the external and internal environments. However, if after time the core values need to be corrected, it is the primary responsibility of the leaders of the organization to introduce these changes.

The next layer surrounding the core values refers to the *shared operational values*, which reflect and are strongly aligned to the core values. Operational values are related to the professional and other specific peculiarities of the

---

*By external environment, we mean all economic, technological, competitive, political, and other forces that may impact the organization's actions.[2]

workplace. This set of values is less robust than the core values and can be shaped with time according to the internal and external environments.

On the same layer that surrounds the core values, we can see a special type of value called *espoused values*. Usually these values are not based on the organization's experience but rather are brought in from outside as a result of following a modern wave of management. Because of this, a discrepancy between what people say they value and what they actually value may occur. However, if the espoused values are reasonably congruent with the organization's culture, they may gradually be transformed to operational values or even become core values. This transformation depends on the organization's learning and experience, which is ongoing in relation to the espoused values.

The next layer is related to *behavioral standards and business ethics*. Again, they are a strong reflection of both core and operational values. This is the part of the system in which culture becomes more visible. We cannot see or measure values, but we can easily see how people behave according to these values. The ethical part of behavior tells us what is right and wrong in relation to the established values in the organization.

And, finally, the peripheral layer is the most visible part of the system. These are the *symbols and symbolic actions*, which consist of histories, heroes, rites, rituals, and myths, etc. These artifacts are an important part of the cultural pattern of the organization and can motivate people to work according to the value system. As soon as you walk into the lobby of an organization, you will begin to see all these artifacts. However, even though they are easily observable on the surface, they are very difficult to decipher—to determine the real culture of the organization just on the basis of artifacts.

The dashed lines of the circles symbolize that there is not an actual separation of layers. The values can move from one layer to another, which means that values influence behavior and behavior keeps values alive. All elements of the system interact with each other and have a very complex interrelationship. Therefore, the real culture is not so much reflected in the elements of the system as it is reflected in the interaction of these elements. This is why some organizations can articulate the same values but still not have a strong culture and prosper. The difference lies in the form of interaction, and the form depends on the people who work in the organization. This is why we say that people interacting with each other create a culture that is not transferable, and this differentiates one organization from another.

All the elements described above are interconnected and interrelated and work together as a whole system, which is subject to influence from the external environment. This means that for a culture to survive, it needs to be dynamic and capable of adapting itself to the external and internal environments.

The following chapters of this part will be dedicated to a detailed description of all the elements included in the system of the organizational culture.

References

1. Edgar H. Shein, *Organization Culture and Leadership*, Jossey-Bass Publishers, San Francisco, CA, 1992, p. 12
2. Arie DeGeus, *The Living Company*, Nicholas Brealey Publishing Limited, London, UK, 1997, p. 35

Chapter 17

Managing the Core of the Organizational System

Just as the health of a human being depends on the strength and relationship of all the body's organs, so does the health of an organization depend on the strength and relationship of all its components: purpose, vision, mission, values, goals, strategies, objectives, etc. All these components work together, complement each other, form the core of the organization's strength, spirit, and keep it alive. As in the human body, if any of the components are not developed properly or do not work as needed, it will create a problem for the whole body of the organization. In this and the following chapters, we will describe the core components of an organizational system (see Figure 16-1) and will show the relationships between them and with other components of the system.

"Values, Vision? Can't live without 'em in today's world, which won't tolerate those hefty manuals.
Can't live with 'em either."[1]

Tom Peters

17.1 THE POWER OF VISION

Most of us have envisioned things since we were children. As a student you probably dreamed about what your career would be after you graduated. After receiving an entry-level position in an organization, you probably envisioned becoming a manager or a senior executive. When you were dating, you may have envisioned building a family, etc. All our lives we dream about the future. The power of our visions always moves us forward and makes our lives more interesting and better. Jerry Sanders, chairman and CEO of AMD, remarked that, "Great things happen when people dare to dream."[2] Walt Disney also stated that, "If you can dream it, you can do it."[3] These statements can easily be applied to both personal and organizational vision because they have the same roots. In both cases, we need imagination, we need to know how to observe and use external information (what is going on around us), we need to use our gut feelings, we need to have entrepreneurial skills, and we need to be willing to take rational risks. The only major difference between personal and organizational vision is that in addition to the personal qualities mentioned above, the leader of an organization needs to have special skills and the knowledge to encourage the employees in the organization to follow him or her on the journey toward the new vision and share in the dream—this talent is necessary to become a successful leader.

A classic example of a great vision that is very popular is Henry Ford's vision of a widely affordable automobile. He said, "I will build a motor car for the great multitude.... It will be so low in price that no man making a good salary will be unable to own one—and enjoy with his family the blessing of hours of pleasure in God's open spaces.... The horse will have disappeared from our highways, the automobile will be taken for granted...."[4] Try to penetrate the meaning of this powerful statement. It inspires, it is concrete and easy to understand, and it concerns all the American people. Did Ford know at this time how he would make it happen? Certainly not, but he had a strong belief in his vision. He found a way to articulate it properly and make it a shared vision for all the people involved in the process of making the vision a reality.

A more recent vision that resembles Henry Ford's vision is the vision of Microsoft Chairman/CEO, Bill Gates, "A computer on every desk and in every home, all running Microsoft software."[5] Again, it is clear and catches the attention of a large population of people.

The most powerful and compelling vision, which impressed the whole world, is John F. Kennedy's vision of landing a man on the moon: "... achieving the goal, before this decade is out, of landing a man on the moon and returning him safely to the earth."[6] To understand the value of this vision for the American people and be able to apply the vision principles derived from it, we developed a case study that you will find at the end of this chapter.

AMD's vision statement indicates that the company sees itself as a leading supplier of critical enabling technology for the information age and an organization that is empowering people everywhere to "lead more productive lives" (see Fig. 17-1).

These examples of great vision statements are all different, but they have one major thing in common: they are inspiring, they touch the hearts and minds of people, they are shared, and because of this they are powerful.

Today, the importance of vision as a component of organizational success is globally recognized by many leaders in a survey reported by Korn/Ferry International, where CEOs from 20 different countries were asked to describe the key traits or talents important for a CEO in the year 2000. Ninety-eight

> **AMD'S VISION STATEMENT**
>
> We at AMD share a vision of a world that is enhanced through information technology, which liberates the human mind and spirit.
>
> AMD is a leading supplier of critical enabling technology for the Information Age. In concert with our customers, we empower people everywhere to leadmore productive lives by creating, processing, and communicating information andknowledge. We are our customers' favorite integrated circuit supplier.
>
> With a strong commitment to our core values and mission, we anticipate and respond quickly to changing customer needs while preserving a culture that brings out the best in each of us.

Figure 17-1 AMD's Vision Statement

percent of respondents described that a "strong sense of vision" is the most important trait and talent for a CEO."[7] This, however, does not mean that a vision should always come only from the CEO.

The originator of the vision may be the leader of the organization, but it may also be a team from the organization. In other words, a vision may also be a product of collaborative work. In any case, to be accepted in the organization, the vision must be shared rather than imposed from the top. AMD's vision was truly shared with all groups and divisions and every employee because it was articulated by a large team of representatives from all levels of the corporate organizational structure and then communicated to all employees through special dialogues in which the content of the vision was discussed.

17.2 VISION AND TRANSFORMATIONAL LEADERSHIP

There are many qualities we usually expect leaders to have: we want them to be honest, inspiring, competent and certainly forward-looking. We demand that our leaders have a vision for the future. In particular, when an organization is in crisis, we want a leader that can provide a new vision that permits the employees to see a way out of the difficult situation. But in reality, this does not always happen. Sometimes a new leader, who is brought in to turn the organization around, starts not with a vision of the future, but with a strategic plan to fix the present. The leader needs time to better understand the specifics of the organization, assess the existing environment, determine the organization's core competencies, and then create a vision. In other words, sometimes the leader needs to make sure that he knows the current reality so that he can formulate the desired future state.

In his book, *Organizational Culture and Leadership*, Edgar H. Schein promulgated two absolutely different leadership models for transformational leadership. One is known as the *Strong Vision Model*, and the other as the *Fuzzy Vision Model*.[8]

In the *Strong Vision Model*, the leader has a pretty clear vision of how the organization will look after the transformation and in the long-range future. The leader clearly sees how to make the vision a reality. He articulates the vision to encourage people and rewards those who actively participate in its realization. This model is used when (1) a visionary leader who is familiar with the strengths and weaknesses of the organization is available, and (2) the future of the organization is predictable and the leader has a strong sense of its future. The leader articulates a strategic vision that is in sync with the internal and external environments—a vision that will inspire people to achieve the necessary transformation. If either of the two conditions mentioned above are not available, the second, *Fuzzy Vision Model*, is more appropriate.

The *Fuzzy Vision Model* is usually applied when (1) the new leader recognizes that the organization is in crisis and fast measures are needed to save it, and (2) the new leader is not very familiar with the strengths and weaknesses of the organization, and he needs some time to better estimate the future pattern of the organization. In this case, instead of articulating a long-term strategic vision, the leader focuses on introducing effective short-term measures to save the company. In this situation, the articulation of a strategic vision will take place only when the organization is out of crisis and back under control. You can see an example of the application of the *Fuzzy Vision Model* in Chapter 20, where we describe IBM's story of cultural development.

17.3 PURPOSE—THE GUIDING STAR OF AN ORGANIZATION

Over time, organizations, like people, develop their personality and character. They shape their ideology and culture, which has an influence on the organization's behavior. To live and grow, organizations need a purpose or mission—a reason for being. A purpose that will shape the organization's identity is the first step in forming an enterprise.

A purpose statement is a source of direction, a special instrument that allows stockholders, employees, customers, and other stakeholders to know what the organization stands for and where it is directed. A properly formulated and well-articulated purpose statement gives the organization a strong sense of identity and gives the employees a good reason to come to work every day—a reason that is stronger than just making money.

Articulating the purpose statement becomes even more important in the period of globalization. It serves as a touchstone and provides a sense of importance. The purpose is the foundation of the organization's future.

17.3.1 The Purpose of an Organization

What is the main purpose of an organization? For many years, philosophers, economists, and practitioners have been providing different answers to this question. Today, even though there are still different opinions on this subject, there is a tendency toward forming a common understanding of the role of a contemporary organization. Looking back in time, we can see that the meaning of purpose has gone through a number of transformations.

Before the creation of the limited company, the purpose of work was closely tied to religion. People were seen to be working to maximize their contribution to the world. During the Industrial Revolution, the religious connection with the world became less relevant and was replaced by work as a social duty. The purpose of organizations was to produce products for the society. Gradually, the purpose of organizations has become unclear. For many, work has become a way of creating wealth; for others, work has become a creative activity or a way of achieving success. In general, work has become a much more selfish activity connected with personal gratification. In short, in the 1970s, the role of business in society became totally unclear. There was a need to clarify the role of business, and this generated the publication of articles and books on this subject.

In his book *Capitalism and Freedom*, Milton Friedman railed against the mood in business for what was called in the 1970s, "corporate social responsibility," which means that businesses needed to do charitable work for the benefit of the community. He argued that the purpose of business is to create wealth, which is best done by maximizing profits. In this regard, he wrote, "...There is one and only one social responsibility of business—to use its resources and engage in activities designed to increase its profits so long as it stays within the rules of the game, which is to say, engages in open and free competition, without deception or fraud."[9] Under Friedman's definition of purpose, it is assumed that the maximizing of profit in the capitalist system will result in a greater contribution to the overall system. Today this view of purpose—maximizing profits—is still shared by many organizations. However, there has been a shift toward a larger meaning of why an organization exists. In their book *Built To Last*, James C. Collins and Jerry I. Porras defined purpose as "the organization's fundamental reasons for existence beyond just making money—a perpetual guiding star on the horizon; not to be confused with specific goals or business strategies."[10] This definition became popular in many organizations, especially in the high-technology industry. For example, Advanced Micro Devices (AMD) doesn't exist just to make money by producing and selling semiconductors; its purpose is "to empower people everywhere to lead more productive lives." A sense of purpose beyond just making money has the capability of guiding and inspiring people throughout the organization and remaining relatively fixed for a long period of time. However, while recognizing the importance of articulating the organization's purpose, it should not be forgotten that a purpose statement does not work by itself.

The purpose should be used as part of the operational system, which combines purpose with the organization's vision, mission, values, strategies, and other elements. In this section we will provide some examples of how purpose works in relationship with other instrumental elements.

17.3.2 Purpose and Vision

In one of our AMD seminars, a manager brought up some questions. He said, "We have goals, and we have a vision. Isn't this enough to inspire people to do their job? Why do we also need a purpose?" What is the difference between vision and purpose?

In his book *The Fifth Discipline*, Peter M. Senge answers these questions. He said, "Purpose is similar to a direction, a general heading. Vision is a specific destination, a picture of a desired future. Purpose is abstract. Vision is concrete."[11] As we mentioned above, the classic example of a concrete vision is John F. Kennedy's "man on the moon" vision. It was concrete because it had an end destination, but at the same time it was a "fuzzy" picture because at that time no one knew how it was to be done. This vision generated a lot of enthusiasm and energy to pool together all the country's available resources and make the vision a reality. But this vision would not have taken place if the United States had not had the purpose—the direction—to maintain superiority on the ground and in space to serve as the peacemaker of the world.

As you can see from this example, the vision had an end and to maintain the purpose, a new vision should be articulated, but a properly defined purpose has a longer life span, possibly forever. This example illustrates that nothing happens until there is a vision—a picture of the future with a specific destination. It also illustrates that a vision with no underlying sense of purpose is just a good idea, which loses its calling, its creative tension, as soon as the vision is accomplished.

A purpose allows an organization to maintain a constant creative tension. In contrast, a vision loses its inspirational power when it comes close to its realization. For example, AMD's purpose is "We empower people everywhere to lead more productive lives." To live its purpose, AMD needs continuous growth. Years ago, when Jerry Sanders visualized AMD as a ten billion dollar company, it was just a desired picture of the future. Today, this is still a challenge that builds creative tension for AMD, but as time goes by and the company approaches its vision, the tension will be released. However, the purpose will remain, and this will require a new, challenging vision to create new tension. So visions are created and achieved in accordance with the main purpose of the organization. In other words, a purpose without a vision makes no sense. The vision creates the tension and empowers people to make the vision a reality. The purpose serves as a guiding star and shows the direction in which to go. They work together to serve the purpose (see Fig. 17-2).

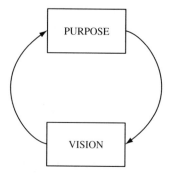

The vision cannot be understood in isolation from the idea of purpose, and the purpose makes no sense without a vision.

Figure 17-2 The Relationship Between Purpose and Vision

17.3.3 Purpose and Core Values

According to James C. Collins and Jerry I. Porras, purpose together with core values form a core ideology that should not change over time if an organization wants to be a visionary company. In *Built To Last*, they wrote, "Over time, cultural norms must change; strategy must change; product lines must change; goals must change;... ultimately the *only* thing a company should *not* change over time is its core ideology—that is, if it wants to be a visionary company."[12] They define visionary companies as institutions that in most cases not only are successful but also "are the best of the best in their industries, and have been that way for decades."[13] General Electric, Hewlett-Packard, IBM, Motorola, and Sony are examples of the long list of visionary companies. Not only does this type of company have great culture, but something much stronger is at work, which Collins and Porras call "cultism," or "cultlike"—a set of characteristics that play a key role in presenting the core ideology. As the authors of *Built To Last* suggest, in most visionary organizations with a cultlike atmosphere, we can observe that these organizations:

- Have *stronger indoctrination* into a core ideology through the history of the company
- Show *greater tightness of fit*—people tend to either fit well with the company and its ideology or not to fit at all ("buy in or get out")
- Show *greater elitism* (a sense of belonging to something special and superior)

One of the major tasks of management in developing a visionary organization is to make the purpose and core values work together and, based on this, develop and enhance the organization's core ideology.

17.3.4 Purpose from a System Perspective

As we have already mentioned, ordinary machines serve the purposes of other, but have no purpose of their own. A computer system, for example, serves the user's purposes by doing complicated computations, communicating through the Internet, and many other things, but the computer has no purpose of its own. This is why thinking of an organization as a machine for making money would suggest that the whole organization is a machine that only serves the purpose of the stakeholders who own the machine. Therefore, using a machine as a metaphor of an organization does not conform to the requirements of a contemporary organization. Thinking of an organization as an organism would get you closer to the meaning of organization because unlike a machine, any organism has a purpose of its own—survival. But this is still not a complete image of a contemporary enterprise, because the parts of an organism don't have a purpose of their own. For example, the stomach, brain, heart, lungs, etc. serve the purpose of a human being but have no purpose of their own.

As we know, no metaphor will ever fully reflect the subject in question. This is why when we say that an organization is a living organism, we need to remember that metaphorical expression has its limitations. A real organization has different levels of purpose. In his book *Creating the Corporate Future*,[14] Russell L. Ackoff suggests that when we focus on organization we are concerned with three levels of purpose: the purpose of the system, of its parts, and of the system of which it is a part, the suprasystem (see Fig. 17-3). For example,

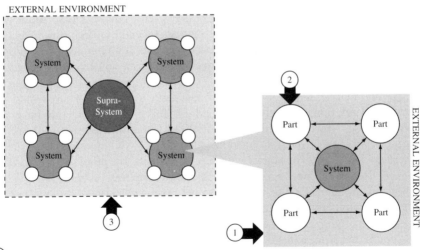

① The purpose of the system.
② The purpose of the system's parts.
③ The purpose of the supra-system of which the system is a part.

Figure 17-3 The Three Levels of Purpose

if we take a division of a corporation, this system has a purpose of its own, which is aligned with the purpose of the corporation (the suprasystem) of which it is a part.

At the same time, the organizations that are a part of a division also have their own purpose. This is to say that every element of an organization has a purpose of its own, which is aligned with the purpose of the higher system of which it is a part.

Seeing the purpose of an organization in this way allows people to see their place and role in serving the main purpose. This, however, does not mean that a corporation should have a series of articulated purposes on all levels of the organization. What it does mean is that everybody should be involved in a continuous dialogue of how to serve the organization's purposes. The organizational purpose should be a living thing that requires a continuous discussion and sharing of thoughts on how to live the purpose as best as possible. Limiting ourselves only to the formally articulated purpose statement and putting plaques in the entrance of every building of the organization will not help to make the purpose an important part of the organization's ideology.

17.3.5 Creating Partnership Through the Exchange of Purpose

Peter Block sees the exchange of purpose as an important component in building partnerships. In his book *Stewardship*, he writes, "Each party has to struggle with defining purpose, and then engage in dialogue with others about what we are trying to create."[15] We share this thought. Every organization needs to engage in partnerships with other institutions to materialize their purpose. Block explains, "Let people at every level communicate about what they want to create, with each person having to make a declaration. Let the dialogue be the outcome. The same process holds for relationships with customers, suppliers, and other stakeholders. Each has a voice in discussing what they want the institution to become."[16] Can you imagine the power of purpose that would be released by using the process described above? There would be a network of relationships between individuals and organizations that have their own purpose, which can only be materialized in relationship with others. For example, AMD sees its main purpose of being as "We empower people everywhere to lead more productive lives." To live its purpose, AMD needs to engage itself in a network with customers, suppliers, subcontractors, and other stakeholders. This will allow it to get products to the end users that will help them have more productive lives. This reasoning can be applied to any level of the organization, including the individual. Young engineers, who have just graduated from college, need the company as much as the company needs them to achieve its purpose. Peter Block suggests that the purpose being defined through dialogue and building relationships between customers, suppliers, and other stakeholders is more attractive and powerful than just the practice of formulating the purpose statement at the top and then enrolling people in a discussion to support and share the organization's purpose.

17.3.6 The Personal Purpose

So far we have emphasized how important it is for an organization to have an articulated purpose statement, but what about a written personal purpose statement? While employees share the organization's purpose, they have, at the same time, their own purpose, which is aligned in part with the organization's purpose. A personal purpose does not have to be formally written, but as Richard J. Leider suggested in his book *The Leader of the Future*, "a written personal purpose statement reduces anxiety in times of change."[17] He wrote, "Although some leaders do have an overwhelming sense of purpose, many don't. Nevertheless, it's important to continually ask the big question, 'Why do I get up in the morning?' How would you answer that question for today?"[18] You probably know from your experience that verbalizing your thoughts and assumptions allows you to think more clearly and to make your implicit assumptions explicit. It allows you to focus your thoughts and to have a better understanding of who you are, where you are, and where you want to go.

17.3.7 Being True to Our Purpose

In one of our workshops at AMD on purpose, vision, mission and values (PVM and V), we initiated a discussion on elaborating the differences and commonalities between these concepts. We started the discussion with an example from AMD. For many years this company, and in particular its chairman and CEO Jerry Sanders, shared the vision of becoming a leader in the design, production, and global marketing of the highest-speed microprocessors in the world. Today this vision has become a reality. We asked the workshop participants a question: Why were we at AMD working so hard, and risking a lot of capital, effort and time? Many of the participants responded, "Because we want AMD to become more profitable." High profitability is an important goal, but we challenged the participants to think of a more intrinsic reason, and we asked, "Why choose high annual profits?" The responses can be boiled down to "to expand our company's capabilities, make it stronger, and keep AMD alive." This is also correct. We then asked, "Well then, why is it so important to keep the company alive?" One of the managers quickly responded, "Obviously, to have a place to work." We continued to ask a lot of "why?" questions, and all the responses were legitimate but could not completely satisfy all the participants. Almost at the end of the discussion, someone said, "To stick to the company's purpose, which is, 'We empower people everywhere to lead more productive lives.'" The participants supported this and expanded on this statement. Being true to the company's purpose has the greatest intrinsic significance to the people who work there. All the rest—profit, survival, etc.—are a means to the end. These means might change in particular circumstances, but the main purpose will remain.

Achieving the highest profits, lowest cost, and best quality and many other goals are very important components for success, but they are still secondary goals. The main and ultimate goal of an organization is being true to its purpose.

17.4 THE MISSION OF AN ORGANIZATION

In literature and in practice you will see that the terms "purpose, "vision," and "mission" are frequently used interchangeably when it comes to formulating the organization's reason for being. Some organizations articulate only mission statements, others articulate only purpose or vision statements, and still others use purpose, vision, and mission statements together to articulate a broader statement. In this section, we will elaborate on the difference between mission and purpose and how they complement each other. We will also discuss the relationship between mission and vision. At the end of this section, we will describe the requirements of a good mission statement and describe an example of how all the fundamental organizational principles such as vision, purpose, mission, and core values can be linked together.

17.4.1 The Mission Statement of a Contemporary Organization

Despite the existing differences in opinions about an organization's mission, it is possible to observe two major schools of thought. One view, which we'll call the strategic school, sees mission mainly as a concept that is linked to strategy, but at a higher level. In this context, it is assumed that mission defines strategy and strategy defines structure. Because of this, mission is perceived as the first step in strategic management, where management's role is to determine the organization's commercial, rational, and target markets. In the strategic school of thought, a mission exists to answer two fundamental questions: What is our business today and what should it be tomorrow?

The second view, which we'll call the philosophical school of thought, sees mission as the cultural "glue" that enables an organization to function as a collective unity. The glue is represented as the core values and beliefs that influence the organization's attitude and behavior. Here, mission is linked to the cultural aspect of the organization. A classic example of a cultural view of mission is IBM. For many years, the company described its mission in terms of distinct business philosophy, which was based on strong cultural beliefs. IBM stands for three things: Respect for the Individual, Dedication to Service, and A Quest for Superiority in All Things. These three things represent the belief system of IBM, which means that everything else could change, but these three things will not change. When describing IBM's beliefs in his book *A Business and Its Beliefs*, Thomas J. Watson, Jr. wrote, "I want to begin with what I think is the most important: Respect for the Individual. This is a simple concept, but in IBM it occupies a major portion of management time. We devote more effort to it than anything else."[19] A good illustration of how this concept was applied in IBM is described in Stephen R. Covey's book *The Seven Habits of Highly*

Effective People. In his story, Covey demonstrates how the organization's culture and its values work in real life (see Fig. 17-4).

Describing IBM's second belief, Thomas J. Watson, Jr. wrote, "Years ago we ran an ad that said simply and in bold type, 'IBM Means Service.' I have often thought it our very best ad. It stated clearly just exactly what we stand for. It also is a succinct expression of our second basic corporate belief. *We want to give the best customer service of any company in the world.*"[21] We see this as an excellent example of how mission can inspire people with simple and yet powerful words. It is a good example of how the content of the mission statement is used to popularize what the organization stands for. In today's environment, making and selling a good product is not enough. Serving and delighting the customer is an important mission for a contemporary organization.

Describing the third belief, Watson wrote, "The third IBM belief is really the force that makes the other two effective. *We believe that an organization should pursue all tasks with the idea that they can be accomplished in a superior fashion.* IBM expects and demands superior performance from its people in whatever they do."[22] IBM's philosophy was established by Thomas J. Watson, Sr. and followed by his son, Thomas J. Watson, Jr. for many years. We bring this to your attention to demonstrate the importance of the "soft stuff" of a mission statement. While recognizing the importance of the hard stuff—the strategy, technology and economic resources—that should be reflected in a mission statement, paying attention to the human side of mission will make the whole mission statement more powerful. In other words, developing a mission statement that is based on both the strategic and philosophical schools of thought will involve both the minds (structure) and the hearts (culture) of the employees.

Once I was training a group of people for IBM in New York. It was a small group, about twenty people, and one of them became ill. He called his wife inCalifornia, who expressed concern because his illness required special treatment. The IBM people responsible for the training session arranged to have him taken to an excellent hospital with medical specialists in the disease. But they could sense that his wife was uncertainand really wanted him home where their personal physician could handle the problem.

So they decided to get him home. Concerned about the time involved in driving him to the airport and waiting for a commercial plane, they brought in a helicopter, flew him to the airport, and hired a special plane just to take this man to California.

I don't know what costs that involved; my guess would be many thousands of dollars. But IBM believes in the dignity of the individual. That's what the company stands for. To those present, that experience represented its belief system and was no surprise. I was impressed.

Figure 17-4 Stephen R. Covey's Story About IBM

Source: Reprinted from Stephen R. Covey's book, *The Seven Habits of Highly Effective People*, Simon and Schuster, Ltd., 1989, p. 139–140[20]

Organizations today have started to recognize that our workforce has become more professional. The number of knowledge workers is growing rapidly, and this will require a significant change in management style. It is also time to recognize that our employees are emotional and sensitive to the way we treat them. Money is no longer the only factor that makes employees give their best effort to the company. It is time for management to learn more about how to win the employees' hearts. As Gary Hamel and C.K. Prahalad noted in their book *Competing For The Future*, "Many companies are beginning to realize that all their employees have brains. How many companies, we wonder, understand their employees have hearts as well?"[23] Formulating mission statements that include the hard and soft stuff are a requirement for today's organizations.

17.4.2 Mission and Purpose

For an organization to properly formulate its reason for being, it needs to have articulated purpose and mission statements that will complement each other. A purpose statement will state the reason for the organization's existence; it should be fundamental, broad, and inspirational; it should generate a feeling of the enterprise's importance in society, a feeling that if the organization ceased to exist, it would be a great loss for the world. The purpose statement does not need to be entirely unique. Two companies may have many of the same purposes as they may have the same beliefs. What is important is that the purpose be inspiring and generate within employees the willingness to work and contribute to the organization. What is more important here is not the uniqueness of the purpose, but the ability of the organization to live its purpose. For example, many organizations declare a purpose statement to contribute to the society via electronic equipment for the advancement of science and the welfare of humanity, but how many organizations have lived this purpose as completely and consistently as Hewlett-Packard?

If a purpose explains why an organization exists, an effective mission statement translates the organization's purpose into tangible, energizing, highly focused objectives and aspirations that draw the organization into the future. A properly articulated mission statement must create a structural tension that will stretch the organization's capabilities and bring it to a higher level of performance and success. In contrast with the purpose, the mission statement must strongly differentiate your organization from others—it must create the organization's personality. The mission statement must have a clearly established hurdle and time frame for achievement, whereas the purpose statement is relatively enduring.

One more element that differentiates purpose and mission is the longevity of these statements. Although a mission can change with time, the purpose remains forever. Walt Disney nicely described the never-completed nature of purpose by saying that "Disneyland will never be completed, as long as there is imagination left in the world."[24]

As you can see, even though the purpose and mission have a number of similarities that are related to the explanation of why an organization exists, these two concepts have different connotations, but they can work very well together and complement each other.

17.4.3 Mission and Vision

Compared with a vision statement, a mission statement is more concrete and specific. It provides a platform from which you can operationalize your organizational goals. The mission statement usually covers a span of three to five years. However, in high-technology organizations, where there is constant urgency to change, a mission statement may have a shorter life span. Independent of its life span, for a mission statement to be meaningful, it must be incorporated into the working fabric of the organization. This document must be specific and logical, so that it can be translated into the tasks that drive the organization's daily work.

Some managers have difficulty seeing the difference between a vision and a mission statement. If a vision statement usually focuses the organization on the next 5–10 years, a mission statement should be able to respond to the question, "What are we intending to do *right now*?" Another difference is that if the vision statement usually covers the whole corporation, the mission statement can and should be developed on different levels, such as groups, divisions, and other organizational units. This allows the organization to tailor the mission statement to their specific needs.

17.4.4 Testing Your Mission Statement

Jeffrey Abraham's book *The Mission Statement Book*,[25] contains 301 corporate mission statements from America's top companies and has a list of key words along with the number of times each key word appears in these 301 mission statements. By analyzing these key words and the frequency with which they are used, we can assess the vocabulary organizations use to articulate their mission statements. For example, the word "best" is used 94 times, which may imply that 94 companies have a mission to be the best. How in the world can you use this term to assess how your organization is fulfilling its mission? Another word that is frequently used in formulating a mission statement is "quality." In our case, it is used 169 times among these 301 companies, which means it appears in more than every other mission statement.

However, today quality is a must. You cannot stay in business very long with a low-quality product. For an organization to survive and attract new business, it must have a mission to deliver something beyond the traditional definition of quality, something beyond the customer's expectation. So what is a "good" mission statement? Below are five requirements for a mission statement that was based on Russell L. Ackoff's development described in his book

Management in Small Doses.[26] These requirements can be used as criteria to test your mission statement. They are:

1. A mission statement should contain a formulation of the organization's objectives that enables progress toward them to be measured. The mission statement must have the power to influence the organization's change and transformation. If it is not capable of changing the organization's behavior, it has no value.
2. A mission statement should differentiate it from other companies. The statement should establish the individuality, if not uniqueness, of the organization. An organization that wants the same thing as many other organizations want wastes its time in formulating a mission statement. Show me, for example, one organization that is not planning to grow or optimize their profit. You probably would not find one. This is not what a mission statement should contain.
3. A mission statement should define the business that the company wants to be in, not necessarily is in. This certainly doesn't mean that an organization's mission statement should contain a total switch from one industry to another. Contrarily, it should find a unifying concept that enlarges its view of itself and brings it into focus; for example, 3M considers itself an organization that is in the sticking business, enabling objects and materials to stick together, it has the opportunity to innovate and enlarge itself by creating new business while still using this core competence.
4. A mission statement should be relevant to all the organization's stakeholders. These include its customers, suppliers, employees and other shareholders. The mission should state how the organization intends to serve each of them. Some organizations formulate their mission statements focusing on satisfying only the needs of their shareholders. Others focus on their customers. A more general approach would be to focus on all major stakeholders, including the employees.
5. A mission statement should be exciting and inspiring. It should motivate and inspire all those who can make the mission a reality.

Try to test the mission statement of your organization against these five criteria and see how much it differs from these major requirements.

Important Note: Offering the test described above does not mean that every organization's mission statement should be structured so that it totally satisfies the test requirements. Every organization is unique, and because of this, the structure of the mission statement should also be unique. For some organizations, it is more important to connect the mission statement with their core values; for others the focus should be on customers, etc. However, comparing your mission statement with these five requirements will allow you to think more about the meaning and requirements of a great mission statement.

17.4.5 The Process of Forming a Mission Statement

To make a mission statement a source of employee inspiration, it should be formulated with the participation of key employees and shared with the whole organization. It must go through a cascading process of employee familiarization, with the current state of the organization and the vision of the future. It must form a sense of urgency and a creative tension that will motivate employees to accomplish the mission.

We looked for a case story that would describe the process of forming a mission statement and would fit the requirements described within this chapter. We found Pfizer's experience on this matter deserved attention, and we hope you will find their case story interesting.

Case Story 17.1   Pfizer's Process of Forming a Mission Statement[27]

In a meeting in New York in June 2001, Dr. Henry A. McKinnell, the Chairman and CEO of Pfizer, Inc., challenged 400 of the company's top executives to join him in leading an "internal revolution" in the way Pfizer does business around the globe. That goal is reflected in the new mission statement Pfizer unveiled at that time (see Fig. 17-5).

In 1997, Pfizer launched a mission of becoming the world's premier pharmaceutical company by 2001. The company declared "mission accomplished" on that goal two years ahead of schedule. The theme of the 2001 Global Leaders' Meeting—the first in Pfizer's history—was "Beyond #1." "We all know of companies that made it to the top and then declined," McKinnell told the gathering. "We must learn what it takes to go beyond number one. That journey includes a self-imposed transformation."

The process of formulating the new mission statement had begun the previous November, when McKinnell asked 25 senior leaders throughout Pfizer to identify and address the most significant issues facing the company. The need for a new mission quickly emerged as an area of focus. Lou Clemente, The Executive Vice President of Corporate Affairs, was tapped to lead a Mission Working Group (MWG), which included a number of the company's senior leaders and communications people. "We had a general idea of where we wanted to take things," Clemente said. "But we wanted to validate our views with a broader group of colleagues."

Focus groups (FG) were conducted at a number of Pfizer sites, during which colleagues were asked to "think out loud," not only about a new mission for Pfizer, but also about the company's purpose—its very reason for being.

Participants expressed a desire to broaden the company's purpose and mission, both of which previously focused almost exclusively on human pharmaceuticals. They also suggested formulating the statements in a concise form that would be easy to remember.

(continued)

344   MANAGING THE CORE OF THE ORGANIZATIONAL SYSTEM

> **Case Story 17.1**   (*continued*)
>
> Clemente's team synthesized all of these comments and produced drafts of a new Pfizer purpose and mission. It was here that a crucial idea was first set down on paper—that Pfizer would no longer measure itself solely against others in our industry, but against all other companies, regardless of industry. "As hard as sustaining our industry leadership will be, it just didn't seem like an inspirational enough goal," said John Mitchell, President of Pfizer Global Manufacturing and a member of the MWG. "That's when we started thinking about a mission that would move beyond the Mercks and GlaxoSmithKlines of the world to the Microsofts and GEs of the world."
>
> To ensure that the mission was truly global in scope, it was tested on focus groups made up of colleagues from around the world, including those from Australia, Brazil, Germany and Japan. This was a necessary step before the mission statement could be finalized for presentation at the global leaders' meeting in New York City, June 4–5, 2001. At the conclusion of that meeting, the new Pfizer mission appeared on a 20-foot-high by 40-foot-wide screen to wide applause from the people who are now putting their talent, enthusiasm, and effort into achieving it.
>
> Many attendees asked McKinnell whether he really thought that Pfizer could fulfill its new mission of becoming the most valued company in the world. "I don't know," he said, displaying the kind of candor that was shown throughout the meeting. "I believe that we should never take on a mission we absolutely know how to do. The fun comes in proving to ourselves that we can do it—that we can move beyond number one, and build a company like no other."

As you can see, the process was developed in such a way that more and more people got involved in formulating the organization's purpose and mission. We want to emphasize the idea of comparing Pfizer to the best companies regardless of their industry. This is probably the best way to protect a company that has made it to the top in its industry from declining, and also help it continue to perfect itself.

The People of Pfizer

Our Purpose: We dedicate ourselves to humanity's quest for longer, healthier, happier lives through innovation in pharmaceutical, consumer, and animal health products.

Our Mission: We will become the world's most valued company to patients, customers, colleagues, investors, business partners, and the communities where we work and live.

Our Values: To achieve our Purpose and Mission, we affirm our values of Integrity, Leadership, Innovation, Performance, Teamwork, Customer Focus, Respect for People, and Community.

Figure 17-5   The People of Pfizer

17.5 SETTING MOTIVATIONAL GOALS

17.5.1 Goals and Tension

Musicians know how important it is to have the right tension on the strings of an instrument to produce the right sound. The same thing applies to goals. If you want a goal to be value added and a source of excitement, it needs to be set in such a way that it will create the right tension. In his book *Corporate Tides*, Robert Fritz describes tension as a key element of organizational competence. He wrote, "There is a structural relationship between the actual state (where we are) and the desired state (where we want to be). The difference between these two conditions, the discrepancy, creates a tension. This tension generates a tendency to move from the actual state to the desired state—our goals. The organization will thus resolve the non-equilibrium between the actual and desired states by achieving its goals."[28] We'll illustrate this by using an actual example. In the internal AMD magazine, *Dialog* (June/July 2001), at that time AMD's President and Chief Operating Officer, Hector Ruiz, said, "There are four things from a semiconductor perspective that drive information technology: microprocessors, memory, DSP (digital signal processing) and communication-intensive products. We are in the fastest growing and the largest segments—flash memory devices and microprocessors. It is relatively easy for me to see AMD becoming a $10 billion corporation just exploiting these two core competences."[29] According to the 2001 results AMD was a $3.9 billion company, which means that there is a discrepancy of $6.1 billion between the actual and desired states. This creates a structural tension (see Fig. 17-6), which requires resolution.

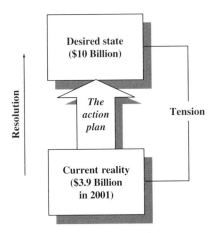

Figure 17-6 Creating Structural Tension to Achieve the Goal (Using AMD as an Example)

Source: Used with permission of the publisher. Based on *Corporate Tides*, copyright © 1996 by Robert Fritz, p. 25 Berrett-Kochler Publishers, Inc., San Francisco, CA. All rights reserved. 1-800-929-2929[30]

Once structural tension is established, the organization is in a phase of forming action plans—the actions that will resolve the tension between the points of discrepancy and move the organization toward its goal (becoming a 10 billion dollar company). The concept of building structural tension is not just a form of illustration. It has its own power, which helps to create momentum. When we establish, with a relatively high degree of accuracy, the desired and current states, we create an opportunity to determine more precisely what we need to do to achieve the goal. Please note that the distance between the two states is not just empty space that needs to be filled with something. Rather, it is tension, as a dynamic force, that requires resolution. This motivates us to develop action plans to resolve the tension. In other words, the structural tension becomes a *creative tension* (a term coined by Peter Senge in *The Fifth Discipline*) that motivates people toward creativity and innovation. Here energy is generated from the system, which then enables people to take advantage of the tension. The management's role here is to facilitate the implementation of the action plans and, at the same time, be ready to introduce new goals to keep the momentum going. As musicians tune their instruments to make them capable of producing the right sounds, managers need to make sure that the organization has the right tension to produce the right results.

17.5.2 A Goal from a Systems Perspective

Whether you are in charge of a large corporation or you are responsible for a small team, you are dealing with a system and you are responsible for making the system work properly. Systems are built to accomplish goals. As Dr. Deming said, "A system must have an aim. Without an aim, there is no system."[31] In setting goals for a system, we sometimes have a problem in differentiating goals from necessary conditions. For example, are "customer satisfaction" or "world-class manufacturing" goals, or are they conditions necessary to achieve other goals? According to Webster's Dictionary, a goal can be defined as "a result or achievement toward which effort is directed" and a necessary condition is defined as "a circumstance indispensable to some result, or that upon which everything else is contingent."[32] According to these definitions, we can say that there is a dependent relationship between the goal and the necessary conditions—you must satisfy the necessary conditions to attain the goal. Eliyahu M. Goldratt, author of *The Goal*,[33] suggests that the relationship between the goal and the necessary conditions is actually interdependent and that because of this, it doesn't matter what factor you choose to be the goal because all other factors will become the conditions necessary to achieve the goal. For example, the CEO may decide that "to become a global company" is the company's goal (see Fig. 17-7). In this case, "superior quality, "advanced technologies, "competitive products," "cost leadership," and "profound knowledge" may be all necessary conditions. However, at the same time, the CEO could choose "customer satisfaction" to be the company's main goal. In this

Figure 17-7 A Goal must be Surrounded with Necessary Conditions

case, all the remaining components in Figure 17-7 would be considered to be necessary conditions to achieve the goal.

From a systemic point of view, if we have a system of interdependent components, it is not important what component we single out as the goal as long as the remaining components act as necessary conditions. However, from a psychological point of view, it makes more sense to focus on an element that will catch people's attention. In our example, customer satisfaction is probably more appropriate than becoming a global supplier because it focuses the attention on the customer and not on the supplier who will serve the customer.

The main point here is that when setting a goal we must recognize all necessary conditions and act accordingly. Just concentrating on the aim without reacting to the necessary conditions will make goal fulfillment more difficult and sometimes unachievable.

17.5.3 Hard and Soft Goals

Every company has its own goals that come from its mission, but most organizations include mainly hard components in their goals such as return on sales, return on assets, market shares, earnings per share, and other traditional elements related to financial and operational performance. However, when organizations are viewed as living systems, it is important to include components in the goal statement that reflect the soft part of the organization's goals. In this regard, Hewlett-Packard's experience is interesting. Their statement of corporate goals consists of two parts: (a) hard goals and (b) soft goals (see Fig. 17-8). The two sets

HARD GOALS

- *Profit*: to achieve sufficient profit to finance our company growth and to provide the resources we need to achieve our other company objectives.

- *Customers*: to provide products and services of the highest quality and greatest possible value to our customers, thereby gaining and holding their respect and loyalty.

- *Fields of interest*: to participate in those fields of interest that build upon ourtechnologies, competencies and customer interests that offer opportunities for continuing growth, and that enable us to make a needed and profitable contribution.

- *Growth*: to let our growth be limited only by our profits and our ability to develop and produce innovative products that satisfy real customers needs.

SOFT GOALS

- *Our people*: to help HP people share in the company's success which they make possible; to provide them employment security based on performance; to create with them an injury-free, pleasant and inclusive work environment that values their diversity and recognizes individual contributions; and to help them gain a sense of satisfaction and accomplishment from their work.

- *Management*: to foster initiative and creativity by allowing the individualgreat freedom of action in attaining well-defined objectives.

- *Citizenship*: to honor our obligations to society by being an economic, intellectual and social asset to each nation and each community in which we operate.

Figure 17-8 Hewlett-Packard's Statement of Corporate Objectives (Goals)

Source: Excerpted from the Hewlett-Packard corporate objectives document, at http://www.hp.com/hpinfo/abouthp/corpobj.htm[34]

of goals complement each other and together form a force that drives the organization's success.

In their book *Strategic Analysis and Action*, Joseph N. Fry and Peter Killing came up with a table that describes the structure of hard and soft goals (see Fig. 17-9). This table can be used to formulate an organizational goal on any level: corporate, division, plant, or team.

17.5.4 How to Make the Goal Motivational

To make the work interesting and motivational, the goal should be set high enough so that it will build a creative tension for those who are involved in accomplishing the task. David C. McClelland and John W. Atkinson have demonstrated in their research that the degree of motivation and effort rises until the probability of success reaches 50% and then begins to fall even though the probability of success continues to increase (see Fig. 17-10). In relation to goal setting, this means that people are not highly motivated if the goal is

Hard Goals

- Profitability
  - Return on sales, return on net assets, return on equity
  - Earnings per share
- Market Position
  - Market share
  - Rank in industry
  - Diversity of product line
- Growth
  - Increase in sales, assets, earnings
  - Increase in earnings per share
- Risk
  - Liquidity
  - Ratio of debt to equity
  - Fixed charge coverage

Soft Goals

- Management
  - Autonomy
  - Status
- Employees
  - Economic security
  - Opportunities to advance
  - Working conditions, quality of working life
- Community
  - Control of externalities
  - Contribution to welfare, cultural life
- Society
  - General benefits through innovation and efficiency
  - Preservation of environment
  - Responsible political involvement

Figure 17-9 Business Strategy Goals

Source: Adapted from J.N. Fry and J.P. Killing, *Strategic Analysis and Action*, Pearson Education Canada Inc., copyright © 1986, p. 9. Used with permission by Pearson Education Canada Inc. [35]

perceived as almost impossible or when there is a zero risk in failing to achieve the goal. So, when setting goals, we must make sure that they build enough creative tension and at the same time make sure that the tension is not overstretched to the extent that people think they will never make it. Finding the right creative tension in goal setting is a very important component for success.

17.5.5 Goals in a Multicultural Organization

These days, when globalization is speeding up, working in multicultural teams has become a way of life. People from different countries come together towork on a global project, and as soon as the project is finished the team is usually disassembled and ready for another cross-functional project. This is a way of life in many highly technological global organizations. In this kind of environment, the team members hardly have time to get used to each other before they need to work together in a new environment to start a new project. This requires not only professionalism but also strong interpersonal skills from the team members. To achieve the team's goals, management must make sure that everyone in the organization learns how to work in teams with members from different countries, with different cultures and attributes.

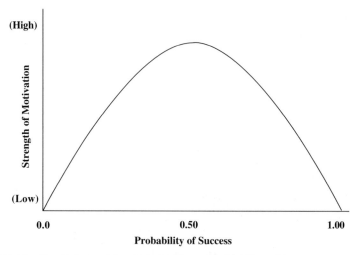

Figure 17-10 The Relationship of Motivation to Probability of Success

Source: Reprinted by permission of Harvard Business Review. From J. Sterling Livingston, "Pygmalion In Management," *Harvard Business Review*, September/October 1988, p. 127. Copyright © 1988 by the Harvard Business School Publishing Corporation, all rights reserved.[36]

Although we acknowledge the importance of professional skills in the realization of goals, our studies of multicultural and cross-functional teamwork have shown that interpersonal skills are even more important. Team members can tolerate it if someone on the team lacks professional skills, and will be willing to teach them, but they have a problem tolerating inappropriate behavior in the group. As one engineer noted, "If you know how to communicate with people, you learn faster. Everybody is willing to help you. But if interpersonal skills are missing, and at the same time you are weak in professional skills, you are in deep trouble." It is natural to have disagreements on the way to achieving a goal. They can be resolved in open dialogues. Our experience has shown that the vast majority of conflicts in multicultural teams are not personal, but job related. However, if these job-related conflicts are not properly resolved in time, they can create personal conflicts. This is why management needs to pay proper attention to the problems of conflict resolution and the development of interpersonal skills.

To reduce the opportunity for conflict to occur, it is very important that the goals of individuals and teams be nested within the goal of the organization as a whole (see Fig. 17-11). In other words, the goal of an individual or a team must be in line with the goal of the higher-level business unit—whether it's a department, division, or company—in which that individual or team is embedded. This also means that all teams on the same level must complement each other. This allows all members of the organization to know what is going on, not only in their group, but also in the whole organization.

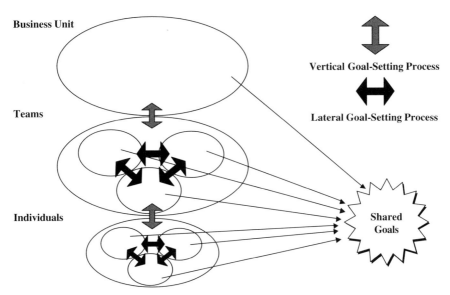

Figure 17-11 Goal Setting as a Vertical and Lateral Process

Source: From Susan Albers Mohrman, *Designing Team-Based Organizations*, p. 174. Copyright © 1995 by Jossey-Bass, Inc. Reprinted by permission of Jossey-Bass, Inc., a subsidiary of John Wiley & Sons, Inc.[37]

17.5.6 Creating Learning Goals

Our success in achieving the goals in an organization strongly depends on the knowledge and skills of the people who work there. Most of the time, new goals require new knowledge. Managers need to make sure that there is no gap between the knowledge needed to achieve the goal and the knowledge that people actually possess. If there is a gap, the manager must take action to eliminate it or at least reduce it to a minimum. The leader of the organization should provide learning opportunities where people can receive the necessary skills and knowledge to fulfill the goal. One example of this is Motorola's Six Sigma Institute, where all employees can receive the necessary education in the six-sigma concept and its methodology.

The process of creating knowledge should be developed concurrently with the process of achieving a goal. This is why it is important that learning goals should be developed in parallel with other business goals. In their book *The Learning Edge*, Calhoun W. Wick and Lu Stanton León developed a useful methodology for setting learning goals. In AMD's Manufacturing Services Group, we used their methodology to develop and implement learning goals that are aligned with our manufacturing goals and objectives. Calhoun and Lu's methodology is applicable to all layers in the organization, including individual learning goals. To be effective, a learning goal should satisfy both

Table 17-1 The Learning Value Index—An Aid in Evaluating Learning Goals

Learning Goal	Learning Value Index				
	Low				High
Value to Business	1	2	3	4	5
Value of Self	1	2	3	4	5
Your Commitment	1	2	3	4	5

Source: Reprint with permission from C. W. Wick and L. S. León, *The Learning Edge*, McGraw-Hill Companies, p. 51. Copyright © 1993 by Wick and Company.[38]

the business and the individual needs. Table 17-1 can be used to evaluate your learning goals by determining the learning value index.

When developing individual or group learning goals, it is important to consider current and future needs. To do this, it is recommended that open dialogues be held in which people can establish their individual and group needs for knowledge required to achieve a goal. When forming learning goals, it is important to benchmark the internal knowledge with the outside world to see what can be brought in from the outside to expand the existing organization's "container" of knowledge. Another important component of learning goals is to develop new knowledge in management and leadership that fits the new organizational environment. Finally, the learning goals should be aligned to the customer's requirements of doing business (e.g., knowledge of the methodology for introducing the ISO-9000 series of standards). As soon as the learning goals are established, learning plans should be developed. Learning plans should include the sources from which the new knowledge will be adopted, what forms of education and training will be used, what the expectations are, what resources are needed, and how to measure the learning results. The main purpose of this section is to bring to the reader's attention that knowledge-creating goals are an important component of business goals.

17.6 CREATING A GREAT STRATEGY

Having defined its purpose, vision, mission, and core values, the organization must have a great strategy to close the gap between where we are and where we want to be. It is the leadership's responsibility to create a great strategy that will satisfy the requirements of the competitive environment. What is a great strategy? How do you measure it? Strategy is conceptual and depends on the specifics of the organization. A great strategy must precisely designate what the organization will do better than its competitors to be the choice of customers. A great strategy must inspire people, generate creativity and innovation, and stimulate changes. It should create a learning environment and be able to absorb the best practices available in the world and, based on this, create a result that makes the difference. For a strategy to succeed, it must be accepted by those who will materialize the strategy, and to be accepted, it must be achievable and have an element of risk taking. Strategy is directly related to change, and there is no real change without risk, without taking people out of their comfort zone. Creating a strategy that raises the organization up to a higher, more competitive level is the leadership's responsibility.

An organization, especially a high-technology enterprise, must have core competences that differentiate it from others. Before developing strategy, we must define the organization's core competences. In its simplest definition, a core competence is what an organization does particularly well. Figure 17-12 shows core competences of a variety of leading corporations.

Figure 17-12 describes core competences that are on the surface and easy to recognize. For example, everyone associates 3M with tape, Intel with Pentium®, and AMD with the AMD Athlon™ microprocessors, but there are many other important core competences that need to be brought to the surface and recognized. This is not an easy task to accomplish. In his book *Leading Corporate Transformation*, Robert H. Miles suggested some guidelines to determine core competences. He said, "As an analytical tool, we suggest that strategic relevant core competencies should be:

- Recognized by the customer because they are essential to creating the product and service features that lead to superior customer satisfaction and loyalty
- Respected by peers because they drive the business system design that produces 'best of breed' performance in cost, speed, and quality
- Responsive to external trends because they embody the capabilities of adjusting to and exploiting changes in the external environment
- A riddle for competition because they are the roots of an often invisible competitive advantage
- Relished by the organization because the shape the vision, beliefs, and culture that attract outstanding people
- Rewarding for shareholders because they drive growth into new markets, protect above-average profitability, and are a source of resilience"[41]

These guidelines can help you to analyze the organization's potential and have a better understanding of its core competences. Please note that core competences are not something you wish to have; they are competences that the organization has already developed and that have demonstrated they can bring success to the organization. Management can enhance these core competences, and can find new areas and ways to apply them, but they cannot adopt a new competence from its competitor overnight. Management can certainly work on developing new competences, but to become a core competence for your organization, it must be inherent in your organization. A full understanding of the organization's core competences allows management to create strategies that properly utilize and develop them.

3M Adhesives, substrates, advanced material
AMD Microprocessor and flash memory technology
Black & Decker Small electric motors
Canon Optics, precision mechanics, microprocessor controls
Casio Microprocessors, material science, display systems, precision castings
Citicorp Online/interactive global MIS (for internal operations)
EDS System integration
Federal Express Logistics management
GE General management
Hewlett Packard Measurements, computing, and communications
Honda Engines, power trains
Intel Microprocessor and flash memory technology
JVC Videotape
Microsoft Computer software expertise
Motorola Wireless communication
NEC Electronic components, central processors, communications
Philips Optical media
Sony Miniaturization, videotape
Vickers Fluid power, electric power, electronic controls
Wal-Mart Logistics management

Figure 17-12 Core Competences of a Variety of Leading Corporations

Source: Based on Robert H. Miles, "Leading Corporate Transformation," a Jossey-Bass Inc., a subsidiary of John Wiley & Sons, Inc., 1997, p. 54,[39] and Gary Hamel and C. K. Prahalad, "Competing For The Future," Harvard Business School Press, 1994, p. 199.[40]

17.7 THE MOON VISION: A CASE STUDY

"I believe we should go to the moon."[42]

John F. Kennedy

17.7.1 A Vision Statement That Becomes a Model

John F. Kennedy's inspirational announcement that we needed to land a man on the moon before the end of the 1960s has become a model of a vision statement. Why? Is it because of the uniqueness of this subject? Maybe. However, there is much more to it. The American people shared the moon vision and made it a reality only because they understood its importance to the nation and themselves, because they understood its urgency, and because it was clear and concrete. We selected the moon event as a case story to study what can be applied in any organization and, at the same time, to demonstrate that to have a great vision and make it come true, we need to use the systemic approach that is proposed in Part V of this book.

Let's start analyzing this case story with some historical facts.

- On May 25, 1961, the President of the United States, John F. Kennedy, proclaimed, "...This nation should commit itself to achieving the goal, before this decade is out, of landing a man on the moon and returning him safely to the earth...."[43]
- On July 20, 1969, the American astronauts Neil A. Armstrong and Edwin E. "Buzz" Aldrin occupied and powered up the lunar module Eagle and deployed its landing legs. That same day, there was a transmission from the moon's Sea of Tranquility, "Eagle has landed."[44]
- On July 21, 1969, after 22 hours on the lunar surface, the astronauts headed for home. On July 24, 1969, the astronauts splashed down in the Pacific Ocean. President Nixon greeted them on the aircraft carrier Hornet.[45]
- President Kennedy's vision came true. The Apollo program achieved its objective five months and ten days before the end of the decade.

The attractiveness of this case story is that it has a concrete time frame with an inspiring beginning and a successful end. It began with a picture in a leader's head and ended with an accomplishment that shook the world. The moon vision statement deserves to be a model for any organization in any industry. Now let's see what really happened in this time frame and what we as organizations can learn from these facts.

17.7.2 How Realistic Should a Vision Be?

We usually say that a great vision should be a realistic one. How realistic was John F. Kennedy when he proclaimed the moon vision? In other words, how confident was he that this would be one of the greatest successes in the world? How prepared was the nation for such a scientific and technical breakthrough? How much courage does a leader need to proclaim something such as this to the whole world? Russia, for example, always announces space successes after the fact. It is less risky. We are talking about the prestige of the United States, about the risk of people's lives, about billions of taxpayers' dollars. Is it really necessary to take this kind of risk?

In 1961 when John F. Kennedy articulated the moon vision, the space program was in its infancy. So, what drove the leader to make such a vision statement and ask people to share it with him? Logic suggests that there had to have been some strong forces in motion to make the President say, "There's nothing more important"[46] [than the moon project].

17.7.3 Looking for Answers by Analyzing a Higher System

We will not find all the answers to the questions raised above by looking inside the system—the United States. As we mentioned above, to fully understand a system, we need to study the higher system of which it is a part—in this case it would be the whole world. In other words, we need to understand the external environment. Here are some historical facts that describe the environment at the time (1961) when John F. Kennedy proclaimed his moon vision. In one of Kennedy's letters, he wrote. "We are in a strategic space race with the Russians, and we have been losing."[47] Then the President provided some facts that demonstrated the Russians' superiority in the space race, "The first man-made satellite to orbit the earth was named Sputnik. The first living creature in space was Laika. The first rocket to the moon carried a Red flag. The first photograph of the far side of the moon was made with a Soviet camera."[48] Concluding, he observed, "If a man orbits earth this year his name will be Ivan."[49]

This is an excellent example of an objective assessment of "where we are" and how a leader should describe the existing situation, which builds a sense of urgency and helps people share the vision.

Recognizing that Russia was ahead and that the U.S. was losing superiority in space, President Kennedy asked, "If somebody can just tell me how to catch up...."[50] He was looking for information that would help him make the right decision. He realized that in this situation, anything less than landing a man on the moon would not be acceptable. This example demonstrates the point that sometimes the environment creates heroes and influences a leader to make challenging vision statements like this. As we know by now, the country pooled all the necessary talents and resources to make the vision come true. However, without this vision statement, we might not have had the opportunity to create and demonstrate our talents and capabilities. So great vision creates great opportunities and great successes.

Now how important was it to be number one in the space race? President Kennedy explained, "If the Soviets control space they can control earth.... But we cannot run second in this vital race. To ensure peace and freedom, we must be first."[51] Listen to these words—how simple and clear they were and how deeply they touched people's hearts and minds. There was no need for days of meetings to elaborate the importance of achieving the challenging vision or to gain acceptance and support from the Congress and the American people. It became a real, shared vision on its own strength. Some organizations articulate vision statements like "We will be number one" but do not take the time to analyze and share with the employees what will happen if we do not become number one. To make the employees follow the leader, we need to take the time and explain where we are going and why we need to go there.

There were obviously more components related to the environment at that time, but we will limit ourselves with just these examples, which demonstrate that assessing the external environment, studying the competitor's strengths, and elaborating to the employees where the organization currently stands are important components of the visioning process.

17.7.4 Building an Environment for Risk Taking

In some organizations, the existing culture creates norms of behavior that have less risk. In this environment, goals and visions have a high probability of success; and because of this, the creative tension is less, the opportunities are fewer, and obviously the results are less attractive. In the moon example, President Kennedy said, "For while we cannot guarantee that we shall one day be first, we can guarantee that any failure to make this effort will make us last."[52] In other words, he was not guaranteeing success. He said, "I believe we possess all the resources and talents necessary."[53] With this statement, the President was saying that he had full confidence in the people's talents and the country's strength. It was only a matter of pulling together the talents and resources and using them properly to achieve the desired results. From an organizational point of view, possessing all the resources and talents does not necessarily mean that we have everything required to achieve the vision. It actually means that the organization possesses the necessary talents and core capabilities, which can be used as a foundation to build on the new capabilities, knowledge, and talents needed to make the vision a reality. If you are trying to have 100% of what you need to achieve a vision, what kind of vision is it?

We need to create a culture that will motivate people to make "unrealistic" goals, and this can happen only when people know that the winner is not the person or department who just achieved the goal, but those who have achieved greater results. We need a cultural environment in which people will not be scared to take rational risks. Jack Welch, expressing the need to set the bar far beyond what is thought realistic, wrote, "For example, how do you get to ten when you are now at three? We believe you should compare yourself to the best, set the bar at ten, and then have a culture that says, if you get to six, you won't be punished."[54] If we introduce this way of thinking into our culture, we will motivate people to take more challenging goals and organizations to create more challenging vision statements.

17.7.5 Pulling the Resources Together

Sometimes an organization has all the resources needed to achieve a challenging vision, but those resources may be spread all over the organization. What is important here is to have the right strategy and structure that will allow the organization to pool and further develop all its strengths and focus on the most important goals. For this, we need great leaders, which by itself can be considered as one of the organization's core competences. The moon example illustrates this point. To improve the space program, President Kennedy nominated James E. Webb to be the administrator of the space agency. At first, Mr. Webb was not confident enough to accept this position. He felt that this position would be more appropriate for a scientist or engineer; but when he found out that President Kennedy was seeking a policy maker who could manage scientists and engineers, he accepted the position.[55] This episode illustrates that sometimes the organization's core technical capabilities are much stronger than they appear on the surface. It is only a matter of strategy and structuring. Structure creates behavior. The readers can probably remember examples where, only when the organization has an obvious problem do they start reshuffling the staff and pulling the strengths from different parts of the organization to "fix the problem." This reactionary approach in management is much less effective and sometimes too late to achieve the desired results. Creating an attractive vision and upgrading the capabilities, culture, and strategy to match the needs for achieving the vision is a better path to success. This is reflected in our Organizational Principles System.

17.7.6 Strategy and Culture

While analyzing the history of the space program, we can observe an interesting relationship between such elements as power, knowledge, success, cooperation, and competition. In a document entitled, 'Competition Versus Cooperation: 1959–62, we read, "and Kennedy continued to espouse the cooperative theme in his State of the Union address on 30 January 1961. The President invited all nations, including the U.S.S.R., 'to join with us ... in preparation for probing the distant planets of Mars and Venus, probes which may someday unlock the deepest secrets of the universe.' "[56] At that time, the U.S.S.R.'s space policy and, in particular, the behavior of the Soviet Premier, Nikita Khrushchev, was unpredictable. One assessment that characterizes this period states: "There appeared to be two Khrushchevs: one, a 'coexistentialist' eager for enhanced intercourse between the U.S. and the U.S.S.R., ... the other, a militant Communist and bully ready to cash in on each and every weakness and hesitation of the West ... "[57] With Major Yuri Gagarin's *Vostok I* April flight, the tone of the Soviet statements on cooperation in space significantly changed. Phoning Gagarin his congratulations, Khrushchev crowed, "Let the capitalist countries catch up with our country!"[58] Khrushchev stopped playing a dual role, enjoyed his country's success, and lost any interest in cooperation (if there was any desire for this in the first place). On the evening of April 14, President Kennedy held a meeting in which he asked, "Is there any place we can catch them? What can we do? Can we go around the moon before them? Can we put a man on the moon before them?"[59] As we can see, the time for cooperation was over. To get ahead, the U.S. needed to compete.

In our opinion, this is an excellent example for demonstrating the relationship of core values and external environment. Some leaders argue that now is the time for cooperation; others still hold competition as a core value for success. What we observe is that you can cooperate when you keep a required level of power in comparison with the other party, a required level of knowledge and demonstrated achievement, which will attract the other party to cooperate. When these organizations cooperate there should be, at least in the beginning, an observable win-win situation from merging the core capabilities. Later, it depends on how much every party is capable of learning from the cooperative effort.

Core values need to be seen as the organizational strength, which needs to be maintained even when it is difficult to do so. An excellent example to illustrate this would be the Tylenol crisis experienced by Johnson and Johnson in 1982. "The deaths of seven people in the Chicago area revealed that someone—not an employee—had tampered with Tylenol bottles, lacing them with cyanide. J & J immediately removed all Tylenol capsules from the entire U.S. market—even though the deaths occurred only in the Chicago area—at an estimated cost of $100 million and mounted a twenty-five-hundred-person communication effort to alert the public and deal with the problem."[60] J & J made this decision with reference to their organization's credo. Even then, we could imagine the difficulty and dilemma involved in making this major decision. On the one hand is responsibility to the community we serve; and on the other hand is profit and the interests of the stockholders. However, the importance of core values, as we mentioned, is the organization's strength. In times of test, they remain as the backbone, the guidepost of decision-making.

At the same time, we are not saying that core values are stagnant and should always be considered unchangeable. If serious changes in the environment are observed, and one of the organization's core values no longer works for the organization's long-term benefits, this core value should be revised. As the core of an organizational culture, values are as dynamic as the culture as a whole is dynamic. We will elaborate more about core values in Chapter 18.

17.7.7 How Do We Create a "Moon" Vision?

People may say that the most attractive part of the moon vision is the moon itself, because reaching the moon was the greatest technological achievement our imaginations could realize. This is true, so how can we relate the moon vision statement to organizations that are not involved in placing a man on the moon? Twelve years ago, the Manufacturing Services Group in AMD decided to adapt Motorola's concept and achieve six-sigma quality for all critical manufacturing processes. For us, it was equivalent to going to the moon. Maybe not as costly, risky, and attractive as the moon project, but it was a great technological stretch that required a lot of talent and effort; and when we achieved it we had a great feeling of accomplishment and satisfaction. Hence, even in smaller visions, we can see the attractiveness if we build the creative tension properly. If a process engineer who is responsible for a particular process takes a goal to reduce the process variation around the nominal to a quality level that is higher than the best existing standard in the industry, this is a "moon project." If a pizza maker has a vision to deliver pizzas to customers in the shortest possible time, while the pizza is still hot and tasty, this can also be considered a moon vision.

17.7.8 What We Can "Take Home?"

Below is a summary of what we can learn from the moon story.

1. A vision statement is not a vision. A vision should have substance. Even if it is "soft" you should be able to "touch" it. If you cannot touch it, it is difficult to perceive it, and if you cannot perceive it, you cannot share it. A "dry" formal vision or mission statement is not enough to stimulate progress. An organization needs to define its destination in broad, but easy to grasp terms. A compelling vision, mission, or goal is needed to focus the organization's efforts and give the employees a sense of direction, of where the organization wants to go and what it wants to be.
2. Consider the whole system. The vision, mission, or goals should be interconnected with the organization's core capabilities, strategy, and culture, and they should work together as a whole system to achieve the desired state.
3. Benchmark with the best practices. The vision, mission, or goals should be based on comparing the organization's practices to other best practices.
4. Consider the external environment. The process of developing vision, mission, and goals should start with an assessment and analysis of the external environment.
5. Know where you are. To develop the right map of the journey to the desired state, a realistic assessment of "where we are" is crucial. Determining the existing organizational capabilities is an important step in determining the future state.
6. Achieve a shared vision. To make the vision, mission, and goals attractive and to encourage the employees to participate in achieving the desired state, we need to make sure that all employees share the same picture of the future.
7. Pool all the strengths. The strength of a laser beam is its ability to focus its energy. This also applies to an organization. It should be able to concentrate its talents, core capabilities, and financial resources to achieve the desired state. Visioning is the focusing instrument.
8. Have superiority over the competitor. A vision should have an element of superiority over the competitor. If a vision does not contain elements of superiority, not only is it unattractive, but it makes the organization a follower and not a leader, even when the vision has been achieved. The moon vision gave a strong signal to the competitors that the U.S. was the strongest country in the world. Putting a man on the moon is a great example of superiority over the competitor.
9. Have an aggressive time frame. Even though it is not always possible to assess the exact time needed to reach the desired goal, a vision becomes more attractive when it has a time frame of accomplishment. This mobilizes people to achieve the target by the right time. Sending a person to the moon, by itself, was a challenging task; but if the American astronauts had gotten to the moon and found the Russians already there, the symbolic strength of superiority would have been lost. Make the best

product one month too late, and the market will be almost gone. Superiority in a competitive business is very important. Becoming number one sounds great, but when? This is the question.

10. Take rational risks. A vision or mission should have an element of risk. Calculated risk creates the necessary "adrenaline" for great performance. If organizations can calculate the risks of achieving the corporate vision or mission, and articulate them so that the employees can relate to it, the vision or mission will be perceived as more real and tangible.

11. Take a stretch to a breakthrough. A vision should not be an incremental improvement that can give a 5–10% increase in productivity. It should motivate a radical change with a great improvement. This is what we mean by an attractive vision. Today we have so many technological and scientific breakthroughs that would not be happening without creating the right structural tension.

12. Involve people. President Kennedy's vision became a mission for NASA. A large number of organizations and subcontractors developed their goals in context with NASA's mission. A corporate vision needs to be translated into its organization's vision, mission, and goals, which are in context with the grand vision of the corporation. Just as small rivers combine to flow toward the ocean, vision, mission, and goals of the organizations, teams, and individuals should flow together to achieve the desired state.

13. Find the right language. When President Kennedy communicated the vision to the Congress and to the American people, the way it was presented was simple and easy to understand. Leaders of organizations should learn how to communicate complicated issues in a simple form to their employees. If Einstein could find a proper language to explain gravitation, we can find a language to explain the corporation's purpose, vision, mission, and core values in such a way that they will energize employees to do better and make them feel that they're a part of the whole organization.

References

1 Tom Peters, *Liberation Management: Necessary Disorganization for the Nanosecond Nineties*, Alfred Knopf, Inc., New York, NY, 1992, p. 617
2 Jerry Sanders "A Tale of Two Dreams" http://amdonline/info/speeches/fab30dedication.html
3 C. Patrick Lewis, *Building A Shared Vision—A Leader's Guide to Aligning The Organization*, Productivity Press, Portland, OR, 1997, p. 13
4 Robert Lacey, *Ford: The Men and the Machine*, Ballantine Books, New York, NY, 1986, p. 93
5 Lewis, *Building A Shared Vision—A Leader's Guide to Aligning The Organization*, p. 10

6. John F. Kennedy, *Special Message to the Congress on Urgent National Needs*, May 25, 1961, Section IX: Space, paragraph5
7. Lester B. Korn, "How the Next CEO will be Different," *Fortune*, May 22, 1989, p. 175
8. Edgar H. Schein, *Organization Culture and Leadership*, Jossey-Bass, Inc., San Francisco, CA, 1992, p. 330
9. Milton Friedman, *Capitalism and Freedom*, The University of Chicago Press, Chicago, IL, 1982, p. 133
10. James C. Collins and Jerry I. Porras, *Built To Last*, HarperCollins Publishers, Inc., New York, NY, 1994, p. 73
11. Peter M. Senge, *The Fifth Discipline*, Doubleday and Currency, New York, NY, 1990, pp. 148–149
12. Collins and. Porras, *Built To Last*, p. 82
13. Ibid.
14. Russell L. Ackoff, *Creating the Corporate Future*, John Wiley & Sons, New York, NY, 1981, p. 23
15. Peter Block, *Stewardship*, Berrett-Koehler Publishers, Inc., San Francisco, CA, 1993, p. 29
16. Ibid.
17. Richard J. Leider, *The Leader of the Future*, Jossey Bass Publishers, Inc., San Francisco, CA, New York, NY, 1993, p. 197
18. Ibid.
19. Thomas J. Watson, Jr., *A Business and Its Beliefs*, McGraw-Hill Book Company, Inc., New York, NY, 1963, p. 13
20. Stephen R. Covey, *The Seven Habits of Highly Effective People*, Simon and Schuster, Ltd., London, 1989, pp. 139–140
21. Watson, *A Business and Its Beliefs*, p. 29
22. Ibid., p. 34
23. Gary Hamel and C.K. Prahalad, *Competing For The Future*, Harvard Business School Press, Boston, MA, 1994, p. 134
24. "Disney's Philosophy," *New York Times Magazine*, March 6, 1938, New York, NY
25. Jeffrey Abrahams, *The Mission Statement Book*, Ten Speed Press, Berkeley, CA, 1999
26. Russell L. Ackoff, *Management In Small Doses*, John Wiley & Sons, New York, NY, 1986, pp. 39–41
27. The case story is a result of collaboration with Pfizer, Inc., and is being used here with their kind permission.
28. Robert Fritz, *Corporate Tides*, Berrett-Koehler Publishers, San Francisco, CA, 1996, pp. 22–23
29. AMD magazine *Dialog*, "Interview with Hector Ruiz," June/July 2001, p. 10
30. Fritz, *Corporate Tides*, pp. 22–23
31. W. Edwards Deming, *The New Economics*, Massachusetts Institute of Technology, Center for Advanced Educational Services, 2nd ed., 1994, p. 50
32. *Webster's New Universal Unabridged Dictionary*, 1989

33 Eliyahu M. Goldratt, *The Goal*, 2nd ed., North River Press, Great Barrington, MA, 1994
34 Hewlett-Packard website, http://www.hp.com/hpinfo/abouthp/corpobj.html
35 Joseph N. Fry and J. Peter Killing, *Strategic Analysis and Action*, Pearson Education, Canada, Inc., Don Mills, Ontario, Canada, 1986, p. 9
36 J. Sterling Livingston, "Pygmalion In Management", *Harvard Business Review*, September/October 1988, p. 127
37 Susan Albers Mohrman, *Designing Team-Based Organizations*, Jossey-Bass, Inc., San Francisco, CA, 1995, p. 174
38 Calhoun W. Wick and Lu Stanton León, *The Learning Edge*, McGraw-Hill Companies, New York, NY, 1993, p. 51
39 Robert H. Miles, *Leading Corporate Transformation*, Jossey-Bass Inc., San Francisco, CA, 1997, p. 54
40 Hamel and Prahalad, *Competing For The Future*, p. 199
41 Robert H. Miles, *Leading Corporate Transformation*, pp. 230–231
42 John F. Kennedy, *Special Message to the Congress on Urgent National Needs*, May 25, 1961, Section IX: Space, paragraph 12
43 Ibid., paragraph 5
44 *Spaceport News*, 23 July 1969
45 Ibid., 30 July 1969
46 Daniel J. Boorstin, *The Americans: The Democratic Experience*, Vintage Books, New York, NY, 1974, p. 595
47 John F. Kennedy, "If the Soviets Control Space...They Can Control Earth," *Missiles and Rockets*, 10 Oct 1960, pp. 12–13; and in the same issue, Clark Newton, "Kennedy's Stand on Defense and Space," p. 50
48 Ibid.
49 Ibid.
50 Boorstin, *The Americans: The Democratic Experience*, p. 595
51 Kennedy, "If the Soviets Control Space...They Can Control Earth," pp. 12–13; and Newton, "Kennedy's Stand on Defense and Space," p. 50
52 John F. Kennedy, *Special Message to the Congress on Urgent National Needs*, May 25, 1961, Section IX: Space, paragraph 3
53 Ibid., paragraph 2
54 Lynne Joy McFarland, Larry E. Senn, John R. Childress, *21st Century Leadership—Dialogues With Top 100 Leaders*, Linc Corporation & Senn-Delaney Leadership Group, Inc., Long Beach, CA, 1994, p. 147
55 *Competition Versus Cooperation: 1959–1962*, www.hq.nasa.gov
56 *Public Papers of the Presidents of the United States, John F. Kennedy, 1961*, Washington, 1962, pp. 1–2, 26–27, and 93–94
57 *Competition Versus Cooperation: 1959- 1962*, www.hq.nasa.gov
58 Boorstin, *The Americans: The Democratic Experience*, p. 594
59 Ibid., pp. 594–595
60 Collins and Porras, *Built to Last*, p. 60

Chapter 18

Values, Behavioral Standards, and Business Ethics

18.1 A GENERAL CONCEPT OF VALUES

Values are the most important elements in a culture. Before we discuss organizational values, we would like to shed some light on the issue of values in general. As we know, there are different types of values: local and family values, spiritual and religious values, material and achievement values, positive and negative values, etc. Coming from different sources, these values form our attitudes and influence our behavior. They tell us what we should do and what principles we are supposed to follow.

Usually we accept values from somebody or from some group, and we follow them because we trust the leader or we fear group disapproval if we do not. However, if these values work positively for us for a reasonable period of time, they will indeed become our own values.

18.1.1 Values and Beliefs

Is there any significant difference between these two terms—values and beliefs? In literature and in practical application, you will find that these two terms are used interchangeably. For example, in the book *Father, Son & Co.—My Life at IBM and Beyond*, Thomas J. Watson, Jr., wrote, "Eventually I was able to distill into a simple set of precepts the philosophy Dad had followed in managing the business for forty years:

- Give full consideration to the individual employee.
- Spend a lot of time making customers happy.
- Go the last mile to do a thing right.

I thought that to survive and succeed, we had to be willing to change everything about IBM except these basic beliefs."[1] If we used these precepts today, we would probably rephrase them as "core values."

You can find another example in the book *In Search Of Excellence—Lessons from America's Best-Run Companies*. The authors, Thomas J. Peters and Robert H. Waterman, Jr. state, "Led by our colleague Allan Kennedy, we did an analysis of 'superordinate goals' about three years ago. (... Since then we have changed the term to 'shared values;' but although the wording has changed, we have always meant the same thing: basic beliefs, overriding values.)"[2] If we had to give short descriptions of these two terms to differentiate them from each other, we would say that values mainly spell out "what is important within the organization" and beliefs mainly refer to "the way we are supposed to be around here."

There are many examples that suggest that even though there are some differences in defining beliefs and values, these two terms can be used interchangeably. In this book we will mainly use the term "value," but in some situations we may use the term "belief" if we feel that it is more appropriate to the subject.

18.1.2 Recognizing Different Values

As Dr. Edward De Bono said, "You need to have in mind a wide range of 'positive-values' in order to recognize them. It is like bird-watching. You will not recognize that bird unless you have had previous experience with that shape...."[3] From the whole variety of values, we need to learn how to recognize different values, those that drive the situation and those that are used mainly for assessment. We also need to recognize that not all values are equally important and that sometimes the importance of a value depends on time and other situations. For example, when the customer is expecting a shipment from a supplier at the end of the quarter, or when there is a high demand for a product, the value of "On-Time Delivery" is probably more important than the cost of the shipment. At other times, the customer will look more closely at the quality and cost.

It is important to note here that in these cases we are not talking about the core values that are usually robust in any situation; we are talking about operational values that can and should be changed for the organization to adapt to the environment and move forward.

As we will describe later in this section, organizations tend to concentrate on a set of core values that are expected to be the organization's blueprint of behavior. Articulating the core values of the organization is very important; however, it is also important to be aware of all the different values involved in a decision-making process or in other situations. We need to keep in mind a wide range of values to be able to recognize them and select the right values for the right situation.

The family, spiritual, and other values mentioned above may sound unrelated to manufacturing life, but all of them exist and work in an organization because an organization consists of people who carry all these types of values. So when we work on the development of a value system for a corporation, or for a smaller group, we need to consider that we are developing a value system that should reflect not only the organization's interest but also the interest of every individual.

18.2 BUILDING SHARED VALUES IN AN ORGANIZATION

Values have the greatest power only when they are shared, which helps the organization develop a common language and a better understanding between its members. However, we also need to recognize that a company is composed of individuals, and not all people have the same needs at the same time, nor do all people recognize and appreciate the same values. For example, whereas an engineer places more value on innovation, a salesman will probably place more value on the art of negotiation. This is especially true when we are talking about an international organization with a diversity of cultures. For example, whereas Thai employees may value meekness more than open confrontation, American employees probably feel the opposite.

In analyzing the variety of values, we can always observe two extremes: essential values and luxury values. In between these two extremes there is a spectrum of other values. We need to learn how to recognize all the values along this spectrum. Below is a story that may illustrate the point.

In one of the AMD offshore plants, a dialogue was conducted in which the issue of openness and trust was brought up. Local engineers and representatives from the United States participated in the dialogue. "We are talking about openness as an important organizational value," said one of the U.S. engineers, "but how come you guys, knowing that the proposed solution would not work, didn't speak up?..." "We couldn't at that time," interrupted one of the local engineers. He continued, "This was the manager's decision." Without describing all details in relation to this issue, just from these two comments we can see two extremes. The American representative recognizes openness as the most important value. He will do everything to get the job done, even if he may hurt someone's feelings. In contrast, the Asian engineer considers protecting the manager's "face" as the essential value. Is the Asian engineer less honest? Absolutely not!

Given that there is a spectrum of values between the extremes, it is important to find the right set of values that work for each particular situation. The art of management is to develop, together with the employees, a value system that will satisfy the major needs of the organization and, at the same time, satisfy the needs of the individuals within the organization. So it is not as simple as it sounds when we say that we need to have a system of shared values.

18.3 CORE VALUES

For a long time, many authors, scientists, and practitioners observed the relationship between success and core values. In *In Search Of Excellence*, it was observed that, "Virtually all of the better-performing companies we looked at in the first study had a well-defined set of guiding beliefs."[4] This was written in 1982. Twelve years later (1994), in another successful book, *Built To Last*, James C. Collins and Jerry I. Porras wrote, "We've found that most companies benefit from articulating both core values *and* purpose in their core ideology, and we encourage you to do the same."[5]

We could continue with examples that demonstrate the importance of core values. There must be something almost magical about core values, because they attract the attention of so many people. Contemporary management recognizes the significance of defining and articulating the organization's core values, because they serve as a blueprint of the organization's behavior and time has proven that core values work. However, it is also important to note that core values, no matter how well they are defined, articulated, and lived, cannot create success by themselves. A company can go down the tubes if the core values are not connected with a set of operational values and behavior that make the company flexible and adaptable to the continuous and rapidly changing environment. Some organizations limit themselves with the articulation of a set of core values and assume that if these values have made the company "stay in business for the last 20 years," there is no need to change the existing culture. This is a dangerous way to think. To preserve the core values of an organization, there should be something to support them that makes the organization flexible and adaptable (see Fig. 16-1).

18.3.1 Where Do Core Values Come From?

There are some innate human values, such as respect for people, a sense of justice, freedom of expression, dignity and worthiness, etc., which are shared by most people. This type of value can become a core value for any organization. However, the leader of the organization has to make these values work.

Values also come from deep learning. When a new enterprise, a department or a team, is formed, the leader brings in his values, beliefs, and assumptions about what and how work should be done. With time, these values are tested, accepted, and shared. Otherwise, the leader must leave. This obviously will not happen overnight. Time is required to transform the leader's personal values and beliefs into a system of the organization's shared values.

In 1969, W.J. "Jerry" Sanders III and seven other co-founders launched Advanced Micro Devices, Inc., a company that has since grown to be one of the largest U.S.-based merchant semiconductor manufacturers. In the initial period, Jerry Sanders might not have had a clear vision of what the company would look like in 25 years, but what he did have was a set of personal values and beliefs that he was able to share with the group of co-founders. These values, tested by time, have gradually become AMD's core values, which are:

- Respect for People
- Initiative and Accountability
- Integrity and Responsibility
- Knowledge
- Competition
- Our Customers' Success

On the basis of these core values, AMD continues to make its major decisions, deriving its norms of behavior and codes of ethics. These values are the foundation of AMD's culture. They interact with each other and work together as a whole. For example:

- To be competitive, we need knowledge, integrity, and responsibility.
- To contribute to the customers' success, we need to take initiative and assume accountability; etc.

Generally, core values are the most stable elements in the organizational culture system (see Fig. 16-1). This is not to say that if an organization successfully develops itself on the basis of a set of core values, they must stick with them forever. What we want to say is that the set of core values, which is usually small in number (3–5), is the most robust part of the culture system. However, if an organization feels that some of the core values no longer serve their purpose, the leader of the organization must initiate an activity that will allow them to introduce a change in the set of core values. For example, 10 or 15 years ago you would rarely find an organization that would consider "knowledge" as a core value. But now that more organizations and products are knowledge based, knowledge has become a core value in many organizations.

While recognizing a leader's influence on the formation of an organizational value system, it is also important to recognize that when a leader forms a new group, he is dealing with people who have their own individual values. Furthermore, when a leader joins a well-established organization, with a solid culture that has a set of shared values, he needs to use his talent to transform his own values and all the employees' values into a shared value system that will serve the organization's purpose.

We need to emphasize the point that changing core values is a very difficult thing to do. It takes time, patience, and special skills to make a smooth transition to a new or revised set of core values. If there is not enough time to do it gradually, the organization must be prepared to encounter a lot of conflicts and resistance, which may lead to an increase in turnover and other changes.

18.4 OPERATIONAL VALUES

Recognizing the importance of core values, which work as a blueprint for behavior in the organization, we should also realize that they do not ensure the desired results. A core value should be viewed as a central point in a cluster of other related values. This point of view can be illustrated with an example of a mind map (see Fig. 18-1) where the core value "knowledge" has a set of related values. Looking at it this way, we can think of the organization's value system as a planetary system that contains a small number of core values, each of which is clustered with other operational values.

Some organizations spend a lot of time defining and articulating a set of core values, assuming that this is enough to give direction toward the right behavior. However, to ensure organizational success, we need to derive a major set of operational values from and aligned with the core values (see Fig. 16-1). This set of operational values can include teamwork, continuous improvement, innovation, creativity, self-improvement, etc., which will help the employees, particularly those who work on the front line, better understand how to relate the core values to their everyday working life.

The arsenal of operational values is gradually changed by adopting new values or upgrading the existing values. For example, creativity and innovation are gradually taking an important place in the value system because they mobilize employees' imaginations and foster unconventional thinking. Empowerment and teamwork is another set of values that many organizations have recently adopted. The power of these values has already been recognized by organizations that move toward employee involvement. Many organizations have also started to give more credit to values such as cooperation, collaboration, and coordination because they are in line with today's business environment. Competitiveness has become more important than competition because most organizations have recognized that to become a winner it is not necessary for someone to become a loser.

Because we are talking about the system of values, we must mention *speed* as one of the dominating values, because it is related to the fast adaptation of new

Figure 18-1 An Example of a Set of Organizational Values Clustered Around a Core Value

technologies, new science, and a quick reaction to change, including cultural changes. In providing these examples, we do not intend to provide a complete list of new values that will gradually take place in the value system. We find it important enough to emphasize again that the value system is dynamic and that we need to be open to adopt any new value if it can contribute to the organization's progress.

18.5 ESPOUSED VALUES

We mentioned that core values should come from prior learning. They are values that time has proven to serve our purpose. However, sometimes we are tempted to adopt some values that either reflect the contemporary way of thinking or are frequently adopted by some other successful organizations. One such example is teamwork. When this concept became popular, some executives and managers became strong believers in it and rapidly adopted teamwork as one of the organization's core values. What can happen in this case? After some work is done on the surface, people will claim that they work as a team, but when difficult situations arise, they will act in their former way.

Core values must come from our past experience and learning. If not, organizations and their employees will find themselves in a dilemma because they have adopted values that are difficult to implement in an unprepared organizational environment. Chris Argyris called values that are not based on prior learning, "espoused values."[6] These types of values are usually not patterned and may sometimes be mutually contradictory. Sometimes, espoused values are also inconsistent with observed behavior (see Table 18-1).

Table 18-1 An Example of Espoused Values that Are Inconsistent with the Observed Behavior

Espoused Values	Behavior
Innovation	Yet when it comes to the implementation of an employee's idea, which may or may not produce a desired result, the manager says no because it has never been done before.
Teamwork	Yet there exists a multilayer organizational structure in which all the decisions come from the top.
Cooperation and Collaboration	Yet there exist "walls" between departments, and nobody knows what is going on in other areas.
Knowledge	Yet there is no time dedicated for dialogue and knowledge sharing.

As you can see from Table 18-1, we may espouse values as in the left column but may act as in the right-hand column. If the organization does not introduce a structure that will support the espoused values, the situation will not change. However, if espoused values are reasonably congruent with the organization's structure and culture, they may gradually transform into real operational values or, with time, even become core values.

A Note of Caution

If espoused values "hang on" in the value system a long time without being transformed into operational values or core values, they may destroy the belief in the whole system. Because the incongruence is likely to develop or be perceived as organizational contradictions, we need to make sure that the espoused values receive a base—a structure—that supports them.

18.6 BEHAVIORAL STANDARDS

Culture is one of the most influential factors on people's behavior in an organization. In this section, we will describe the relationship of values and behavior, show how to develop behavioral standards; and show how values and these standards work together to improve the bottom-line results.

18.6.1 What Comes First: Values or Behavior?

As we mentioned above, once an organization survives for a relatively long time under the leadership of the founder, the founder's values will gradually become embedded in the organization's culture, which will be shared with all employees. However, while the organization continues to grow, some people who have worked in the company from its inception will eventually leave and be replaced by new employees. How then do we preserve the culture? By telling the newcomers to behave according to the organization's values? This will probably not help. For example, you cannot tell people that "you must be open and trusting" just because one of the company's core values is openness and trust. What an organization can do is to make sure that employees behave according to its behavioral standards and code of ethics. For example, a rule or a statement like "No commitments that we cannot keep" can be observed, measured, and managed. If values cannot be measured, behavior can and should. So what comes first, values or behavior? The reader will probably recognize a "chicken and egg" type of question here. On one hand, there should be an entrepreneur or leader who carries some values and has the talent to convince people to follow him, so that when he shares his values with others he becomes a true leader. In this case, values support the creation of a set of particular behaviors. On the other hand, when an organization has an established system of behavior, people act according to the system's requirements;

and when they recognize that the system works, they start to value this set of behaviors, which support a particular value.

We'll illustrate this with an actual example that took place in AMD about 10-12 years ago. Our employees started to apply statistical principles because it was required by many specifications. At that time, people believed in inspection as a form of protecting the customer from poor quality. It took many years of management effort and a lot of learning before the employees started to value statistical principles as a form of making quality products. Finally, the application of statistical principles has become an important element of our value system. Employees apply statistical methods not just because it is required by specifications but because they believe that it will bring good results.

In conclusion, values and behavior support and generate each other. To inculcate the desired behavior, we need the right values, and to keep the values alive, we need a structure that will support the desired behavior.

By analyzing values and behavior, we can observe a strong relationship between them. Let's take *trust* as an example. To develop trust between people, they need a safe environment in which they can express their feelings and thoughts. When employees feel threatened and unsafe, they usually become defensive and their behavior will have a protective character. To develop an environment in which people can trust each other, we need to create a condition in which they will feel themselves a part of a larger group. Any behavior that implies rejection will destroy the feelings of trust. One example of this is criticism. We all know the unwritten rule of "compliment in public and critique in private." Nevertheless, how many times do we consciously or unconsciously violate this rule? Table 18-2 represents an outline of behavior that encourages or hinders trust.

Behavior that supports the development of trust focuses on openness. In contrast, behavior that hinders the development of trust continuously and increasingly generates defensiveness. Similarly, we can find the relationship of other values that have influence on behavior. With this relationship, we can demonstrate the link and influence of the values on the bottom line: The right values create the necessary behavior to achieve the desired results (see Fig. 18-2). We can also view this relationship from another angle—the achievement of the desired results influences the organization's values, which continually reinforces more desirable behavior.

18.6.2 Developing Core Behavioral Standards

Even though operational values are closer to the worker's "skin" than core values, this is still not enough. What are needed are established behavioral standards that come from and support the operational values (see the system of organizational culture in Fig. 16-1).

The value system of an organization can have an impact on the operational results only when it affects the employees' behavior. The system sends signals to

BEHAVIORAL STANDARDS

Table 18-2 Development of Trust

Behavior That Helps Develop Trust:	Behavior That Hinders the Development of Trust:
Accept ideas.	– Ignore people.
Tolerate faults.	– Embarrass someone in front of a group.
If you disagree, criticize the idea, not the person.	– Fail to keep a confidence.
Clarify to make sure people understand each other.	– Avoid eye contact.
Include others in your activities.	– Withhold credit when it is due.
Offer to help.	– Interrupt when others are talking.
Ask others for feedback.	– Withhold information important to a decision.
Play together (Ping-Pong, volleyball, darts, etc.)	– Break a promise.
Work with others on a project.	
Tell others information that checks out to be accurate.	
Follow through on something you agree to do.	

Source: Adapted with permission from *Groups at Work* by Mink, Mink, and Owen, Educational Technology Publications, Inc., 1987, p. 65[7]

employees about how things should be done and helps them decide what to do and what not to do. These signals remain ineffective until they are translated into standards of behavior and are reflected in actual performance.

We usually say that the organization's culture is "the way we do things around here." This is true only if the value system is connected with the behavioral system and they interact with each other. So how do we achieve this type of interaction? Should we formalize a set of behaviors that reflect the value system, or is it more appropriate for people to determine themselves how to behave as long as they share the organization's values? In our opinion, it is

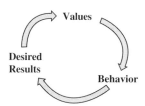

Figure 18-2 The Relationship between Values, Behavior, and Results

not necessary to tell people exactly how to behave. With some methodological help, they can develop guiding behavior themselves. For example, if teamwork is recognized as an operational value, we need some guiding principles, which will build a team environment where people will work together to accomplish their tasks. These may be:

- Acting as a team and respecting the individuals' needs
- Continuously sharing their knowledge with their peers
- Being willing to sacrifice short-term personal benefits for the long-term benefits of the organization, etc.

Our experience in AMD has shown that all these guiding behaviors can be developed by the team.

However, there are some core behavioral standards that need to be developed and articulated by management to capture the essence of an organization's culture. By making these standards explicit, we can focus our employees' attention on the general direction of the organization's culture. This will project a strong message to all employees of what management's expectations are in relation to organizational behavior, help people translate the shared values into concrete behavior and actions, allow the managers and supervisors to facilitate compliance with these behaviors, and help develop a method of measuring the improvement in the employees' behavior. All this will create a strong link between the value and behavioral systems.

Hewlett-Packard (HP) provides a good example of how to develop and articulate core behavioral standards in their document, "The HP Way," in which they articulate a set of behavioral standards that reflect their corporate culture (see Insert 18-1).

Their experience of developing and implementing effective ways of motivating the right behavior has been adopted by many organizations. One example is Management By Walking Around (MBWA), which has become a widely accepted concept.

Influenced by HP's concept, AMD has developed an analogical behavioral standard called "Learning By Walking Around (LBWA)." Getting out of the office and walking around inside and outside the organization is an effective way to learn and share knowledge. Individuals or small groups consisting of employees from different layers of the organization, and sometimes from different plants, walk around, observe, have short dialogues with employees from different areas, and then put together a report of what they observed and how their observations could be deployed throughout the entire organization. The same approach is used when the group, or an individual, walks around outside the organization's boundaries. The major goal is to absorb and deploy knowledge. LBWA has become even more important because we are planning to develop knowledge workers and build knowledge-based organizations.

Insert 18-1

An Extract from" "The HP Way"

Strategies and Practices

Management By Wandering Around. An informal HP practice, which involves keeping up to date with individuals and activities through informal or structured communication. Trust and respect for individuals are apparent when MBWA is used to recognize employees' concerns and ideas.

MBWA might look like:

- A manager consistently reserving time to walk through the department or be available for impromptu discussions.
- Individuals networking across organization.
- Coffee talks, communication lunches, hallway conversation.

Management By Objective. Individuals at each level contribute to company goals by developing objectives which are integrated with their manager's and those of other parts of HP. Flexibility—and innovation in recognizing alternative approaches to meeting objectives—provide effective means of meeting customer needs.

MBO is reflected in:

- Written plans which guide and create accountability throughout the organization.
- Coordinated and complementary efforts, and cross-organizational integration.
- Shared plans and objectives.

Open Door Policy The assurance that no adverse consequences should result from responsibly raising issues with management or personnel. Trust and integrity are important parts of the Open Door Policy.

Open Door may be used:

- To share feelings and frustrations in a constructive manner.
- To gain clearer understanding of alternatives.
- To discuss career options, business conduct and communication breakdowns.
- ***Open Communication*** At the core of our practice of open communication is the belief that when given the right tools, training and

(continued)

Insert 18-1 (*continued*)

information to do a good job, people will contribute their best. Open communication leads to:

- Strong teamwork between HP people, customers, and others.
- Enhanced achievement and contribution.
- Customer relationships built on trust and respect.

Source: Reprinted with permission from HP website.[8]

It is important to note that behavioral standards should not be confused with technical "standards" in which we give exact requirements, such as what the size of a product should be. "Behavioral standards" is a much broader term that includes the means of helping people behave in a desirable manner. Behavioral standards are developed under the assumption that people always want to do a great job. What they really need is support and facilitation. This thought is captured in a statement made by Bill Hewlett, the co-founder of HP, "What is the HP Way? I feel that in general terms it is the policies and actions that flow from the belief that men and women want to do a good job, a creative job, and that if they are provided the proper environment they will do so."[9]

Summary

Core values exert a tremendous power. To release this power and make it work for the bottom line, we need to align the organization's core values with operational values and behavioral standards. Together, they represent a value system that is the foundation of the organizational culture.

18.7 BUSINESS ETHICS

18.7.1 Understanding the Concept of Ethics

Our experience with the introduction of the ethical system in AMD showed that some managers, even though they apply ethics in their day-to-day work, need more elaboration on the concept and terminology of business ethics. In our workshops regarding this issue, when we asked the participants to describe what ethics meant to them, they usually related this term to feelings about what was right or wrong, to the Golden Rule, to their religious beliefs, etc. All this certainly has some relation to the subject in question, but we need a more complete definition of ethics.

In literature we can find a variety of definitions on this subject. For example, in her book *Fundamentals of Organizational Communication: Knowledge, Sensitivity, Skills, Values*, Pamela Shockley-Zalabak defined ethics as "...the standards by which behaviors are evaluated for their morality—their rightness or wrongness."[10] However, no matter how clear it is, a definition by itself does not give people as thorough an understanding of the concept as they can get from case stories related to the subject. Taking this into account, when we conducted our workshops on ethics, we asked the participants to divide themselves into teams and come up with a real or made-up story that would illustrate situations in which morals and ethics are at work. It is interesting to note here that the participants could better describe their understanding of ethics and morality through stories than by simple definition. We selected one of these stories to use here as an illustration that will help us better understand the subject of this section.

18.7.2 The "Testron" Story

A small company, let's give it the fictitious name of "Testron," that makes specialized equipment for testing electronic devices received an order from a large semiconductor corporation to design and produce a sophisticated tester. The technical requirements, engineering complexity, and time limitation made this order a challenge for Testron, but it also provided a great opportunity to become a preferred supplier for a major electronics corporation (who was planning to place future orders for a dozen or more testers at a price of several million dollars each). A special team was formed, on which every member was given a unique assignment. As usual in this type of situation, the schedule was tight. If Testron had had at least one more month to make adjustments, it would have been just great; but there was no extra time. Everything was needed in a rush.

Testron pushed its suppliers to deliver the parts on time, and finally the tester was ready for a capability study. Joe Smith (a fictitious person at Testron) was assigned to perform the study. Considering the requirement for special skills and features to perform the capability study, it was understandable that the customer would not duplicate the test and would just rely on Testron's data when the test results were ready.

Unfortunately, Joe found that the tester failed to perform with the accuracy and precision required by the customer. Late at night, he made a call to his supervisor to report the results. In response, he was told, "I'm not surprised, because the parts we used to build the tester won't perform any better, but I have to get a positive report written regardless of your results." Joe hung up the phone and knew he had a dilemma: Should he change the data on his data sheet (which means he would be participating in writing up a false report), or should he leave the data as they were and let his superior decide what to do? Joe was pondering this while he drove home. He ran a red light and was beeped at a few times, but thoughts kept popping into his head, "This is my first job after college. I'm finally doing what I like to do. My boss treats me well. I was promised a promotion at the end of this year. The company is doing well. My wife and I are expecting our third baby in the fall. We just bought a house. My wife won't be working after she delivers the baby. My reputation is great. I am a team player...." Gradually the abrupt thoughts started to take a more complete form: "If I don't participate in the false report, I will have to resign or will probably be forced to resign; the tester will go to the customer anyway, because somebody else will provide the falsified data. However, I will have the satisfaction of knowing that I have not participated in the falsification. My father would be proud of me acting according to my beliefs, but my mortgage payments won't be paid by personal satisfaction. Ethical principles are important, but the bills need to be paid...." Nearing the garage of his new home, he concluded that first thing in the morning he would rewrite his report. He did this and then walked into his supervisor's office and said, "Good morning. Last night I made some new calculations and fortunately I found some mistakes. The test results are just great." The boss looked at the report and then straight into Joe's eyes, said, "I like your calculations, and I like your attitude. It looks like we can work together as a team."

Even though the decision was made, Joe continued to think, "When the customer installs the tester, it will pass nonconforming parts. These parts will go to other companies who will use them, and they will have a high rework rate. People will receive products that will not perform properly. There will be a chain of problems, and the origin will be my report." He could not keep all this inside, so he finally gathered his courage, and went to the President of the company with his problem. The response was, "You think too much for someone in your position. You should do your engineering job and we'll do the thinking. "It would have been better if I had not discussed this with him," thought Joe as he left the President's office. In the meantime, the new tester was shipped to the customer, and the company received a new order for another, even more complicated tester from the same customer. Later, Joe resigned voluntarily. The story is obviously exaggerated, but with some variation, it could take place in some organizations. In the remaining part of this section, we will use this story as an illustration of the concept of ethics. But before we understand ethics, we first need to understand the meaning of morality.

18.7.3 The Meaning of Morality

As we can see from the earlier definition, ethics is related to morality, which can be described as the standards that a person or an organization has about what is right and what is wrong or what is good and what is bad. In the Testron story, Joe's belief that it is right to tell the truth and wrong to hide the information about the tester's capability and his beliefs that integrity and responsibility are good and that covering-up is bad are typical examples of moral standards. *Moral standards* include the *behavioral standards* that we believe are morally right and morally wrong. Moral standards also include the *value standards* that we believe are morally good and morally bad. Moral behavioral standards can usually be expressed as general rules or statements; like "always tell the truth," or "it is wrong to withhold information." Moral value standards can be expressed as statements that describe the worthiness of actions, for example, "honesty is good," or "discrimination is bad" (see Fig. 18-3).

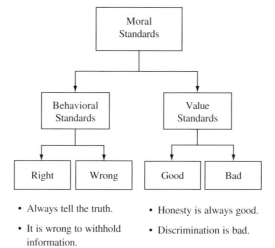

Figure 18-3 An Illustration of the Meaning of Moral Standards

18.7.4 Where Do Moral Standards Come From?

Usually, we learn about moral standards while we are young from our family, friends, school, religion, television, movies, books, etc. When we start to work, we face a different environment and our moral standards go through a kind of "life" test. Some of the standards will survive and others may be discarded or replaced by new moral standards.

In the right environment, through their maturing process people will develop standards that allow them to deal more intelligently with real life. However, there are situations in life in which we do not always live up to the moral standards we hold. In other words, we do not always do what we believe is morally right or good. As Joe stated later in his letter of resignation, "I grew up in a family where telling the truth was our living standard. The school that I attended taught us the same principle. In our church I always heard that lying is wrong. I don't know what forced me to go against my moral standards. It was wrong and I am sorry about it. I don't know how much my truth would have helped at that time. However, I know now that by not participating in the falsification, I would have kept my principles."

18.7.5 The Meaning of Ethics

Now that we have some understanding of the term "morality," we can better understand the definition of ethics, which was elaborated on above. In general, people start to behave ethically when they adopt the moral standards absorbed from their parents, friends, school, and other environmental sources. Then they ask themselves, Do these standards really make sense? Which should I follow? Why do I believe in these standards? Are there any better standards?

Let's go back to our Testron story. As we mentioned above, Joe was raised to accept the moral standard that a person should always tell the truth, so he felt that it would be wrong to write a false report on the test results; but we may ask whether writing the false report was really wrong in this particular situation. He had several important financial obligations and obligations to other people. If he had not behaved as he did, he would have been fired and not been able to fulfill these obligations. Now these questions arise: Why did the moral obligations toward himself and his family outweigh his obligation not to change the data in the test results? Why was the obligation to tell the truth greater or lesser than a person's obligation to himself and his family? Now consider Joe's obligations toward his employer: Doesn't an employee have a moral obligation to obey his employer? Is the obligation to obey one's employer greater than the obligation not to write a false report? Let's also consider who would be responsible if the tester caused a lot of trouble when in use in the customer's facility. The management who decided to ship the tester would take the responsibility—not Joe, who was just an engineer. Did Testron have the moral right to send out the tester knowing that it was not built to the customer's requirements? And, finally, if Joe had not agreed to falsify the data, someone else would've done it, and the tester would have been shipped to the customer anyway. Did he really have a moral obligation to refuse falsification? In other words, does one have a moral obligation to do something that will make no difference except for "shooting yourself in the foot?"

The reader has probably noticed the type of questions the Testron story leads us to ask. These are mainly questions about whether it is reasonable to say that one moral standard is more or less important than another and questions about what reasons we must have to hold these standards. When a person or organization asks these kinds of questions about moral standards, it means the person or organization is applying ethics. In short, ethics is the study of moral standards to determine whether these standards make sense. The other important aspect of ethics is development of a body of moral standards that we, as a person or an organization, feel responsible to maintain.

18.7.6 The Meaning of Business Ethics

Now that we have conveyed an idea of what "ethics" is, we can move on to the description of what is behind business ethics. As the term suggests, business ethics is a specialized study of moral rightness and wrongness in a business environment. Here we concentrate our efforts on moral standards as they apply particularly to business systems, processes, policies, procedures, instructions, and behavior. In short, we can say that business ethics is a form of ethics applied directly in organizations, services, and other institutions.

We can distinguish at least three different types of issues where business ethics apply: systemic, corporate, and individual issues. Systemic issues in business ethics include questions raised about economic, political, legal, and other systems. Using the Testron story, an example would be to question the system of testing finished products at Testron.

Corporate issues on ethics are ethical questions raised about a particular company. This includes questions about the morality of the activities, policies, practices, or organizational structure of a company as a whole. An example would be questions about the morality and organizational culture of Testron reflected in the story.

Finally, individual issues in business ethics are ethical questions raised about a particular employee, a team, or other business units within an organization. This may include questions about the morality of the decisions made and actions taken by groups or individuals. In our story, this would be the questions related to the behavior of the President of Testron, Joe's supervisor, the other people who work in this company, and Joe's own behavior.

Classifying ethical issues by types may help in analyzing ethical issues, which means when an issue is raised it is helpful to determine whether it belongs to systemic, corporate or individual issues. This can help to avoid confusion when complex ethical issues are raised.

18.7.7 The Distinction Between Values and Ethics

Our personal and organizational values significantly influence our ethical behavior. In other words, on the basis of our values, we determine what is ethical and what is not. However, not all values can be viewed as ethical standards of rightness and wrongness. For instance, values such as courage, teamwork, cooperation, competition, hard work, discipline, etc., cannot be viewed primarily as ethical standards. We cannot say that if someone is not a team player, he is unethical. However, value standards such as honesty, fairness, truthfulness, integrity, etc., are usually applied in making ethical decisions of rightness and wrongness in human behavior. In developing or enhancing the value system of the organization, management needs to make sure that ethical values are at work.

As you can see, values and ethics have a close relationship, but they are different. The major difference between them is that values are our measure of importance whereas ethics represents our judgment about rightness and wrongness. Together, values and ethics are a powerful combination that strongly influences our behavior and shapes our culture.

18.7.8 Codes of Ethics

A code of ethics is a collective agreement about the responsibilities of individual performance. It serves as an important component in creating an environment in which ethical behavior becomes the norm. When developing a code of ethics, it is recommended you involve people in a dialogue on this subject to make sure that the code is not perceived as a document coming from the top. If properly prepared, the code can serve as a direction in decision-making. When using the code as a major guide to resolve complex problems, we need to remember that it is not a cookbook that gives you the exact recipe for a particular dish. Ethical codes should serve as motivational tools to inspire ethical behavior rather than as a basis for proceeding against unethical behavior. In other words, an ethical code should have a positive connotation to encourage desirable modes of conduct. A code of ethics, unlike a set of laws, must be based on a generally accepted set of ethical principles, leaving enough space for interpretation in the light of general moral values.

References

1 Thomas J. Watson, Jr. and Peter Petre, *Father, Son & Co.—My Life at IBM and Beyond*, Bantam Books, New York, NY, 1990
2 Thomas J. Peters and Robert H. Waterman, Jr., *In Search of Excellence*, Harper & Row Publishers, Inc., New York, NY, 1982, p. 280
3 Edward De Bono, *Textbook of Wisdom*. Penguin Books, New York, NY, 1996, p. 91
4 Peters and Waterman, *In Search of Excellence*, p. 281
5 James C. Collins and Jerry I. Porras, *Built to Last*, HarperCollins Publishers, Inc., New York, NY, 1994, p. 70
6 Chris Argyris, *Overcoming Organizational Defenses*, Pearson Education, Inc., Upper Saddle River, NJ, 1990, p. 60
7 Oscar G. Mink, Barbara P. Mink, and Keith Q. Owen, *Groups at Work*, Educational Technology Publications, Inc., Englewood Cliffs, NJ, 1981, p. 65
8 HP Website, p. 2
9 Ibid., p. 1
10 Pamela Shockley-Zalabak, *Fundamentals of Organizational Communication: Knowledge, Sensitivity, Skills, Values*, Longman Publishers, 1995, p. 435

Chapter 19

Symbols, Symbolic Actions, and Metaphors

"We are symbol-inventing, symbol-using animals."

John W. Gardner[1]

Symbols and symbolic actions are an important part of the whole organizational culture system. They consist of the artistic creation of buildings and products, the professional language used in communication, the style embodied in dressing, manners of address, emotional display, myths and stories told about the organization, the display of purpose, vision, mission, and values, heroic sagas, observed rituals and ceremonies, etc. If we consider shared values as the driving force (a motor), then symbols and symbolic actions are the devices that start the motor. It is not enough to articulate core or operational values; we need a way to translate these values into day-to-day actions. We need heroes, stories, and symbols that will influence the employees' minds. Below we will describe some of the symbols and symbolic actions that are frequently used in an organization.

19.1 THE FOUNDER AND LEADERS AS SYMBOLS OF THE ORGANIZATION

When you walk into the lobby of AMD's main building, you can see the brass sculpture of Jerry Sanders, the founder of AMD. His personal style, experience, and behavior provide extremely potent symbols not only for the employees but also for the public. His values have become the roots of AMD's culture. The personalities of the founders and their corporations are often inseparable; one is frequently an extension of the other. Jerry Sanders symbolizes AMD, and AMD symbolizes Jerry's character.

Rich Previte, the previous President of AMD, also made a strong impact on the corporation's culture by his symbolic actions. An example of this was his way of asking questions in casual meetings with the employees. It did not matter what topic triggered the conversation, he always ended it with questions related to profit. When a certain type of question is frequently asked by the top executives and becomes a central part of meeting agenda, it will eventually influence the shared values and the whole culture of an organization. Rich Previte symbolized a leader who was interested in the bottom line. He was always interested in "Where's the beef?"

19.2   STORIES AND MYTHS

Stories and myths about how organizations have dealt with complicated situations in the past, how creative the leaders were in accomplishing goals, how the organization shows respect for the people, and so on, form the organization's pattern of culture. Sometimes your actions can result in a story that reflects the organizational culture and makes your behavior a symbol of the organization forever. To illustrate this point, we will tell a short story that took place in AMD Penang in 1997.

It was the end of the quarter, and the business was tense. At this critical time for the plant's performance, Penang had an electrical cable problem and lost power. C.H. Teoh, the managing director of AMD Penang, used his personal business connections to find and utilize portable electric generators and therefore fulfilled the quarterly schedule of shipments on time. This episode made him the hero of the quarter, and he gained the reputation of a man who could run manufacturing, even "without electricity." C.H. Teoh has a history of getting things done and is a man of action, but sometimes it takes the "last straw" to become a hero, and this episode in 1997 was the last straw. His management style is a symbol of the organization's culture.

19.3   THE ORGANIZATION'S HISTORY

Every organization has a history. Some companies spend time accumulating facts that will preserve the history of organizational development, even though they are busy with the present and with the concerns of the future.

The Spirit of AMD, written by Jeffrey L. Rodengen in 1998, captures the highlights of the company history starting with its birth in 1969. This book contains facts and stories about people who contributed to AMD's development. It also describes other moments that demonstrate how AMD became a global supplier of integrated circuits. This book became possible due to the accumulation of periodic publications from all parts of the organization. AMD currently has an internal publication, *Dialog*, which reflects the life of the company. As the name of this publication suggests, it captures an ongoing dialogue among the people who work here and make things happen. Some groups and divisions also have their own publications that preserve the events that take place in the various locations of the multinational organization. For example, the Manufacturing Services Group has its own biannual publication titled, *InFocus*. This magazine focuses on capturing all the activities that occur in the different countries where AMD has its manufacturing plants.

Having publications in different parts of an organization is not only an effective form of motivating people, but also a form of creating an organization's memory. It creates the possibility to go back in time and see how the corporation progressed and how the culture was shaped. All the symbols and symbolic actions, stories about engineers, technicians, operators, secretaries, and other interesting people who have built the company is a treasure that should be preserved. When new people start in a company, they want to know what the company is all about, and they want to know its history. Having a book on the company's history is important. It is a part of the organization's culture.

To preserve traditions, some organizations also use rituals that involve the time capsule concept. Below is a story from MSG that describes this concept.

For example, 10 years ago in one of MSG's operator's symposiums, four 18-year-old operators who represented the Malay, Indian, Chinese, and Thai nationalities—the major nationalities that make up the workforce of AMD's offshore plants—went up on stage. They read a report of AMD's accomplishments and then a vision for the future that was to be sealed up in our time capsule (to be opened in the year 2000, when they would be 28 years old). Many of the participants probably thought that this would take forever. This year, these four operators will have the opportunity to report the accomplishments for the last 10 years at the operator's symposium. A ritual like this also belongs to the arsenal of symbols and symbolic actions.

At times when employee turnover is relatively high, it is very important to ensure that potential employees are familiarized with the organization's culture. Rituals of working life, from the selection process to retirement dinners, help to shape the culture.

All the various locations of AMD have developed an orientation package for new employees. It includes conversations with executives or senior management who will describe the organization's culture and business ethics, short video tapes that show the history and today's life in AMD, plant tours through AMD's facilities, etc. Familiarizing the newcomers with attributes that symbolize the organization's culture is an important orientation activity that helps the newcomer adapt more quickly as an employee who shares the organization's values.

19.4 THE ORGANIZATION'S RITUALS IN WORKING LIFE

Rituals are a type of symbolic action, and they play an important role in shaping the organizational culture. They are visible and therefore can give people a direct impression of an organization's culture. If properly managed, these rituals can be a very powerful method to enhance employees' spirits and hence, the organization's culture. Below are some examples of rituals that are included in AMD's culture system.

19.4.1 Meeting with Leaders

Breakfast with Jerry It has become a ritual for the AMD management to meet on a regular basis with Jerry Sanders, the Chairman of the Board and Chief Executive Officer (CEO). This meeting takes place in the morning with a continental breakfast; therefore, it is called "Breakfast with Jerry." Every organization probably conducts similar meetings. However, what we want to emphasize here is not the content of the meeting but rather the social part of it. First of all, Jerry always comes to the meeting early so people can have a personal talk with him before the formal meeting starts. This short, informal conversation with Jerry one-on-one, or in small groups, has a lot of value. You would need to be there to recognize and feel the culture aspect of this breakfast meeting. On the good days, Jerry has the talent to give the credit to the people who made the success happen. People feel good about it. On the bad days Jerry knows how to encourage people not to give up but to stretch that extra mile to achieve a goal. The "Breakfast with Jerry" is one of the ways that a leader builds comradeship. On this kind of occasion, Jerry uses his charm and charisma to motivate people and share his confidence. His humor and simple form of communicating complicated issues give people confidence and make them feel good. Jerry's presentation and his responses to the questions that arise in this meeting are always shared with the rest of the employees. This is a way to help keep employees informed and on the right course.

'Breakfast with Jerry' has become a tradition for AMD. This is why when Hector Ruiz was appointed the new CEO in 2002, he continued this tradition and is giving 'Breakfast with Hector' a new flavor.

19.4.2 National Rituals

The AMD manufacturing plants, located in different countries, are rich with national rituals and folklore. An interesting process of bringing these rituals into our organizational culture can be observed. Without special planning, some of the rituals have gradually become a part of the organization's culture.

To highlight, in 1998 we had the opportunity to participate in AMD Thailand's New Year's celebration, which included the water festival (Songkran). On a sunny Saturday morning, the employees came to the plant to celebrate this event. Some of them came with their parents, children, and other relatives. Although the Managing Director did not mention much about work in his speech for this occasion, this ritual can be considered an important element of the organizational culture. The employees who participated in various festival activities this day were from different levels and departments of the organization. People dressed in beautiful national costumes participated in different celebration activities, such as a beauty contest, dancing, singing, and games, and the festival activities culminated in everyone throwing water on each other at the end of the day. All this fun helps build a stronger bond between employees, improves comradeship, and enhances communication, and, besides, everyone had a lot of fun. Almost all national rituals, if properly prepared for and supported by the organization, can become an integral part of the organizational culture.

19.4.3 Groundbreaking Ceremonies

We have also developed an interesting scenario to celebrate the expansion of our capacity. Every time we build a new plant or expand an existing facility, a special ritual that reflects the folklore of that national culture is held to make the groundbreaking ceremony a memorable celebration. In 1998, we had a groundbreaking ceremony in Suzhou, China. The ceremony combined elements adapted from the Chinese culture and AMD's culture. To increase awareness of this groundbreaking event, videotapes of the event were made. Expanding AMD's capabilities is a sign of the organization's growth and prosperity. It makes employees feel happier and more secure. This is why it is an important event to celebrate and announce to all employees.

19.4.4 An Organization's Anniversary

Like a person, an organization has a birthday, which needs to be celebrated. In AMD, it has become a ritual to celebrate not only the corporate anniversary, but also the birthdays of its plants. We celebrated the 25th anniversary of AMD Penang, which was held in a stadium with the capacity to hold all 3,000 employees and their guests. This celebration will be remembered for a long time. Again, combining AMD's traditions and the national culture in Penang created a special scenario that contained old and new rituals. Employees who had been in AMD Penang from the beginning shared their life stories of 25 years at AMD, which became the building blocks of AMD's culture. Young employees who had just joined the organization shared their dreams and goals. The Chinese Dragon Dance concluded the celebration, which went on all night.

19.4.5 Humor as a form of Symbolic Action

As we mentioned above, myths and stories are important elements of organizational culture, but what about humor? How many times have you, as a leader, relieved tension or the pain of disappointment by telling a funny joke in a formal presentation, a debate, a difficult situation, or a friendly conversation?

For most of us humor is a part of our nature. As Mark Twain said, "A sense of humor is the one thing no one will admit not having."[2] Our advice to the knowledge manager is to use it more often and not to be surprised if your subordinates also make you laugh. There is a saying that if you cannot take a joke then you will have to take the medicine. So it is more healthy and economical to take a joke. In AMD, friendly humor has become a part of our culture. When we are engaged in lengthy seminars or under a lot of pressure, we use humor to relieve the tension.

In our Singapore plant, we have a ritual to welcome new hires. All new employees will attend the 1-day Orientation Program, which gives an overview of AMD. The Program is followed by a lunch with the managing director, T.S. Tan, and his staff. This lunch acts as the first informal way for the management to welcome and to get to know the new hires. This actual mingling also gives the new employees a much deeper impression of who's who in the various functions, compared with to trying to recognize and memorize the faces seen on the organization chart.

The warm welcome does not end here. During the quarterly employee communication sessions, or the biannual TCPI[2] conference (whichever comes earlier), respective managers will accompany their new hires to stage. This is a light-hearted moment. First, the manager will give a brief introduction of the new hire. Next, the latter will take a couple of minutes to say something about him/herself. This is sometimes followed by questions from the enthusiastic audience, who is usually interested to know their age and availability. Note that a large audience, normally in the range of 200 people, attends these communication sessions or conferences. Despite the business nature of these conferences, this short segment of introducing the new hires, usually at the beginning of the conference, is always one highlight of the day. This can be seen from the abundant laughter and applause for those on stage. It sets the mood for the conference.

We have provided just a few examples of the large spectrum of rituals that are a part of AMD's cultural system. Organizations should put forth more effort and attention to the development of all kinds of rituals of working life. The rituals should include such events as hiring new employees, celebrating large project accomplishments, recognizing people who receive patents, retirement dinners, festive parties, etc. All these are elements of a large spectrum of celebrating work and accomplishments. This is an important part of the organizational culture.

19.5 INTERPRETATION OF SYMBOLS

Because symbols are on the outermost layer of the organizational culture, they are often tangible; but at the same time, they can be very difficult, if not impossible, to properly interpret. For example, can we conclude that an organization has a high level of communication just because they have an open office layout (cubicles) for all levels of management or that they have a bureaucratic attitude just because the executive offices have thick doors that are always closed? The interpretation of symbols and symbolic actions becomes even more difficult to decipher when you work in an international organization or are visiting different countries.

We recently visited a large Japanese supplier of AMD. When we entered the conference room to present an educational program to their staff, we noticed that a stranger would have difficulty determining who was the President and who was an engineer. All the participants were dressed in company uniforms, all of them were smiling slightly, and all of them looked ready to get down to work. Can we conclude from these observations that the organization we visited was very democratic and that there was simple communication between all levels of the organization structure? Later, when we moved to the discussion part of the delivered material and in other conversations, we also observed that an engineer who sat in the conference room with the higher-ranked executive would seldom freely participate in asking questions or giving comments. He would do this only if the higher-ranking managers asked him to do so.

19.6   A NOTE OF CAUTION

The examples above illustrate that symbols and symbolic actions are part of the organizational culture. They are sometimes very impressive but are sometimes also difficult to use so that they reflect the culture of the organization. The interpretation of all these symbols in relation to the organization's culture depends on our perception and may or may not convey the right impression. These are the artifacts that are easily seen but difficult to decipher correctly.

However, because people tend to believe what they see and draw conclusions based on their impression, it is very important for an organization to learn the art of demonstrating their symbols and symbolic actions to create a positive image. It is also very important to create an environment in which all the employees will have a similar interpretation of symbols and symbolic actions. This will significantly improve employee communication and shape the organization's culture.

19.7   A METAPHOR: THE TREE OF CULTURE

Using a fruit tree as a metaphor to describe an organizational culture can help us to achieve a deeper understanding of how its components work together as a whole system (see Fig. 19-1). The skeleton of the tree by itself can be considered as the structure, which supports the organizational culture system. It holds all the components together and allows them to interact and complement each other. The roots of the tree can be considered as the core values, which influence the organization's behavior. Real core values are usually those that are invisible. We do not "touch" them, we do not debate about them, we just act according to their requirements.

390 SYMBOLS, SYMBOLIC ACTIONS, AND METAPHORS

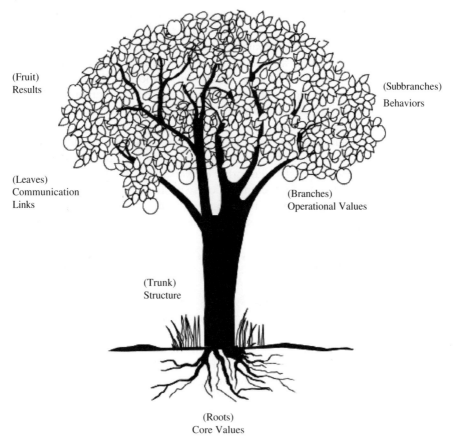

Figure 19-1 The Tree of Culture

The major branches of the tree can be compared to the operational values, which are close to the actual performance. They carry the subbranches, which may be compared to our behavior that holds and delivers the results (the fruits). The leaves can be considered as communication links of the organizational culture system with the external environment. Without these links, the system will die. As you can see, we need all the components to make the system work properly and achieve the desired results.

With time, and influenced by the environment, the branches and the roots of the tree may change their form and sometimes dry out. The same thing is true with the core and operational values; they may also change under the influence of the environment. Here we need to note that core values are less exposed to change than operational values.

As we mentioned above, some organizations develop and articulate the core values and consider this as a job completed. Core values without proper structure, operational values, and behavior will not influence the bottom line no matter how good they are. We need the whole tree to enjoy the fruits.

One can ask, Where is the organization in this metaphor? Thinking culturally, we may say that the organization is its culture (the whole tree). The organization and its culture are inseparable. There is no organization without a culture, and there is no culture without an organization. The culture *is* the organization.

Note: We did not include the espoused values here because they do not come from the roots (the core values).

For some types of trees you need to be very familiar with them to predict what type of fruit they will produce; it depends on the form and patterns of the leaves or other elements of the tree. An undereducated layman would have a problem predicting the type of fruit. The same thing is true with organizational culture. You need to be *in* the organization to recognize its culture. From the posters on the wall, the booklet of core values, and other attributes, it is very difficult and sometimes impossible to guess the culture. All phenomena that one sees, hears, and feels when one visits an organization are called artifacts. They are easy to observe but difficult to decipher.

Sometimes we wonder how the same type of tree can produce fruits that look and taste differently. The same question is legitimate to ask about organizational culture. With almost the same set of core values, why does one organization stay in business successfully and the other organization fail? In relation to the tree, the answer is simple, it depends on the treatment that the tree receives: fertilization, watering, trimming, and other components that allow the tree to unleash its natural capabilities and produce its fruits.

Translating this to an organization we could say almost the same thing: It depends on how you treat the people who work there, the customers, the suppliers, and all the other stakeholders. Core values are just a blueprint. It depends on how you use it as a guide to behave—this makes the difference.

In AMD, we have "Respect for People" as one of our core values. By that, we mean "Honoring their rights as individuals—is the cornerstone of our culture. We respect individual differences and diversity as qualities that enhance our efforts as a team. We believe in treating each other fairly. Fairness is based on what is right, not who is right."

Many organizations have "Respect for People" in their set of core values, too, which may be interpreted in different ways (see Table 19-1). However, only actual behavior in relation to "Respect for People" makes the difference. It is the same with other core values.

So articulating a set of values and posting them on the wall does not require a lot of effort. Releasing the power of core values is an art and a science, and this makes the difference. The roots of the tree are very important, but what will happen to the roots if you cut off all the other parts of the tree? Core values are very important, but they are still only one element of the whole culture system of the organization.

Table 19-1 Different Interpretations of the Value "Respect For People" in Some Famous Companies[3]

3M	Respect for individual initiative and personal growth
Ford	People as the source of our strength
Hewlett-Packard	Respect and opportunity for HP people, including the opportunity to share in the success of the enterprise
IBM	Give full consideration to the individual employee
Johnson & Johnson[4]	We are responsible to our employees, the men and women who work with us throughout the world
Marriott	People are number 1—treat them well, expect a lot, and the rest will follow
Motorola	Treat each employee with dignity, as an individual
Philip Morris	Opportunity to achieve based on merit, not gender, race, or class
Procter & Gamble	Respect and concern for the individual
Sony	Respecting and encouraging each individual's ability and creativity
Wal-Mart	Be in partnership with employees

Important Comment: Even though they are out of sight, the roots determine the quantity and quality of the fruit (the results). What if the environment is significantly changed with time—can the roots support the tree and still bear fruits of quality? Probably not. The same question can be asked about an organization. What if its core values no longer reflect the current internal and external requirements? In this case, instead of major values being the driving force, they may sabotage the organization's ability to address current business realities.

In this rapidly changing business environment, it is very important to learn how to differentiate between requirements and traditions and whether it is necessary to correct the whole value system.

References

1 John W. Gardner, *Morale*, W.W. Norton & Company, Inc., New York, NY, 1978, p. 25
2 Glenn Van Ekeren, *The Speaker's Sourcebook: Quotes, Stories and Anecdotes for Every Occasion*, Prentice Hall, Englewood Cliffs, NJ, 1988
3 James C. Collins and Jerry I. Porras, *Built to Last*, HarperCollins Publishers, Inc., New York, NY, 1994, pp. 68–70
4 John R. Childress and Larry E. Senn, *In The Eye Of The Storm: Reengineering Corporate Culture*, Linc Corporation & Senn-Delaney Leadership Group, Inc., Long Beach, CA., 1995, p. 83

Chapter 20

Understanding an Organization's Behavior

Every organization, corporation, group, division, plant, or team has its own cultural pattern. However, just as any small individual floret of broccoli preserves the dominant shape of the whole head, the cultural pattern of any part of the organization reflects the whole corporate culture. To understand the behavior of any organization in a corporation, we must understand the mental model of the organization, which is based on its system of values and beliefs.

The mission of our engineering group is to continuously develop and implement the most progressive packaging solutions for AMD's integrated circuit devices. This group, based in the U.S., consists mainly of engineers and technicians and coordinates its work with other engineering departments located in four Asian countries. To decipher the behavior of this organization, we need to understand the underlying value system that influences their behavior. As Figure 20-1 suggests, there are four major shared operational values that work in this organization, all interconnected.

1. New ideas come from experimentation and risk taking.
2. A team environment makes the work more productive.
3. Incremental improvement and innovation complement each other.
4. We learn from each other and complement each other.

As you can see (in Fig. 20-1) all four operational values come from and support the corporate core values (see Section 18.3). The belief that new ideas come from experimentation and risk-taking (1) reflects and complements the core value "knowledge." Similarly, we can find the relationship between other

393

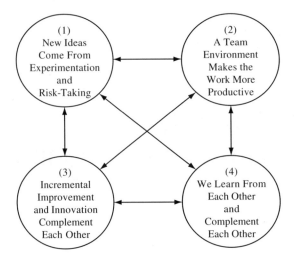

Figure 20-1 The Shared Mental Model of MSD Engineering

operational and core values. The four operational values also have a strong interrelationship, and this is the power of the value system. If the need to have a team environment (2) is not recognized, it will be difficult to learn from and complement each other (4).

By trying to understand the individual elements of the mental model, we may not find out how the organization behaves. Only by seeing the combination and interaction of the operational values can we understand the everyday behavior of the organization. Although every department, plant, division, or group may have its own pattern of a mental model, it is important to ensure that all of them are aligned with the corporate core values. Over time, some elements of the pattern may begin to change, and this can impact the whole pattern, although, as long as they match the corporate philosophy, they are totally acceptable. We may also see that in some larger organizations, in addition to the general mental model pattern, some "submodels" may occur. In this case, we need to study the reason for this occurrence. For example, we have a Supply Management Division that shares AMD's values. However, this division also has a set of operational values that reflect the peculiarities of this division as a pure service organization that works with its suppliers to satisfy AMD's needs as a customer. The Supply Management Division has its own cultural pattern and at the same time, reflects AMD's cultural pattern because it is a part of us.

We presented this simple example to demonstrate the point that although every team, department, or plant may have a system of operational values that are aligned with the corporate set of core values, these are also based on the internal and external forces of this particular group.

20.1 SHOULD ORGANIZATIONAL VALUES CHANGE?

Should an organization's values change, or should they stay the same forever? This is not a question that has a simple answer; the answer depends on a lot of things.

In their book *Built To Last*, James C. Collins and Jerry L. Porras wrote, "Visionary companies tend to have only a few core values, usually between three and six...for only a few values can be truly *core*—values so fundamental and deeply held that they will change or be compromised seldom, if ever."[1]

Core values, if they are really "core," stay with the organization almost forever. It is difficult to imagine that a core value like "Respect For People" could ever become less important than it is now. However, every core value has a cluster of other supporting values (operational values) that may and should change over time. The same thing is true with norms of behavior and standards. As John W. Gardner noted in his book *Morale*, "The whole structure of values, beliefs, laws, and standards by which a society lives must be continuously restored, or it disintegrates. It will survive and flourish only if people continuously renew the values and reinterpret tradition to make it serve contemporary needs."[2] Here the author is referring to a society, but this thought can be related to an organization simply because an organization is a society—a social system. You may recall a time when "bigger is better" was a widely accepted thought. Now we do not value this view so much, and those who have continued to live with this belief have a lot of problems. You may also remember a time when organizations strongly believed in the need to produce all parts of their products themselves. Now, outsourcing and subcontracting is a widely accepted way of working. We could continue to give examples to demonstrate that if we want to preserve a culture, we need to do everything we can to preserve the core values and change our operational values and norms of behavior to match the ever-changing external environment. And when it is time to change even some of the core values, we will need to do that. The whole value system is dynamic, which is why it is so powerful.

20.2 SHOULD WE PRIORITIZE VALUES?

When an organization is working to put together a set of core values, the question that frequently comes up is, How important is the prioritization of these values? Usually it would take a lot of effort to select a set of most important values from a long list and further reduce them to a set of three to five values. We would feel that all of the values are so important that it would be difficult to prioritize them. For example, we all understand that the external customer is the most important stakeholder, but we also feel that our employees are our most important assets. So, who is first? If the people in the organization are not satisfied, you will probably not have satisfied customers either. On the other hand, if you do not have external customers, how do you satisfy your employees?

For every particular organization, there is always a feeling that some values must be at the top. For example, Disney Corporation's four core values for their theme parks are, in order of importance, *safety*, *courtesy*, *the show* (performing according to the requirements of your particular role), and *efficiency*.[3] Who can argue that safety should be one of the most important core values for Disney, and, of course, it is. However, at the same time, this should be a given factor, as sanitation is for a restaurant and as prescribing the right medicine to the patient is for a doctor. There was a time when the quality of the product was one of the major core values. Now we do not put this in our core values—it is the entry ticket to the marketplace. What Disney is actually selling is not safety but entertainment. Of course, this is a matter of opinion.

Jack Welch, the Chairman and CEO of General Electric stated, "Values are key to our success at GE. When I meet with groups at our training center, I put these values up on the board and talk about what they mean to our leaders. You can argue about what's first and what isn't, but not about the importance of our values."[4] When we look at the organization's values as a holistic system, there is no need to argue about what is most and least important. All values are equally important because they are interrelated and work together as a whole, and the greater their interaction, the more important the values become by themselves. So we need to move away from the habit of straight-line thinking to closed-loop thinking, in other words, move toward thinking systemically.

20.3 CULTURE AND CHANGE

Usually, when the term "culture" is used, people will relate it to something stable and unchangeable; something that has been gradually built up for years. When saying "This is our culture," people usually mean something good that we would like to preserve. In contrast, when the term "change" is used, people presume something related to transformation, doing things differently, something dynamic.

So what is the relation between these two terms? How do they work together? In its simplest form, culture means "It is the way we do things here." When we say organizational change, we mean changing the way we do things in an organization. We have already accepted the notion that we need to continuously change the way we do things to survive. Does this also mean that we need to continuously change the culture? What about the external environment? How much does it impact the organization's culture? How much can (or should) a culture be preserved? How much do the personal characteristics of a leader impact the organization's culture? How much do new leaders carry on the culture of their predecessors?

To shed some light on these questions, we will use IBM's culture as an example. We will go through a full cycle, from its inception through development, maintenance, transformation and change. We have selected IBM for the following reasons. First of all, IBM is one of the oldest companies in the United States—it was founded in 1911 so it has a culture that is more than 90 years old. Secondly, IBM is one of the largest companies in the world—it started as a four million dollar business with 1,200 people in 1911 and became a global corporation that has more than 319,000 employees located in over 100 countries. Third, IBM has a number of leaders that have worked long enough to have an influence on the organization's culture. Since IBM has survived for almost a century, we can observe and obtain answers to these questions of cultural development. However, the main purpose here is not so much to respond to these questions but rather to generate new thoughts for the readers and provoke new questions.

20.4 LEARNING FROM IBM'S EXPERIENCE

20.4.1 The Timeline of IBM's Leaders

We would like to start this section with a brief description of the timeline of IBM's leaders because of its uniqueness. IBM has a long history that contains a start-up period, a family leadership period, a number of leadership transitions, a crisis, and a successful transformation. Even though we are only able to provide a short description of this timeline, we think that the readers will get the feeling of IBM's pattern of leadership for the last 91 years.

From its beginning in 1914 to 1971, IBM was led by the Watson family. Later, a number of other presidents and CEOs have been appointed to lead IBM from 1971 to the present time. They all contributed to the development of the organization's culture. To simplify things, we have divided the whole time frame of IBM's leadership into seven periods (see Fig. 20-2).

The first period started with George W. Fairchild, who was the first president of Computing-Tabulating-Recording Company (C-T-R). In February of 1924, it was renamed International Business Machines (IBM). George W. Fairchild became the chairman of the renamed company for a very short time. He died in December of that year.

The second period is the time of Thomas Watson, Sr., the first CEO of IBM. He laid the foundation and gradually built a strong corporate culture.

The third period is when Thomas Watson, Jr. followed in his father's footsteps. He further developed the organization's culture and put a lot of effort into maintaining the already-established core values.

The fourth period is special because it started in 1971, when the Watson family leadership came to an end (after 57 years). In a transition period like this, it is common to see a frequent change of leadership because of the work of two types of forces. On one hand were the internal forces of a strong culture that had existed for 57 years, and on the other hand were the external forces of time that necessitated major changes. This was also a period in which new leaders tried to introduce changes according to their philosophies but meeting strong resistance from the well-established culture. All this created a lot of

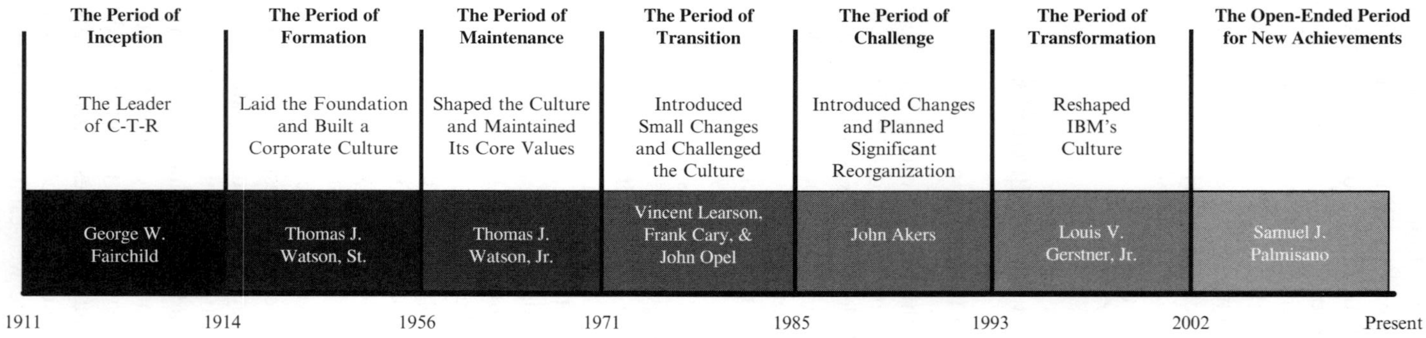

The Period of Inception	The Period of Formation	The Period of Maintenance	The Period of Transition	The Period of Challenge	The Period of Transformation	The Open-Ended Period for New Achievements
The Leader of C-T-R	Laid the Foundation and Built a Corporate Culture	Shaped the Culture and Maintained Its Core Values	Introduced Small Changes and Challenged the Culture	Introduced Changes and Planned Significant Reorganization	Reshaped IBM's Culture	
George W. Fairchild	Thomas J. Watson, Sr.	Thomas J. Watson, Jr.	Vincent Learson, Frank Cary, & John Opel	John Akers	Louis V. Gerstner, Jr.	Samuel J. Palmisano
1911	1914	1956	1971	1985	1993	2002 — Present

Figure 20-2 IBM Timeline

structural tension inside the organization, which continued for years and needed to be resolved. Vincent Learson, Frank Cary, and John Opel led IBM during this period.

The fifth period could be considered a continuation of the fourth period. In these years, when IBM was in crisis, John Akers struggled to implement different drastic programs aimed at turning the company around.

The sixth period was when IBM achieved a major transformation. In this period, the resistance of the strong culture was the weakest, the people were frustrated, and they were ready for a leader that would bring them out of the crisis. This was when Louis V. Gerstner, Jr. stepped in and did exactly that.

And finally, the seventh, open-ended period occurred with the election of Samuel J. Palmisano as CEO of IBM. What will Palmisano's strategy be? Will he put all his efforts into preserving Gerstner's legacy, or will he do something extraordinary? Neither will be easy to do.

Due to space limitations in this chapter, we will concentrate mainly on two areas of the timetable: the Watson's period, which contains historical information about the formation of IBM's culture, and Gerstner's period, which is rich with transformational changes.

20.4.2 A Little Bit of History

To better understand how IBM's culture was developed, it is probably best to start with some chronological data that reflect the historical period.

IBM was incorporated in 1911 as the Computing-Tabulating-Recording Company, or C-T-R. At that time, this company had 1,200 employees. In 1914, when the company experienced economic difficulties, Thomas J. Watson, Sr., age 40, was invited to join the company as a General Manager. Eleven months later, he became the President of C-T-R. In 1924, C-T-R changed its name to International Business Machines, or IBM, and Thomas Watson, Sr. became the CEO of the company. In 1952 Thomas Watson, Sr., after nearly 40 years as IBM's leader, passed the title of President on to his 38-year-old son, Thomas Watson, Jr. He became Chief Executive Officer just six weeks before his father's death on June 19, 1956, at the age of 82. The 1970s saw the end of more than half a century of Watson family leadership. Thomas Watson, Jr. stepped down as CEO in 1971. Later, he served as U.S. Ambassador to the Soviet Union from 1979 to 1981, and remained a member of IBM's Board of Directors until 1984. He died in 1993 at the age of 79.

20.4.3 The Formation of IBM's Culture

As a tree needs at least three components to grow: a seed, the right soil, and the right environment, the formation of an organizational culture also depends on three major components: an entrepreneurial leader, a group of people who support him, and the right environment.

In a way, the beginning of the formation of an organization's culture is a matter of chance. It depends on who the leader is, where he worked previously, his personal values, and his dreams and motivations. It also depends on the economic, political, and social climate at the particular time of inception. It happened in 1914, by chance, that Thomas J. Watson, Sr., a 40-year-old man, was out of work. It also happened that Charles R. Flint, the founder of what was to become IBM, was looking for someone who could bolster a company that was close to collapse. Flint met with Thomas Watson and recruited him as a general manager of C-T-R, a small conglomerate he had assembled in 1911. We can consider this event to be the beginning of the creation of IBM's culture.

Before proceeding further on the development of IBM's culture, we need to go back by asking these questions: Who was Tom Watson at that time? What did he bring with him as a leader to an organization with 1,200 people that had been formed from three smaller organizations and produced different types of products ranging from commercial scales and industrial time records to meat and cheese slicers and, of course, tabulators and punch cards? Before joining C-T-R, Thomas Watson worked for 18 years at National Cash Register Co. (nicknamed Cash by its employees), where he was the second top man in the company under John Henry Patterson. This was one of the major places that Watson formed himself as a leader with particular values, which further influenced the forming of IBM's culture. "Dad worked for him [Patterson] for eighteen years and learned from him many of the ideas that built IBM,"[5] remembers Thomas J. Watson, Jr. in his book *Father, Son & Co—My Life at IBM and Beyond*. Learning from a leader does not mean copying him, but working with Patterson for 18 years certainly influenced Thomas Watson's style.

We will not fully describe Patterson's leadership here, but we will provide one story that will give you a feeling for his management style. In *Father, Son, & Co.*, Thomas J. Watson, Jr. wrote, "Dad told the story of an executive who showed up for work one morning not knowing Mr. Patterson was angry with him. On the grass in front of the Cash he found his desk and the contents of his office, soaked with kerosene and set on fire. He left without ever going into the building."[6] This episode is probably good enough to illustrate Patterson's character. So when Thomas Watson, after 18 years of growth, success, and contribution to the company, was pushed out by Patterson, he was not really surprised. Patterson always had a tendency of firing his best men. He always worried that somehow somebody would take the company away from him, even though he owned almost all the National Cash Register stock.

Patterson had many good qualities as a leader that were adapted by Watson Sr., but the eccentric and egocentric elements of his characters were predominant. You have probably noticed that when some people who have worked in a secondary position to the leader replace the leader, they not only apply the "good stuff" they learned from the leader but also not apply what they felt was not right. This was what Thomas Watson, Sr. did (and didn't do) when he became the leader of IBM.

It is interesting to note here that even though Thomas Watson experienced a lot of unfairness that forced him to resign, he never complained about Patterson's treatment and always respected him. Watson used to say, "Nearly everything I know about building a business comes from Mr. Patterson."[7] Thomas understood that in the real world you could not expect to find only good attributes in any person. Thomas Watson saw in Patterson one of the greatest entrepreneurs he could find at that time, and he wanted to be an entrepreneur like Patterson. He wanted to have a share in the company's profit, not be simply a hired manager. This was why, when he met Charles R. Flint, he agreed to become the general manager of C-T-R only if he could share in the profits as well as the risk of the company, and his request was granted. This was how Thomas J. Watson became an entrepreneur without investing any capital.

20.4.4 Thomas Watson, Sr., a Great Entrepreneur

What Is an Entrepreneur? Before we describe Thomas Watson, Sr., as an entrepreneur, we will establish what we mean by this term. Some people think an entrepreneur is a person who takes risks and starts his own, new, small business, but this is not exactly right. Not every new small business can be considered entrepreneurial. An enterprise does not need to be small and new to be entrepreneurial. A person who opens an ordinary Chinese restaurant in a new area probably has some risk involved, but because of this, can we consider him an entrepreneur? On the other hand, the leader of a large corporation like IBM can be and is an entrepreneur, especially when the company is continuously moving toward high technology. In his book *Innovation and Entrepreneurship*, Peter F. Drucker quotes the French economist J.B. Say, who defines an entrepreneur as one who "shifts economic resources out of an area of lower and into an area of higher productivity and greater yield."[8] This definition also reflects the meaning of innovation. And both of these concepts—entrepreneurship and innovation—are important elements of an organizational culture.

Listen to the Business Environment One of the important characteristics of an entrepreneur is to visualize the future need for something that will have a high demand. An entrepreneur, like a musician, should have good hearing. He should be able to listen to and hear the voice of the customer and the whole "melody" of the business environment.

When Thomas J. Watson, Sr. joined C-T-R, he was interested in a limited number of products and in particular, the tabulating machine. This was the product that Watson bet the future of the company on. He was not the father of the tabulating machine. This machine was designed by Herman Hollerith to provide statistics for the Census Bureau. Watson visualized that if he had printers that could be added to the tabulators, it would be possible to automate the record-keeping process and totally transform the way people processed information. This human quality of seeing the larger picture, combining things together to get a greater value is another important element of an entrepreneurial leader.

Thomas Watson, Sr. understood that with an American industry growing so fast, there would be a great demand for the automation of record keeping and accounting. Watson needed money to invest in the development of new record-keeping machines, but when he applied to the Guarantee Trust Company for an additional loan of forty thousand dollars, the bank turned him down. They pointed out that C-T-R already owed them four million dollars and that the economic conditions of C-T-R were not attractive enough to justify further loans. To this statement, Watson replied, "Balance sheets reveal the past. This loan is for the future."[9] He was able to convince the bank officials, and this was probably Watson's first step in developing the tabulating machine that allowed C-T-R to expand its market. This fact illustrates how important it is for an entrepreneur to have the talent to convince people to accept his ideas (convincing customers to buy the product, convincing banks to invest in future development).

In manufacturing it is interesting to observe the result of combining two types of talent: the inventor who creates new things and the innovator-entrepreneur who brings things to market. These are two types of people who make progress together. The role of the CEO is to support and facilitate the process of combining these two types of effort and capitalizing on the synergy arising from it.

Creating a Culture Now, after obtaining the funds for the development of new products, Watson's major concern was to put together an organization and create a culture that would comply with the needs of that time. This was a very important issue considering that the conglomerate he inherited was unprofitable, disorganized, and demoralized.

Watson started introducing motivational elements, using some of Patterson's techniques. At that time, when the company had no money, all activities to bring people together were based on emotional and morale grounds. He created slogans and newsletters, introduced songs and schools, installed codes of discipline, etc. Watson worked hard to gain the employees' loyalty by making inspiring presentations and recognizing the importance of the workforce to accomplish the company's goals. From the moment Watson arrived at C-T-R, he made a strong point that he had no intention of firing people. Watson knew that the only way to win their trust was by bolstering their self-respect.

Building a Shared Vision As a successful entrepreneur, Thomas Watson, Sr. had a vision that there would be a great need for record keeping and information. He understood that the fast growth of clerical and accounting jobs would need to be mechanized. As Lou Mobley, who was in the task force team that developed IBM's first computer, remembers Thomas Watson, Sr. saying hundreds of times, "You can punch a hole in a card and that's a five...and you'll never have to write it down again. Machines can do the routine work. People shouldn't have to do that kind of work."[10] At that time, this was the thought of a visionary leader-entrepreneur. He shared his vision and dreams, nurtured values and beliefs, and gradually created the organization's culture, which was aligned with his vision.

Within eleven months of joining C-T-R, Watson became its president. Under his leadership, the company focused its efforts on providing large-scale, custom-built tabulating solutions for businesses, letting the rest of the small office products be produced by other organizations. During Watson's first three critical-for-the-organization years, sales doubled from $4.2 million in 1914 to $8.3 million in 1917.[11] He also expanded the company's operations to Europe, South America, Asia, and Australia.[12] In 1924, to reflect the worldwide expansion, he renamed C-T-R to International Business Machines (IBM). At that time, Watson was 50 years old. He had a vision and he had a plan to build, as the new name of the company implied, a global and multinational corporation. Thomas J. Watson, Sr. led IBM to prominence.

Values and Time Analyzing the almost four decades of IBM leadership under Thomas Watson, Sr., you will probably not find a formalized set of articulated core values or other documents directly related to culture. At that time, this was not the way to create an organizational culture, but Thomas Watson, Sr., up to the end of his working career, spent a lot of his time sharing values in which he strongly believed. In their book *Beyond IBM*, Lou Mobley and Kate McKeown wrote, "In 1954, Watson Sr. was eighty years old. . . . [He] had reached the point where he could no longer make personal visits and speeches to factories and sales offices. It was in the thousands of these meetings with IBM employees that he personally had planted and nurtured the values and beliefs and dreams—the culture, although he would never have called it that—of IBM."[13] In contrast, these days some organizations practice activities related to articulating a set of core values and afterwards stop doing anything in this direction, assuming that something has been accomplished.

While giving importance to three basic beliefs—respect, service, and excellence—Thomas Watson, Sr. also paid a lot of attention to such values as *safety*, *security*, and *belonging*. Why? We won't find answers to these questions by studying IBM. To understand the primary importance of such values as safety, security, and belonging, we must understand the environment during the time of IBM's inception. The Great Depression wrenched people's souls. The major dream of people at that time was to never be hungry again, to never be without a job, to be able to feed their family. People were looking for an organization that would take care of their elementary needs, for an organization they could belong to. Watson Sr. was working hard to provide his employees with an organization that promised the security of a lifelong career and the feeling of belonging.

He had a talent for sharing his spirit, his values, and his belief in others. By doing this, he was building a foundation of corporate culture that reflected his beliefs and the values of society; and even though at that time no one called this "building a culture," this was the real process of creating it. He did it by making speeches, having personal meetings with people, and developing different policies that would support his values and beliefs. This was how, in 1984, he came up with the IBM policy on "Job Security," which at that time meant a great deal to his employees. From this, later came another policy of "Building From Within," which meant developing and promoting people from within the organization and giving them another chance if their position became obsolete. In his book, A Business and Its Beliefs, Thomas Watson, Jr. wrote, "No matter how great the temptation to go outside for managers, we have almost always filled these new jobs from within...."[14] Furthermore, his father, Thomas Watson, Sr., initiated the "Open Door Policy," which demonstrated a democratic way of communication, even though he was considered an autocratic leader. Watson, Jr. recalled, "The Open Door grew out of T.J. Watson's close and frequent association with individuals in the plant and field offices. It became a natural thing for them to bring their problems to him, and in time was established as a regular procedure."[15]

Ethics: A Part of IBM's Culture Buck Rodgers, who was vice president of IBM's marketing for 10 years and retired in 1984, wrote,: "I don't know how many companies have actually had a code of behavior articulated for them, but that's what Thomas J. Watson, Sr., did for IBM when he founded it in 1914."[16] To introduce and formalize ethical principles in IBM as a part of the company's culture, IBM has developed a Business Conduct Guidelines booklet that is given to every employee. This booklet is a code of ethics that describes the ethical aspect of employees' behavior at IBM. Linking the organization to ethical standards means that no explanation, description, or analogy of the organization's behavior can be considered complete without an explicit acknowledgement of this relationship. The existence of this type of link begins with a definition of the obligation of the organization's employees, not only to affirm the moral indebtedness of business but also to accept and act upon it. Business Conduct Guidelines states, "First, there is the law. It must be obeyed. But the law is the minimum. You must act ethically."[17] This means that every employee of the company should obey the law, but at the same time should act according to the ethical principles (see Section 3.10).

Today, organizations do not give the necessary attention to organizational ethics. In our opinion, this issue deserves more attention and should be a part of the cultural system of the organization.

Measuring the Influence of Culture Managers who assess the effectiveness of any achievement are usually looking for a measure of the influence of culture on the organization's success. Different authors have attempted to measure the influence of the culture on operational effectiveness. However, culture is not a method or tool that we can apply and see a difference. It is something broader.

Thomas Watson, Jr. remembers "Dad and his ability to make tens of thousands of people march to slogans like 'A salesman is a man who sells' and 'There is no such thing as standing still.' We were known as a personality cult. I thought this image was bad for Dad and bad for the company.... For anyone who questioned his way of operating, Dad had an impregnable defense: 'Look at the record!' You could argue it wasn't slogans that were making the record, but he'd say, 'How do you know?' "[18] In our opinion, the best and simplest method of assessing the influence of culture is by measuring the organization's bottom line, by looking at overall results reflected in our annual reports, the fact that we are still in business, the record we make, and asking how people feel about working in our organization. For 42 years, Watson Sr. led the creation of an organizational culture that strongly differentiated IBM from other companies. This was a culture that supported IBM's growth and prosperity.

20.4.5 The Second CEO: Thomas Watson, Jr.

The Journey to Becoming a CEO To familiarize you with the personal growth of Thomas Watson, Jr. from a young salesman to a CEO, we will make a brief journey in time. Watson, Jr. was born the same year (1914) that his father took over IBM, so he grew up with the company. As a youngster, he was a witness to the company's struggle to survive and its rocketing to fame and fortune. In 1937, at the age of 23, he joined IBM as a salesman. After a five-year interruption, during which he served in the U.S. Army as a pilot, he rejoined IBM in 1946. Six months later, he was promoted to the position of Vice President.

In 1952, after 38 years of leadership, Thomas Watson, Sr. passed the title of IBM President on to his son Thomas Watson, Jr. Four years later, Watson Jr. became the CEO of IBM, just six weeks before his father's death on June 19, 1956 at the age of 82.

IBM in the Period of Transition Thomas Watson, Sr. handed his son a steadily growing company with a stabilized cash flow. This was achieved after the introduction of the rental policy, which means that instead of selling its machines, IBM rented them to customers. This allowed IBM to generate cash internally and to plan their growth. The introduction of the rental policy made IBM a service-oriented company, which created the need for a stable, highly educated, and skilled staff. This was why the "no layoff policy" became even more important for IBM.

As we mentioned above, Thomas Watson, Sr. started building up IBM by borrowing forty thousand dollars. From that time on, for the next 30 years, IBM was in a constant cash crisis. However, when Thomas Watson, Sr. passed the leadership of IBM on to his son, it was a company with stable growth and a cash flow that made it easier for Thomas Watson, Jr. to lead and maintain his father's heritage.

Moving toward a New Vision As we mentioned above, Thomas Watson, Sr. had a strong vision that there would be a market for tabulating machines. He also saw the importance of adding printers to IBM's tabulators to automate the record-keeping process and revolutionize the way organizations processed information. He put all his efforts into making his vision a reality, and this was a major reason for IBM's success.

However, during World War II, when computers began to replace tabulating machines, Watson Sr. failed to recognize the enormous commercial potential of the computer, and he did not see that this could be a new market for IBM. In the 1950s there was a general belief that the computer market would be limited to government and military organizations. Watson Sr. shared this belief. At that time, IBM's line of electronic calculators was so successful that he did not recognize that a new era of computation had begun. This is a classic example of how leaders sometimes have tunnel vision, which does not allow them to see what is really going on in the world. If the leader of IBM had seen that the company's main purpose was data processing, in whatever form, and not just the mechanization of the clerical and accounting processes, it would have been easier for him to see that the computer would eventually replace the electron-mechanical tabulator.

In contrast, Thomas Watson, Jr., who was at that time the Vice President of IBM and a member of the Board, visualized the importance of investing in computer development. He saw it as a new opportunity for IBM's future growth.

Thomas Watson, Jr. tried to push IBM into computer production, but he did not succeed. Only in 1952, when he became the President of IBM, was he able to achieve positive results in pushing the company into the computer market. At that time, this issue became a matter of urgency because of the need to catch up with IBM's competitors. Thomas Watson, Jr. was a visionary leader who brought IBM into the computer business. This required a lot of work to reshape the organization's culture and align it to the needs of the new industry.

Culture and Structure In Thomas Watson, Sr.'s era, IBM had practically no organizational charts. Thomas Watson, Jr. recalled, "I discovered that Dad had his finger in everything at IBM. A phenomenal number of executives—at one point I counted thirty-eight or forty—reported to my father directly.... We had no organization chart because Dad didn't want people to be so focused on specific jobs that they concentrated only on those jobs...."[19] In December 1956, Thomas Watson, Jr. held the famous "Williamsburg Meeting" with 110 key IBM men, at which they developed the details to decentralize the organization. This was one of his first reorganizations after he became the CEO. This reorganization is remembered as proof that IBM is reorganizable. Thomas Watson, Jr. commented in 1973, "The real achievement at Williamsburg was not that reorganization but the fact that it made IBM readily reorganizable."[20]

This event—the change from a one-man structure to a more flexible collaborative structure—is an important step in the direction of cultural change.

Why? Because it is impossible to introduce any significant change in people's behavior without changing the underlying structure. It is amazing how some organizations try to shape the culture just by articulating a set of new values or by introducing a new code of ethics. This is a waste of time simply because it will not work. It is like trying to make a significant change in the shape of a building without changing its structure.

We are not trying to infer that Thomas Watson, Jr.'s organizational structure was better than his father's structure. Structures are dictated by the external and internal environment, by specifics and size of the organization and other factors. Rather, we wanted to point out that moving from one type of organizational structure to another will certainly influence how people act, behave, and feel. In other words, if you want to reshape the culture, you must consider the organizational structure.

Preserving the Basic Beliefs For almost four decades of leading IBM to growth and success, Thomas Watson, Sr. also led the development of a culture that made this growth and success possible. Then when his son assumed the leadership, it was a question of how much of the legacy of Watson Sr. should be preserved and what need to be changed. As usual, anyone who assumes a new position, particularly that of a CEO, plans to change the organization regardless of how well the company was run before. Watson Jr. was no exception. He had a plan to reshape the organization, and he acted accordingly. However, he was convinced that the basic beliefs instilled by his father should be preserved. In his book, *Father, Son & Co.*, he wrote, "Eventually I was able to distill into a simple set of precepts the philosophy Dad had followed in managing the business for forty years:

- Give full consideration to the individual employee.
- Spend a lot of time making customers happy.
- Go the last mile to do a thing right.

I thought that to survive and succeed, we had to be willing to change everything about IBM except these basic beliefs."[21]

What was the major reason that Thomas Watson, Jr. decided to preserve his father's core beliefs? It is easy to assume that after so many years of leadership, the Watson values were shared and had become IBM's beliefs and, in particular, Thomas Watson Jr.'s philosophy. Another reason is probably because those beliefs worked well for the company.

For the whole period of his leadership as President and CEO (1952–1971), Thomas Watson, Jr. continued his father's efforts in developing IBM according to these core beliefs. This was not an easy task to accomplish because it was a period of searching for further stability, explosive growth, and introducing new products. All this made it very difficult to maintain and run the company according to these three beliefs.

To illustrate IBM's growth in that period, in his book *A Business and Its Beliefs: The Ideas That Helped Build IBM*, Thomas Watson, Jr. wrote, "Within a period of two generations, we have expanded from an awkward combination of three small companies into an international company whose revenues in the United States and abroad exceeded $2 billion in 1961.... During the last forty-eight years, IBM has grown from a company of 1,200 people to one of more than 125,000.... Domestic revenues alone have increased more than 400 times over the $4 million a year the company grossed in 1914. Not once during those forty-seven years did the company fail to make a profit."[22]

Today, in these turbulent times, not a lot of companies could report such growth and, in particular, make a profit so many years in a row. However, it is important to note that independent of the tremendous growth, which resulted in a lot of qualitative and quantitative changes, IBM was able to preserve its core cultural elements. Emphasizing this point, Thomas Watson, Jr. notes, "But in its attitude, its outlook, its spirit, its drive, IBM is still very much the same company it has always been and that we intend it shall always be. For while everything else has altered, our beliefs remain unchanged."[23] Thomas Watson, Jr. was convinced that a company's success came from the power of its beliefs. He said, "I attribute our success in the main to the power of IBM's beliefs."[24]

Although recognizing the importance of core beliefs or core values, we should, however, say that this is not sufficient from a value system point of view. To create a full value system, we need to have a set of operational values also, which are closer to the daily work. This is why in addition to the three well-known core beliefs—respect, service, and excellence—IBM also developed a set of what we call operational values that, together with the core values, reflect the company's culture more accurately. These operational values were published in IBM's house organ (*Think*, 1988, no. 1). It is interesting to note here that values such as knowledge, networking, and teamwork, that have only recently become popular in organizations, were on IBM's list more than 10 years ago (see Insert 20-1).

Practice What You Preach Thomas Watson, Jr. recognized that to keep his father's beliefs alive, he had to behave according to these beliefs and encourage others to reinforce these beliefs by their actions. In his book *The IBM Way*, Buck Rodgers used two examples from his own life that described Thomas Watson, Jr. as a person who practiced what he preached. Buck Rodgers remembered, "I was new to the company and completing the final sales training program.... My wife was pregnant and expecting to deliver at any moment.... I wanted to be with Helen, but I was afraid to ask for the time off.... Watson spoke to the class, and when he was finished he moved around the room, chatting with the trainees. I was wondering if I had time to call home, when he approached me. 'How are things going? What's happening?' Well, he asked so I told him. I hadn't got the words out of my mouth when he

Insert 20-1
IBM Values

1. Customer service
2. Profitability
3. People
4. Competition
5. Autonomy
6. Quality
7. Efficiency
8. Cost-effectiveness
9. Education/knowledge
10. Adaptability
11. Growth
12. Teamwork
13. Production
14. Family
15. Innovation/creativity
16. Information
17. Market leadership
18. Education
19. Expertise
20. Entrepreneurship
21. Challenge/opportunity
22. Achievement
23. Social responsibility
24. Image
25. Technology
26. Change/progress
27. Planning/strategy
28. Career opportunities
29. Fun
30. Employee dedication
31. Communication/networking

Source: Based on IBM 1987 Annual Report and *Think #1*, 1988.[25]

interrupted. 'What are you doing here? You should be home with your wife. Get on a plane and get back to Ohio immediately.' Within the next few minutes, IBM's CEO made arrangements for this trainee to get to Cleveland, and I was airborne that same afternoon. In the lectures, I heard a lot about respect for the individual, but the message was made crystal clear when Watson demonstrated how he placed my family and personal worth above the business. Among the flowers that were delivered to the hospital room after the arrival of my daughter was a beautiful arrangement from Tom Watson. 'Do we know him?' Helen asked."[26] This example demonstrates the whole difference between printed values that are posted on a wall and actual beliefs that are alive, working, and generating the necessary power to create a great culture.

In his second story of describing Thomas Watson, Jr., Buck Rodgers wrote, "To me, Tom, Jr., was a great man—but he wasn't perfect. He had a pretty good temper and could be very impatient with what he thought to be inconsiderate behavior. When I was president of the Data Processing Division, he sent a message to my office asking that I come to a meeting at three o'clock that same afternoon. It was a special session, with little advance notice, and I was already on my way to New Jersey to see a customer. Arriving at the customer's office, I received a call from my secretary: Watson had called a meeting and wanted me there. The customer situation was a difficult one, and I didn't get back to Armonk until six-thirty that evening. As an object lesson, Watson hadn't started the meeting, so everyone was waiting for me.... As I walked into the room, he said, 'When I call for a three o'clock meeting, I expect *everybody* to be there at three o'clock.' I took a deep breath. 'Tom, how many times have you said, 'The customer comes first?' I was with one in New Jersey who had a very serious problem.' Watson's face softened. 'Buck,' he said, smiling, 'you have the right priorities.' There was a brief pause; he turned to the others and said, 'We shall now commence the meeting.' "[27]

We used these two stories to emphasize the importance of "Walking The Talk" if we want the value system to serve the purpose. Articulating core values does not mean publishing them, displaying them on a visible place, or talking about them in our presentations. It means living them on a day-to-day-basis.

There Are No "Small Things" in a Culture A culture of an organization is like the panorama of a mosaic—an extensive picture composed of many small pieces, in different colors, shapes, and sizes. Every piece has its function in the whole picture, and together they make up the composition of the architecture. Take away one piece and the composition suffers. The beauty and strength of an organization's culture depends on small things that together make up its composition. However, sometimes in creating a culture, we ignore the small things and lose a lot. When we read about the history of IBM's culture, we found that because they did not divide the culture into big and small things and paid attention to all cultural elements, they became a company with a great culture.

In *Father, Son & Co.*, Thomas J. Watson, Jr., wrote, "I kept a whole squad of secretaries busy doing things in the same way Dad would have—making sure every letter I received was acknowledged within forty-eight hours, sending flowers to the hospitalized wife of an employee, and making thousands of other small gestures of thoughtfulness."[28]

At that time, there was no E-mail to provide an immediate response. But do we always use the capability of E-mail to respond and communicate on time? Do we have a way of knowing when and where to send flowers? How important is all this? How much does it belong to the culture? Does this not belong to the type of things that are nice to have but are not really necessary? Above we compared culture to a mosaic to show how a small piece can spoil the whole picture, but a mosaic is static and a culture is dynamic. To emphasize the relationship between the cultural elements and their interdependence, it is probably better to compare a corporate culture to a forest, which is an ecosystem in which everything is interdependent and the relationships must be balanced synergistically to form the necessary environment. There are no "small things" in forming a culture.

One of the authors remembers a real story when an executive secretary resigned on her birthday after 10 years of service. With tears in her eyes, she said, "After so many years of working for him (her boss), he doesn't even remember my birthday. He walked in this morning in a bad mood, looked at the flowers on my desk (that I had gotten from another department), and said nothing.... Enough is enough! I don't want to work here anymore." The secretary did not resign only because her boss forgot her birthday; this was probably the last straw. There are a lot of great managers who find other ways to express their appreciation to their subordinates. The point is that small things, especially those that could hurt someone's feelings, will go in a special place in someone's memory, and when they accumulate, they become big things. When a person reacts seriously to a small thing, this is a cumulative response that has sometimes accumulated for years. Managers should pay serious attention to small things in communicating with its employees and other stakeholders. As Stephen R. Covey said in his book *Principle-Centered Leadership*, "If we aren't giving twelve hugs a day to some people, we will soon have a withdrawal state because our deposits are essentially evaporative in their nature."[29] These twelve hugs can be expressed physically, emotionally, verbally, or in any form that indicates to your subordinates your appreciation to them for working with you. Managers should continuously be concerned about their emotional bank accounts so as not to go into emotional bankruptcy.

20.4.6 The Beginning of Gerstner's Era: The Renaissance of IBM

The First Step Towards Organizational Trousformation Louis V. Gerstner, Jr. was named Chairman and CEO of IBM on April 1, 1993. This was the first time in the company's history that a new CEO was brought in from the outside. With no related industry experience, he replaced John F. Akers—who was

IBM's CEO from 1985 and Chairman of the Board from 1986. Akers had served IBM for 33 years.

The situation at that time was very difficult, and IBM's future was bleak. They had lost billions of dollars, there were large drops in market share, thousands of employees lost their jobs, and there were no signs of recovery. It was a situation like that of a sinking ship; it needed fast decisions and transformational changes.

Gerstner described IBM's situation when he stepped in as CEO: "IBM was in critical condition in 1993. Revenues were declining. We lost $16 billion from 1991 through 1993. We lost more than 117,000 jobs, and we took more than $28 billion in restructuring charges."[30] He understood that the first step in a situation like that was to create a sense of urgency, and he did this very well. Gerstner's main task at that time was to find the right way out of the crisis and find the right people to materialize the new plans and strategies.

When the signal of urgency was strong and people inside the organization recognized the need for a fast transformation, the question that popped up was, "how much does the existing culture align with the required transformation?" A culture may serve its purpose for many years, but there may come a time when "pruning" and "hybridization" of the old cultural tree are needed. The new leader can influence the existing culture by gradually embedding his values and beliefs. However this takes time, and if the leader needs to introduce quick changes, there may not be enough time for the new 'branches' to grow and produce new fruit on the trimmed cultural tree.

If an organization is in a crisis and its culture is obviously inhibiting the necessary change, the leader must find the internal strength to do two things. One is to challenge some of the existing assumptions in the organization, and another is to lead in the redefining and reshaping of some (maybe even the central) elements of the culture. Even though Lou Gerstner planned a revolutionary change, he understood that the best way to do this was to build on IBM's existing strengths and good traditions. Many leaders who tried to introduce radical changes by starting with a 'blank sheet'—totally ignoring the best practices and traditions—failed. At the same time, Lou Gerstner also recognized the urgent necessity of challenging some of the old assumptions and practices that led to the situation the company faced at that time. He assumed the leadership of IBM and said, "We will build on [IBM's] traditions, but we will not hesitate to make every change necessary to meet the changes of a very rapidly adjusting marketplace."[31]

How did he plan to do this? How could he preserve IBM's core values while introducing the new principles that were required to lead the transformation?

One vehicle that Gerstner employed to bring about the intended organizational change is eight fundamental operating principles (see Insert 20-2), when he declared 1994 the year when the new IBM would emerge.

As we can see, this new corporate philosophy does not contradict the three basic tenets—respect, service, and excellence—that were at the heart of IBM's

> **Insert 20-2**
>
> **IBM Principles**
>
> - The marketplace is the driving force behind everything we do.
> - At our core, we are a technology company with an overriding commitment to quality.
> - Our primary measures of success are customer satisfaction and shareholder value.
> - We operate as an entrepreneurial organization with a minimum of bureaucracy and a never-ending focus on productivity.
> - We never lose sight of our strategic vision.
> - We think and act with a sense of urgency.
> - Outstanding, dedicated people make it all happen, particularly when they work together as a team.
> - We are sensitive to the needs of all employees and to the communities in which we operate.
>
> *Source*: Reprinted with permission from 'Saving Big Blue' by Robert Slater, McGraw-Hill Companies, 1999, pp. 113–114.[32]

culture for so many years. Quite the reverse: The three core beliefs were just redefined and incorporated into Gerstner's eight principles, which provided a broader philosophy and reflect the needs of the new IBM.

Analyzing the new eight principles further, a question may arise: Can all of them be considered IBM's shared core principles? In other words, do they reflect what the IBM employees think and feel? Is it possible that some of these principles reflect only the values of the new leader, or maybe they are just the leader's proposal at this time? It will take time for some of the principles to become deeply embedded into IBM's organizational philosophy. It will also take time to develop new and modify the existing policies and regulations before the proposed principles will truly become core values. Principles that are not totally based on the organization's prior learning and reflect only the leader's values and desires may be called espoused principles, which will need to go through the test of time to demonstrate that they serve the purpose. Only then can they be transformed into real core principles. For now, they are just principles that are aligned with IBM's needs for organizational renewal today.

IBM's Six Strategic Imperatives

To focus people's energy on the realization of IBM's business and technology strategies, in the 1994 Annual Report Lou Gerstner unveiled what he called six strategic imperatives[33]—the roadmap for IBM's near future.

1. Exploiting our technology.
2. Increasing our share of the client/server computing market.

3. Establishing leadership in the emerging network-centric computing world.
4. Realigning the way we deliver value to customers.
5. Rapidly expanding our position in key emerging geographic markets.
6. Leveraging our size and scale to achieve cost and market advantages.

These imperatives are based on a clear understanding of what's happening in the marketplace. To the investors, Gerstner said, "We are doing everything we can to focus everyone here on what's outside IBM—markets, customers, competitors."[34] These six imperatives serve as an instrument for aligning any other initiatives that take place on different levels of the corporation.

Involving People in Leading the Change

In September 1993, after six months of being with IBM, Lou Gerstner announced that he was creating a special advisory group consisting of himself and ten other senior executives to lead the transformation. He named this core team the Corporate Executive Committee (CEC). In his directive, Gerstner described the purpose of the CEC by naming three functions:[35]

- First, it would make IBM's businesses "cohesive and responsive" to customers;
- Second, it would implement "corporate-wide integration initiatives;"
- Third, it would "advise me [Gerstner] on broad issues of corporate strategy."

This indicated that Gerstner was looking for supporters who could help him focus on and lead the organizational transformation.

Having a committee of ten senior managers that would meet on a regular basis (monthly) would not only help introduce a collective style of leadership, but also help Gerstner better understand and more closely evaluate the members of the team he had inherited from John Akers. The team of ten consisted mainly of old-timers at IBM, except two—Gerry Czarnecki and Jerry York. It was very important for Gerstner to assess who of the old-timers would buy into the new culture, and who wouldn't fit the new environment. As Doug Garr wrote in his book, *IBM Redux*, "York said Gerstner put it this way: 'Even if you make your numbers, if you do not become part of the solution—meaning the new way of business rather than the old way—your position is at risk.' Basically, the new way meant that decision-making was a top-down exercise, and executives who didn't implement or act swiftly had no excuses. In other words, failure to execute resulted in your execution."[36]

While having a leadership nucleus to execute change, Gerstner also felt that they needed to involve people from different levels of the organization. From the e-mails he received, he understood that most of the IBM employees wanted change, and this desire had to be fulfilled in the right direction. As Robert

Slater wrote in his book, *Saving Big Blue*, "During his first few months on the job he [Gerstner] called for 5,000 volunteers to help him bring about change in IBM. The 'volunteers' would be expected to influence thousands more, recruit others, and so on. He did not intend to launch a formal program—it was more of a challenge, a call to arms, Gerstner's way of encouraging an ongoing grassroots campaign within IBM to change the company's culture."[37]

How did the committee work? The CEC met periodically to develop plans and strategies of cultural change. Larger groups of creative people from different areas and layers of the organization developed their own execution plans, which were in sync with the six strategic imperatives. They conducted cross-functional projects that involved most of the employees of the organization. This approach to leading and executing change is like rolling a small snowball that gets bigger and bigger over time. There is nothing new in this approach. From the outside, Gerstner's CEC almost looked the same as IBM's Worldwide Management Council (WMC), which was implemented prior to Gerstner's era. The main difference between these two committees was in the way they operated. The WMC operated in such a way that it needed to have absolute agreement between all members of the committee. Jerry York, the CFO of IBM, put this clearly when he said, "From what I was told, one stinking person out of twenty-eight that had equity in a decision could block it. If the vote was twenty-seven to one, the thing still didn't go. John Akers had abolished the nonconcur process officially around 1990, but the culture still carried on, even though whatever forms they used had long since disappeared."[38] While it is good to have full agreement between all the members of a committee, sometimes the full agreement may be a symptom of members being afraid to express their own thoughts.

The former Soviet Union also had plenty of committees, but they had nothing to do with the collective decision-making concept. One of the authors of this book remembers (and it looks funny now) when before making a serious decision, the leader usually asked, "Kto Za?" which literally means, "Who is for it?" and everybody raised their hand, independent of what they really thought or felt. If you tried to behave differently you may have been rolling a "snowball" in a very remote place.

We are talking here about a democratic way of leadership, where you can have a free dialogue and express your thoughts. But when the decision is made, it needs to be executed. You need to support and participate in this execution. If not, you are a failure, with all the consequences. The model of employee involvement in the process of transformational change is now used in many organizations that are involved in the change process. The effectiveness of the CEC depends on the culture of the organization.

The approach we just described is an effective way to energize people and make them a part of the solution. Sometimes management complains about people's resistance to change. Our response to them is, "Explain your thoughts and intentions; form a strong 'magnetic field;' create a sense of urgency;

and, most importantly, involve people. This will allow you to transform the resistance to a force directed at the desired goal." As Lou Gerstner put it, "We've energized an incredibly talented group of people who were here waiting for the leadership that would bring the company back again. And they've been great. I haven't done this. It's been 280,000 people who have done it. We took a change in focus, a change in preoccupation, and a great talented group of people, merged it together, and changed the company."[39]

Other Major Changes to the "Old Culture"   In July 1993, three months after he assumed office, Gerstner announced a plan to lay off, in one single cut, 35,000 employees.[40] This was a very different move from tradition. In fact, it was a challenge to the old IBM culture of no employee layoffs. "For more than 50 years IBM has practiced 'full employment', promising that once you were hired here, you would not be laid off for economy reasons."[41] This move certainly shook the mindset of lifetime employment in IBM. And besides, 35,000 at one cut was sufficiently awakening to show that no complacency was to be tolerated. Only the talented, and those who continuously strive for performance and improvement and embrace the changes that he brought about, would have a place in the organization. Implemented simultaneously with other changes was also a series of cutbacks,[42] which included:

- A pay cut of executive secretaries from a 6-digit figure to $40,000 per year
- Reduction in longtime employees' early payout; from as much as two years' salary to a maximum payout of 26 weeks' pay
- Reduction in pay increment and cash bonuses
- Cutbacks in health plans
- Discontinuation of company picnics, long service celebration
- Reduction of awards
- Reduction in payback package of layoff employees

As it is always more difficult to play the devil's advocate, this series of cost cutting may seem tough and impersonal, even insensitive. But for a quick turnaround of a huge enterprise, these measures may be necessary. However, no employee would welcome cutbacks. We would expect that the efforts to reshape the financial picture of the company would inevitably result in some bitterness from the rank and file.

During World War II, to save lives, surgeons would amputate arms or legs that under normal conditions would not have been amputated. It was tough for the surgeons to see the suffering of the amputees, but this was the only alternative to save their lives. For many leaders, it is difficult and painful to introduce such "surgeries" as we mentioned above. But for IBM, this was the only alternative to save the company from total collapse.

Small Changes that Tell a lot

One of the changes that affected IBM was the dress code. It used to be well-pressed white shirts that dictated formality and image. When Lou Gerstner took over, the style was gradually replaced with more casual striped or sport shirts for the men and blouses and pants for the women. This was not a change in the policy, but rather an example that came from the top. In his book, *IBM Redux*, Doug Garr gave an example that illustrates the change. He wrote, "An IBM vice president recalled one revealing Saturday morning meeting in Armonk. York had gathered his top finance people together, ostensibly to review the current health of the company and unify a cost-cutting strategy that could be presented to Gerstner. As soon as he walked into the conference room, the manager marveled at the contrasting scene. The IBM finance guys were dressed casually, in slacks and open-collared shirts. York showed up looking like an aging biker from *The Wild One*, the 1950s Marlon Brando movie about a motorcycle gang. He was wearing blue jeans and a tight-fitting black Harley-Davidson T-shirt with a pack of cigarettes rolled up in the sleeve. This, uh, was the CFO of IBM?"[43]

Gradually the formal dress code that was embedded for many years in IBM's culture was replaced by a casual dress code that reflected an atmosphere of informal relationships. The dress code is one of the most visible artifacts of a culture, but it is not always easy to decipher correctly. The moment we walk into a place, it is the dress code that greets us, and more often than not, we would form a certain presumption about the place. The dress code reflects the atmosphere, relationships, and power play of that organization. Returning the flexibility to the people to wear what they feel comfortable, without compromising their productivity, demonstrates the leader's openness, trust and empowerment. It carries the subtle message of respecting individual's taste, as long as results are delivered. No longer does a well-pressed shirt guarantee knowledge, it may well be a facade.

Nevertheless, the bureaucracy does not go away just by changing the dress code. Not all who dress themselves casually are open-minded and friendly, and not all who are dressed formally are bureaucrats. However, changing the dress code in IBM was a good signal that suggested the new leadership was against hierarchy and bureaucracy and was encouraging informal and friendly relationships.

Changes of the Mind

The style of informality was also evident in meetings. Knowing that changes have to be effected top down, Gerstner did not spare his surgery at his senior executive level. Before a meeting, he would require facts, problems, to be presented to him so that he could get straight to the matter during the meeting. Business reports were shortened to ten pages, meetings were no longer lengthy but short and precise, and only those who had a direct reason would attend. No

418 UNDERSTANDING AN ORGANIZATION'S BEHAVIOR

presentation was to be done on foils. Foil projectors were removed from most meeting rooms.

These were by no means small changes, but to practice them was much easier said than done. For the senior executives who were so used to just updating business by foils prepared for them, this was a demanding change. CEOs in the past had not requested them to update business details. Now, they would have to understand the business inside out, in order to verbalize them sensibly without the help of prepared foils and data.

In one of the meetings, a senior executive was approaching the foil projector, preparing to present. As he approached the machine, he was startled to find Gerstner also approaching the equipment. He was more startled to see his CEO snap the equipment off with these blunt words, "If you can't explain your business, of which you are the expert and manager, without all kinds of props, you don't understand your business".[44]

The "no foils" example may appear insignificant to some people. So many things to do, and here Gerstner was fighting against the use of foils. We see this example as an effort of a leader to understand those he was working with, and with whom he should entrust the transformation. Gerstner was trying to penetrate people's minds, know their way of thinking and assess their knowledge and readiness to participate in the change process. Gerstner wanted to make sure that his leadership team deeply understood their business. The "no foils" approach was one way to achieve what he wanted.

20.4.7 IBM: Moving Toward a Strong Strategic Vision

When Lou Gerstner first joined IBM as the CEO, he did not bring along a blueprint of how to reengineer the company. He also did not have a clear vision of IBM's future. Gerstner recognized the importance of an inspiring vision. However, from previous experience he knew that it would be more appropriate to take as much time as necessary to learn how IBM was functioning, to assess the capabilities of the existing systems, and get a better understanding of IBM's culture. This would allow him and his executive team to make a decision of what kind of company IBM should be in the future.

Lou Gerstner's major priority in the first year of leadership was to concentrate on short-term strategies that would stabilize the company and bring it back under control. At a press conference in the summer of 1993, Lou Gerstner said, "There's been a lot of speculation on when I will deliver a vision... The last thing IBM needs right now is a vision. What IBM needs right now is a series of very tough-minded, market-driven, highly efficient strategies in each of its businesses."[45]

Even while making such a statement, we believe that Gerstner may have had some kind of vision, but it was probably too fuzzy to be formally articulated. He was perfectly aware that in last few years IBM had declared different strategic visions in an attempt to visualize how the company would look in the future. But all these attempts failed because they were not compatible with the requirements of the external environment. Lou Gerstner needed a strategic vision that would not only turn IBM around, but also regain its prominence as one of the world's preeminent corporations. In this early stage, Lou Gerstner didn't feel that he could articulate such a strong vision, so it would be better not to have a vision than to have one that would not fit what IBM needed.

What we have described so far in this section matches the Fuzzy Vision Model (see Section 17.2) where the new leader recognizes the importance of having a strategic vision, but he is not ready to articulate a vision statement in the early stage of leadership.

Later, in 1995, when IBM had been stabilized, returned to profitability and growth, reconnected to the marketplace and its customers, and had started to build better products and provide better services, Lou Gerstner was ready to unveil a strategic vision. In the 1995 IBM Annual Report, three years from the time he said that the last thing IBM needed was a vision, he said, "... So it's with an enormous sense of irony that now, almost three years later, I say this: What IBM needs most *right now* is a vision."[46] He pointed out that he didn't mean a slogan or promises. He meant something concrete, something people could focus on. He said, "So this convergence of two powerful forces— customer need and advanced network technology—leads IBM to a strategy, or vision, that is simple and clear, and consistent with our company mission. IBM will lead the transition to network-centric computing by:

- Continuing to create the advanced products and technologies needed to make powerful networks real; and
- Working with our customers to help them fully exploit these networks."[47]

The main focus in this strategic vision was the customers' needs. In the same report, Gerstner emphasized that, "Our job is to help our customers leverage their information to their advantage... We are rapidly adapting our systems and delivering new services that allow customers to move their content to network—and do it with security and reliability they have come to expect from IBM."[48] Moving from an internal focus to a customer focus got IBM out of trouble.

20.4.8 How to Measure the Success of Cultural Transformation

Earlier we described some examples of cultural change at IBM under Lou Gerstner's leadership. We talked about such activities as changes in the way people dress; changes in the way they make presentations; changes in the forms of people involvement, etc. All these are symptoms of cultural transformation, which are on the surface and are easy to observe. These activities are like small stones of different colors that form a new pattern in the cultural mosaic. They help make people think and act differently. However, the fundamental changes in the culture are difficult to see. You can't just walk into an organization and see the process of change, and then apply it to your organization. The power of the organization's culture is invisible and impossible to measure. What can be measured are the actual business results that come from the influence of the culture.

The influence of the culture on business results can be metaphorically compared with a magnetic field where we cannot see the force of attraction and repulsion, but the force is evident in the work it can do. In other words, culture is like an invisible field that influences the way people act. It can be measured only by the actual business results. This is why it makes no sense when some leaders claim that they have a great culture at the same time their organization is losing business. We have heard such statements not only from the leaders of organizations, but also from those who lead entire countries. For many years, the leadership of the former Soviet Union claimed that their country had the best culture, even though the people had nothing to eat. Only when Mikel Gorbachov took the risk of offering a reconstruction for the country did the transformation of a culture that had existed for many years begin.

IBM's cultural transformation is reflected in its business results. Its employees started to think and act differently; they "came out of the black box," and started to listen to the signals from the external environment and act accordingly. Figure 20-3 shows some results of the reshaped culture at IBM.

As you can see, for the nine years of Gerstner's leadership, the revenue grew by 37 percent, the stock price rose by 108 percent, and net income increased from a loss of 8.1 billion dollars to an income of 7.7 billion dollars. There is much more hard evidence that can be used as a measure of cultural change.

The main purpose of this section is to make a point that the best way to measure the cultural transformation is through its business. Certainly there are some other indicators that reflect cultural changes, such as surveys, 360-degree feedback, and other techniques that assess the cultural environment and how people feel in the organization. However, these should be complementary measures. The core indicator of a healthy culture should be the bottom line results.

20.4.9 What Type of Leaders Have Led IBM?

What leadership style is the best for a great culture? The answer is—it depends. It depends on the external environment, on people's needs at a particular time, etc. Here in our case study, we have observed two types of leadership. Thomas Watson, Sr., was considered an autocratic leader and Thomas Watson, Jr., a leader who made IBM more bureaucratic. However, both of them were the right leaders for their time.

In their book *Beyond IBM*, Lou Mobley and Kate McKeown wrote, "Looking back on Watson [Sr.] through the lens of today's values, we label him an autocratic, paternalistic leader. He was. And he was precisely what his followers wanted."[49] In the 1930s and 1940s, people felt comfortable with this

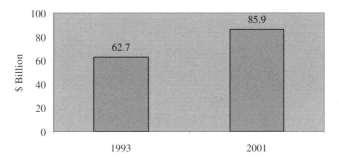
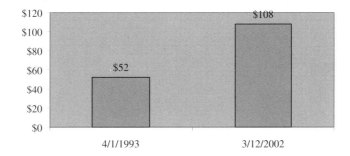
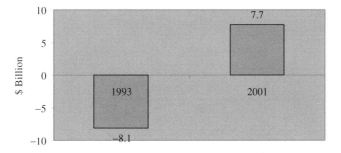

Figure 20-3 The Results of IBM's Cultural Transformation

type of leader. They needed someone who would take care of them. They needed a family environment with a father figure who would tell them what to do, and who would protect their lives when necessary. However, in 1950s and 1960s, when Thomas Watson, Jr., inherited a company of 50,000 employees and the sales at that time were growing by 20–30% a year, they needed to move from autocracy to bureaucracy. The company had become so large—its size and complexity growing rapidly—that Thomas Watson, Jr., could no longer make decisions by himself like his father did. He and his staff introduced fixed organizational structures and all kinds of policies and regulations, which replaced the culture of management in which all decisions were made by one individual.

However willing to move from autocracy to bureaucracy, this does not mean that Thomas Watson, Jr., was a bureaucrat. He considered this move as necessary for proper management, but later, when the time came, he was concerned that IBM might become too bureaucratic. In his article "The Next Challenge," Chris Argyris wrote about Thomas Watson, Jr., "In many meetings with his top group, he led earnest inquiries into the possibility that someday IBM could become too bureaucratized, too rigid, too inflexible, and unable to change as fast as might be necessary. I attended at least a dozen such meetings. Watson's immediate reports told him not to worry."[50]

This was why Watson Jr. made sure that all policies and regulations were aligned with IBM's core values, which protected the organization from becoming too bureaucratic. An example is the "Open Door Policy," which made it possible for every individual to communicate with all levels of the organization's structure. This was the system that allowed every employee to give suggestions on how to manage the organization. The art of leadership is to make sure that the form of behavior is timely and serves people's needs and the organization's purpose.

The above examples are related to the leadership in the past, which certainly had an impact on the behavior of IBM's employees and the organization's culture. But what type of leadership does IBM have today? For the nine years of leadership that Lou Gerstner had at IBM, he demonstrated great capabilities in crisis management. If he had stayed at IBM longer, what should his leadership style be like after the shock?

Below is an extract from an article in *Fortune* magazine written by Betsy Morris, April 14, 1997, in Lou Gerstner said, "*On mentoring*: I do a lot of nurturing in the sense of mentoring. I work hard at helping people improve their skills, grow, get better at what they do. I'm not a big patter on the back. They're very different things. . . . *On being called arrogant*: I'm tough minded. I'm tough in business situations. I'm really focused and have little time. I guess I'm also relatively blunt. So if you want to know if I'm intense, competitive, focused, blunt, and tough, yes. That's fair. I'm guilty. Quite frankly, I am not very comfortable in chitchat. When I go to [IBM] board meetings, I arrive two minutes before, and I leave when it's over. I don't stay for lunch or go early and have coffee. But if arrogance means pride, wanting to take credit for everything, not seeking others' advice, I don't think those are fair characterizations."[51] As you can see from his own analysis of his character, it would be very difficult to attach a simple label of a particular type of leader to him. He is complicated; and this may be exactly the type of leader that IBM would need even today, after the transformation. In the same article, the author wrote, "So now Lou Gerstner, boy wonder from Mineola, has his opportunity to stand next to industry titans like Jack Welch, Robert Goizueta, and, yes, Tom Watson, Jr., as a true-blue company builder."[52] We believe that Lou Gerstner has strongly proven that he can stand next to the industry titans mentioned above.

In January 2002, IBM announced that its board of directors had elected Samuel J. Palmisano as CEO of the company effective March 1, 2002. Mr. Gerstner said, "Over the last decade, Sam Palmisano has taken on a number of IBM's most significant challenges, from building the services business to transforming our server line. In each instance, Sam has done far more than manage operations effectively. He has made it both his personal mission and that of IBM to become the number one competitor in each of these markets. Sam's unique mix of strategic vision, passion and discipline, combined with his intimate understanding of IBM, make him the right person to become IBM's next CEO."[53] Samuel J. Palmisano is not an outsider to IBM. He joined the company in 1973, and for all these years, he had diverse positions where he was proven to be a successful leader. But in this new role, as Lou Gerstner's successor, Mr. Palmisano will need to demonstrate a leadership style and significant results that IBM never had in its 91 years of history. As David Kirkpatrick wrote in the February 18, 2002 issue of *Fortune* magazine, "Palmisano will have to deliver something big, something we can't yet envision. He can't be just a caretaker of Lou's legacy. He has to surprise us as much as Lou did."[54]

20.4.10 Studying IBM's Pattern of Cultural Change

It is logical to assume that most organizations are constantly going through a cultural change because of the continuous introduction of new artifacts, which are usually consistent with the prevailing culture of the organization. For example, an organization can introduce a new performance appraisal system or another program, but these changes may still be consistent with the existing paternalistic culture. Such changes are considered incremental changes in the culture, changes that are generally consistent with the underlying paradigm of assumptions. In this example, however, we are interested in a more dynamic change, which will come from a major organizational transformation.

Here we want to demonstrate a cultural change model developed by W. Gibb Dyer,[55] which was based on the examination of the histories of five large organizations that had experienced significant changes in their culture. To illustrate this model, we will use IBM's cultural change pattern to see how it fits the model (see Table 20-1).

Dyer's model articulates the dynamics of performing major cultural transformations and consists of six stages:

1. The organization develops a sense of crisis because of the declining performance, and the leadership's abilities are in question.

This stage involves questioning the leadership's ability to lead the organization in the current situation. The necessity of questioning comes from recent

Table 20-1 The Model of Cultural Change

Stages	General Symptoms	IBM
1. The leadership's abilities are called into question.	• A crisis situation is recognized. • There is a perception that the crisis cannot be solved by the existing practices.	• A financial crisis was observed in 1992. This provided a strong impulse to begin the process of cultural change.
2. The pattern maintenance starts to break down in the sense of values, policies, and procedures.	• Values, norms of behaviors, symbols, and procedures that support the old system are weakening and cannot support the organization's needs.	• A declining ability of the leaders to reinforce IBM's culture was strongly seen during the recession. • There arose a need for a revised set of values, policies, and guiding corporate principles.
3. A new leader emerges with new values and beliefs.	• There is a questioning of the old leadership abilities and a breakdown of its power. • The organization is still in crisis.	• A new leader and some key executives and managers from outside were brought in.
4. Conflicts develop between the proponents of the old and new cultures.	• There is an increase in turnover. • New people are brought in to support the change.	• There was a reduction of the workforce. • There was also new hiring.
5. The new leadership brings the organization out of the crisis.	• Finances improve, and other symptoms of improvement are seen. • The credit goes to the new leadership.	• There was a significant improvement in 1996. • The credit was given to the new leader.
6. The values system and the organization's culture is redesigned to share the existing culture.	• There is a need for new values, beliefs, and symbols. • Existing structures are not satisfying the need to introduce the transformation.	• A new set of organizational principles was articulated. • A new structure was set up to sustain the new culture.

conditions that created a crisis in the organization that cannot be solved by using traditional forms of leadership. At this stage, a search begins to find alternative solutions and patterns of behavior that will allow the organization to get out of the crisis. In the case of IBM, the process of cultural transformation actually started in 1993, when a major financial crisis forced John Akers, the CEO, to pass the leadership to Louis V. Gerstner, Jr.; in other words, a significant crisis triggered the need to start the process of cultural transformation.

2. Breakdown of the pattern-maintenance in the sense of weakening the value system, policies, procedures, and symbols that supported the old culture.

At this stage, when the leadership's ability to manage is in question, a weakness in maintaining the existing policies that were based on the organization's value system can also be observed. Contradictions occur between the espoused values and actual behavior in the organization. All this is questioning the basic premise on which the organization's culture is based. One example is IBM's Lifetime Employment policy, which had been maintained for so many years, but became untenable in the period of crisis. For instance, in 1993 117,000 employees were laid off.

3. A new leader, usually an outsider, is brought in to deal with the existing crisis.

Questioning the leadership's activity to bring the organization out of the crisis, and challenging the existing values system, policies and procedures, is an important start toward cultural change, but all this is usually not sufficient to get the necessary momentum. Without having a new attractive vision, a strategy to make that vision a reality, a new or revised set of values, and someone or a team who can clearly evaluate the current situation, the organization will usually try to introduce incremental changes based on the old assumptions that contributed toward organizational successful results in the past. In the period of crisis, IBM started to take measures to fix the problems, but real transformation of the organization began when a new leader, Louis Gerstner, Jr., came in and brought with him a team of executives and managers from outside.

4. Conflict develops between the supporters of the old culture and proponents of the new culture.

After the new leadership takes place, a period of conflict can be observed between those who used to live and work under the old leadership and the proponents of the new culture. The people who are the losers in the conflict feel great resentment toward the new leader's philosophy; and therefore, are usually not deeply involved in the process of transformation, and have difficulty accepting the new policies and behaviors that came from the newly introduced organizational principles. Because of this, some people leave the organization voluntarily, while others are forced to go. In this period, there is a process of hiring new people who are closer to the new philosophy. The remaining employees who are used to the old philosophy start to go through a gradual process of self-transformation to adjust themselves to the new internal environment. Usually in this period, a general reduction in the workforce is observed.

When the new leadership took place at IBM, the workforce was reduced from 256,000 in 1993 to 220,000 in 1994. However, in 1994, the year after Gerstner became the IBM CEO, more than 60 new executives were brought in to support Gerstner while he introduced the new philosophy. In 1995, 15,000 new employees were hired, and in 1996, 26,000 were hired. These activities are typical when a conflict between the old and new culture arises, and this is an integral part of the process of cultural change.

5. The new leadership brings the company out of the crisis and becomes a winner in the conflict.

If the conflict is resolved and the crisis is eased, the new leader receives the credit, and the new organizational philosophy takes place. Usually a new set of values or principles is articulated and implemented by a new set of pattern maintenance activities. In other words, new policies and norms of behaviors are introduced to maintain the new set of organizational principles. At the beginning of the transformation, not all people may understand, like, or share the new approach to doing business. However, taking into consideration that they cannot argue with hard facts that prove the organization is out of crisis, they start to understand that something meaningful is going on. Because of this, they start to rethink their position on the process of transformation, and gradually adapt themselves to the new environment.

In 1994, when Gerstner proclaimed that, "IBM is back," this was impressive, but not sufficient. However, in 1995, when he continued to demonstrate new economic results, it was difficult not to give him the credit for resolving the crisis and difficult not to accept his new philosophy and new structure.

6. The value system and the organization's structure are redesigned to establish a new culture.

As the new leadership is formed, it begins to create new patterns, symbols, values, beliefs, and structures to sustain the organization's culture. New structures are often also developed to support the new culture.

In IBM, the new leadership articulated a new set of principles that reflected the new culture. A new vision and strategy of its materialization was also developed and shared with the employees.

So, what is next? It is understandable that the new culture will maintain until some event again calls into question the leadership's ability and activities, and the cycle of cultural change will start anew. However, if the organization uses the capabilities of an open system and corrects itself based on the feedback from the internal and external environments, the cycle will repeat itself less frequently. As for IBM, if the organization operates according to its principles, and continuously introduces incremental changes in the existing culture, it will not be necessary to go through a radical cultural transformation, at least for a long time.

20.4.11 Ten Major Lessons We Can Learn from IBM's Story

We hope that by reading IBM's story that you will find something that you can take home and consider when managing your own organization. However, we decided to draw attention to some aspects of the story that we believe deserve to be emphasized. We will limit ourselves to ten major points.

1. An organization, as any other living entity, is an open system, which to survive and grow, must be interconnected with the external environment. Reading the signals from the external environment (customers, suppliers, competitors and others) and acting accordingly is the only way a company can stay alive.

2. If an organization is in a crisis, it must reshape the existing culture by challenging its assumptions, mindsets and values. Most of the time this is done by bringing in a new leader from outside the organization— someone experienced in managing transformation in a critical situation; someone with a different mindset, who would look at the problems from a different perspective.

3. The first step a leader should take when undertaking a transformation is to create a sense of urgency and demonstrate a strong need for change. People usually move faster when they are on the edge of a catastrophe. A sense of urgency creates a structural tension that enhances people's thinking and allows them to become more creative.

4. To accomplish the transformation, the new leader usually forms a Corporate Executive Committee (CEC) that consists of old-timers and newly hired executives. Having some newly hired executives—usually people with whom the CEO has worked before—will help him accelerate the speed of transformation because they share the same values and can easily understand each other. At the same time, having some selected senior managers on the CEC who have worked in the organization for a long time will allow the newcomers to better understand the specifics of the existing organization and its culture.

5. While the CEC is the core of transformational leadership, it is important to form a larger group of enthusiastic people to represent the CEC on other levels of the organization. They will spread the idea of transformation and lead special projects. Independent of the existing culture in an organization, there are always people who are waiting for an opportunity to express themselves by doing something different. These people usually have less resistance to change. What is needed is to find them, encourage them, and give them challenging assignments.

    The core CEC, together with a large number of Initiative Groups (IG), will reach all layers of the organization and involve them in the transformation. The best way to have people positively accept the change is to involve them in doing the work needed for a successful transformation.

6. The assumption that a culture is something that will endure forever is wrong. Under this assumption, some organizations may fall into the trap of bankruptcy. Compared to the speed at which technologies and products change, culture is more stable; but in general, the cultural pattern of an organization changes over time. It is important to recognize that culture is dynamic. Some cultures may not need to be changed for many years, while other cultures may need to be reshaped, and yet others may need total transformation.
7. Customer focus should be at the core of any organizational philosophy. When you lose that focus, you lose customers, and then you lose the organization. Customer focus helps create customers, which is the main purpose of any organization.
8. A leader should limit himself from putting everything into policies and procedures; for example, dress codes or work schedules. What are important are the results. However, this does not mean that professional ethics are not a part of the culture. Allowing some chaos that is within the frame of the organization's needs is also healthy.
9. Organizational culture is invisible (except for artifacts). It is like a magnetic field that can only be observed by the results it produces. The best assessment of the level of culture is, therefore, through the demonstration of business results. An organization with a great culture should have great results. A great culture should be able to pull the organization out of trouble.
10. It is important to recognize that there is no such thing as a great organizational culture that has had bad operational and financial results for a long time. Bad results are a signal that something is wrong with the existing culture. We are not talking about the natural variations that always exist in any business process. Nor are we talking about situations such as recessions or catastrophes that may influence the whole industry or country, or even the whole world. Rather, we are talking about organizations that cannot sustain competitiveness; that have losses for a long period while their competitors prosper. An organization like this has a culture that needs to be reshaped.

References

1 James C. Collins and Jerry I. Porras, *Built to Last*, HarperCollins Publishers, Inc., New York, NY, 1994, p. 74
2 John W. Gardner, *Morale*, W.W. Norton & Company, Inc., New York, NY, 1978, pp. 23–24
3 Ken Blanchard and Terry Waghorn, with Jim Ballard, *Mission Possible: Becoming a World-Class Organization While There's Still Time*, McGraw-Hill Companies, New York, NY 1997, p. 34

4 Lynne Joy McFarland, Larry E. Senn, and John R. Childress, *21st Century Leadership: Dialogues with 100 Top Leaders*, Linc Corporation and Senn-Delaney Leadership Consulting Group, Inc., Long Beach, CA, 1994, p. 130
5 Thomas J. Watson, Jr. and Peter Petre, *Father, Son & Co.—My Life at IBM and Beyond*, Bantam Books, 1990, p. 11
6 Ibid., p. 12
7 Ibid., p. 13
8 Peter F. Drucker, *Innovation and Entrepreneurship: Practice and Principles*, Harper & Row Publishers, Inc., New York, NY, 1985, p. 21
9 Watson and Petre, *Father, Son & Co. – My Life at IBM and Beyond*, p. 15
10 Lou Mobley and Kate McKeown, *Beyond IBM*, McGraw-Hill Companies, New York, NY, 1989, p. 5
11 Regis McKenna, *Who's Afraid of Big Blue? How Companies Are Challenging IBM—And Winning*, Addison-Wesley Publishing Company, Inc., 1989, p. 17
12 IBM though the years, at http://www-1.ibm.com/ibm/history/history/decade_1910.html
13 Mobley and McKeown, *Beyond IBM*, p. 4
14 Thomas J. Watson, Jr., *A Business and Its Beliefs: The Ideas That Helped Build IBM*, McGraw-Hill, New York, NY, 1963, p. 23
15 Ibid., p. 19
16 Francis G. Rogers and Robert L. Shook, *The IBM Way: Insights into the World's Most Successful Marketing Organization*. HarperCollins Publishers, Inc., New York, NY, 1986, p. 9
17 Ibid., p. 222
18 Watson and Petre, *Father, Son & Co.—My Life at IBM and Beyond*, pp. 147–148
19 Ibid., p. 151
20 Nancy Foy, *The Sun Never Sets on IBM*, William Morrow & Company, Inc., New York, NY, 1974, p. 38
21 Watson and Petre, *Father, Son & Co.—My Life at IBM and Beyond*, pp. 301–302
22 Watson, *A Business and Its Beliefs: The Ideas That Helped Build IBM*, pp. 7–8
23 Ibid., p. 9
24 Ibid., p. 7
25 Charles Conrad (Ed), *The Ethical Nexus*, Greenwood Publishing Corporation, Westport, CT, 1993, p. 60
26 Rogers and Shook, *The IBM Way: Insights into the World's Most Successful Marketing Organization*, pp. 14–15
27 Ibid., pp. 20–21
28 Watson and Petre, *Father, Son & Co.—My Life at IBM and Beyond*, p. 301
29 Stephen R. Covey, *Principle-Centered Leadership*, Summit Books, New York, NY, 1991, p. 254
30 Lou Gerstner, *Lou's Page*, 1997 IBM Annual Meeting of Stockholders, April 29, 1997, p. 1
31 Robert Slater, *Saving Big Blue: Leadership Lessons & Turnaround Tactics of IBM's Lou Gerstner*. McGraw-Hill, New York, NY, 1999, p. 54

32 Ibid., pp. 113–114
33 Lou Gerstner, IBM 1994 Annual Report, Letter to Investors, p. 2. http://www.ibm.com/investor/financials/irfiar.phtml
34 Ibid., p. 5
35 Doug Garr, *IBM Redux: Lou Gerstner and the Business Turnaround of the Decade*, HarperCollins Publishers, Inc., New York, NY, 1999, p. 61
36 Ibid., p. 121
37 Slater, *Saving Big Blue: Leadership Lessons & Turnaround Tactics of IBM's Lou Gerstner*, p. 103
38 Doug Garr, *IBM Redux: Lou Gerstner and the Business Turnaround of the Decade*, HarperCollins Publishers, Inc., New York, NY, 1999, p. 120
39 Thomas J. Neff and James M. Citrin, *Lessons From the Top: The Search For America's Best Business Leaders*, Doubleday, New York, NY, 1999, p. 140
40 Garr, *IBM Redux: Lou Gerstner and the Business Turnaround of the Decade*, p. 62
41 Levering and Moskowitz, *The 100 Best Companies To Work For In America*, p. 193
42 Garr, *IBM Redux: Lou Gerstner and the Business Turnaround of the Decade*, pp. 63, 133
43 Ibid., p. 61
44 Slater, *Saving Big Blue: Leadership Lessons & Turnaround Tactics of IBM's Lou Gerstner*, p. 127
45 Ibid., p. 113
46 Lou Gerstner, "Letter to Investors," *1995 Annual Report* p. 3 (http://www.ibm.com/investor/financials/irfiar.phtml)
47 Ibid., p. 4
48 Ibid.
49 Mobley and McKeown, *Beyond IBM*, 1989, p. 8
50 Chris Argyris, "The Next Challenge," *The Organization of The Future*, Frances Hesselbein, Marshall Goldsmith & Richard Beckhard (eds.), Jossey-Bass Inc., Publishers, San Francisco, CA, 1997, p. 368
51 Betsy Morris, "Big Blue," *Fortune*, April 14, 1997, p. 74
52 Ibid., p. 81
53 IBM News Release, Samuel J. Palmisano Elected IBM CEO: Louis V. Gerstner, Jr. to Remain Chairman Through 2002, January 29, 2002 (http://www-916.ibm.com/press/prnews.nsf/jan/42CE5EC(91D7983B985256B500058 A102)
54 David Kirkpatrick, "The Future of IBM," *Fortune*, February 18, 2002, p. 68
55 W. Gibb Dyer, "The Cycle of Cultural Evolution in Organizations," in R.H. Kilmann et al (eds), *Gaining Control of the Corporate Climate, 1985* and "Culture Change in Family Firms, 1986," Jossey-Bass Inc., Publishers, San Francisco, CA, a division of John Wiley and Sons, Inc., New York, NY

Index

3M, 103, 118, 353, 392
AMD, 4, 5, 14, 15, 16, 19, 36, 58, 59, 78, 113, 114, 116, 121, 123, 140, 141, 142, 143, 181–182, 194, 197, 217, 248, 249, 255, 272, 281, 291, 294, 303, 304, 309, 323, 328–330, 333, 336, 337, 345, 351, 359, 366, 367, 368, 372, 374, 383–389, 391, 393, 394, see also Total continues process improvement and innovation (TCPI^2) macro system
   AMD Athlon™, 15, 16, 90, 178, 186, 353
   AMD Hammer, 113, 186
   AMD-K5, 177
   AMD-K6®, 90, 178, 182, 186
   AMD-K6-2®, 186
   NexGen, 177–178
Acquisitions, 171–179, 201–203, 206–207
Assumption, 192, 251–252

Baldrige award criteria, 227–238
Behavior, 200, 205–206
Behavioral standards, 371–376

Change
   behavioral, 205–206
   concept of, 289–291
   culture, 304–305, 396
   cultural, 203–204, 206
      meaning of, 312–313
      methods and forms for, 314–318
      strategies for, 312–318
   cycles of, 293–309
   management
      force field analysis, 309–312
      three phases of, 298–306
   resistance, 292–293
   second-order, 306–309
   stages of, 308–309
Competitive cost, 187–188
Core competence, 196–197, 353–354,
Creative thinking, 120–122
Cultural
   blending, 203
   change, 423–426
   conditioning, 190–192
   differences
      ignoring, 193
      making a core competence, 196
      minimizing, 194
   influence on mergers & acquisitions, 201–203
   pluralism, 202
   resistance, 203
   strategies, 192–198
   takeover, 203

432 INDEX

Culture
    and change, 396–397
    and strategy, 313–312, 358
    and structure, 406–407
    beliefs, 407–408
    changing during a merger, 203
    creating quality, 219
    definition, 191, 323–324
    global, 192, 195
    organizational, 99, 191, 192, 195, 201, 321–322, 323–326
    shaping in a global environment, 200
    stories and myths, 384
    symbolic action, 387–388
    symbols, 383–384, 388–389
Current reality tree (CRT), 93–99
    a manufacturing example, 96–99
    steps to build, 95–96
Creative tension, 66, 77, 102–103, 144–145, 297, 345–346

Deming
    7 deadly diseases, 42
    14 points of management, 41–43, 47
    chain reaction, 54–56
    incrementalism, 119
    innovation, 114–115, 119
    organizational psychology, 56–57
    plan-do-check-act cycle, 41, 43
    profound knowledge, 33–45
    theory of knowledge, 47–48
    theory of systems, 30–32, 46–47
    theory of variation, 52–54
Disney, 329, 340, 396
Drucker, Peter
    business purpose, 211
    entrepreneur, 401
    knowledge worker, 127, 128, 129, 134, 135
    knowledge-based innovation, 101, 106, 107, 108, 109, 111
    management, 127
    organization, 4, 12
    total quality management (TQM), 273
Due diligence
    cultural aspect of, 176
    process of, 176–178

Entrepreneur, 401
Ethics, 376–382, 404

FASL, 181
Feedback, 24–25, 26–27
Ford, 329, 392
Fujitsu, 78, 181. *See also* Joint venture

General Electric, 102, 204, 273–274, 334, 357, 396
Global
    competitiveness, 184–188
    environment, 199, 200
    manager, 198–200
    mind-set, 49, 185, 188, 199
    organization, 194, 198
Globalization,
    and culture, 190–207
    driving forces of, 154–155
    evolution of, 157–165
    meaning of, 165–169
    subsystem, 38
    theory and practice of, 48–49
    trend toward, 149–153
Goal(s)
    and tension, 345–346, *see also* tension
    from a system perspective, 346–347
    hard and soft, 347–349
    in a multicultural organization, 349–351
    learning, 351–352
    make motivational, 348–350

Hewlett-Packard, 118, 182, 334, 340, 347–348, 374, 375, 376, 392
Holon
    business systems, 281–284
    networks, 281–284
    Portuguese man-of war, 283–284

IBM, 78, 90, 110, 119, 141, 183, 204, 289, 334, 338–339, 392, 396–428
    beliefs, 407–408
    Corporate Executive Committee (CEC), 414–415
    formation of culture, 399–401
    Gerstner, 411–418
    principles, 413
    strategic imperatives, 413–414

strategic vision, 418–419
Thomas Watson, Jr., 405–411
Thomas Watson, Sr., 401–405
timeline of leaders, 397–399
values, 409
Incrementalism, 115–120
Inflection Curve, 15
Innovation, 112–122, 223–224, 231, 265, 266. See also Knowledge-based innovation; Deming, innovation
Innovative environment, 101–105
Intangibles, 215–217
Intel, 113, 155, 177, 178, 353

Janssen's Four-Room Apartment, 294–298, 301
Joint Venture, 78, 171–172, 180–183

Kennedy, John F., 354–361
Knowledge, see also Profound Knowledge; Knowledge-based organization
and mental models, 122–124
freezing model, 66–67
management, 144–146
manager(s), 126–135
T-shape of, 128, 129–132
theory of, 47–48
transfer, 137–143
workers, 126–135
Knowledge-based innovation, see also Innovation
characteristics of, 105–108
convergences of, 107–108
lead time of, 106–107
requirements for creating, 108–112
Knowledge-based organization, 63–64

Learning,
anticipatory, 75–76
double-loop, 71–74, 78–79, 254–256
emotional, 76–77
from partners, 76–77
how to learn, 67–71
maintenance, 74–75
meaning, 65
occur, 65–66
organizational, 65–80
participatory, 75

shock, 74–75
single-loop, 61–73, 78–79
two scenarios, 69–71
Licensing Technology, 182–183
Low ppm environment, 248–253

Management
time and meaning of, 127
Maslow's Hierarchy of Need, 222
Mental model(s), 122–124
Merger, 172, 173, 201–203, 206–207
Metaphor, 4–14, 335–336, 389–392
Microsoft, 329
Mission
and purpose, 340–341
and vision, 341
statement, 338–340, 341–344
Moore's law, 113, 289
Motivation, 57–59
Motorola, 55, 116, 257–260, 272–273, 334, 392. See also Six sigma, Motorola's methodology

Nortel, 182

Organization
architecture, 17–19
as a community, 12–13
as a living entity, 7–11, 14
as a machine, 6–10
as a symphony orchestra, 11–12
as a web, 13–14
behavior, 393
comparison of mechanic and living entity metaphors, 8–11
growth and development, 15–17
history, 384–385
learning, 65–80
life time, 14–15
psychology, 56–59
rituals, 384–385
what is it, 4–5

Peters, Tom
a thirst for change, 292
importance of intangibles, 215
view of innovation and incrementalism, 117–118
Pfizer, 102, 109, 343–344

Problem solving, *see also* Systemic problem solving
  anatomy of, 84–86
  forms of, 88–90
  holistic view, 83–84
  oscillation, 86–88
Profound knowledge, 33–36, 38–59. *See also* Deming, profound knowledge
Purpose
  and core values, 334
  and vision, 333–334
  being true to, 337
  from a system perspective, 335–336
  personal, 337
  of an organization, 331–333
  the exchange of, 336

Quality
  and customer satisfaction, 212–214, 218
  challenging & changing our assumptions, 251
  cost, 214–215
  culture, 219–226
  evolution, 284–286
  improvement, 223–224, 263–264
  initiatives, 262–286
  Kaizen approach, 264–266, 277, 286
  Kano model, 212–214
  managing, 210–238
  meanings of, 214–215
  measuring, 226–238

Reengineering, 275–281, 286
Risk taking, 145, 357, 361

S-Curve, 103, 112–114, 298–299, 304
Senge, Peter
  creative tension, 346
  learning, 65, 106
  purpose and vision, 333
Shewhart, 50–51
Six sigma
  Motorola's methodology, 257–260, 286
  philosophy, 272–275
Sony, 334, 392
Statistical process control (SPC), 25, 256–257, 279, 285
Strategic alliance, 171–172, 180–182
Strategic position, 110–112

Strategy, 352–353
Supplier agreements, 182
Systemic problem solving, 90–93. *See also* Problem solving
  dissolving, 88–93
  proactive, 91
  reactive, 90–91
  resolving, 88–93
  solving, 88–93
System(s)
  a general concept, 21–32
  and analytical thinking, 27–29
  control system, 23–27
  definition of, 22–23
  Deming's view, 30–32
  meaning, 22–23
  planning system, 26–27
  theory and application, 46–47. *See also* Deming, theory of systems
  thinking, 21, 27–29
  three conditions of, 22–23

Taguchi's quality philosophy, 245–248, 287
Theory of constraints (TOC), 93–95, 97, 99
Total Continuous Process Improvement And
  Innovation (TCPI^2) Macro System, 33–39, 44, 88, 90, 255, 280–281
Total quality control (TQC), 265–267, 271–272, 285
Total quality management (TQM), 87, 268–272, 276, 277, 278, 279, 287

Values
  and beliefs, 364–365
  and ethics, 381
  and time, 403–404
  core, 194, 200, 228–232, 358–359, 367–370, 389–392, 395–396
  espoused, 370–371
  general concept, 364–366
  operational, 369–370, 390, 393–394
  shared, 366
Variation, *see also* Deming, theory of variation
  managing, 240–260
  theory and knowledge of, 49–56

Vision
   and purpose, 333
   and transformational leadership, 330–331
   attractive, 297
   case study, 354–361
   power of, 328–330
   shared, 402–403